The Rise and Fall
of the
Japanese Imperial Naval
Air Service

'Those who do not remember the past are condemned to relive it.'
Santayana

'Who can be safe? Guard as we may, every moment is a danger.'
Horace

The Rise and Fall
of the
Japanese Imperial Naval Air Service

Peter J Edwards

Pen & Sword
AVIATION

First published in Great Britain in 2010
by Pen & Sword Aviation
an imprint of
Pen & Sword Books Ltd
47 Church Street
Barnsley
South Yorkshire
S70 2AS

ISBN: 978 1 84884 307 3

A CIP catalogue record for this book is available from
the British Library

Typeset in Palatino by S L Menzies-Earl

Printed and bound in England
by CPI

Pen & Sword Books Ltd incorporates the imprints of:
Pen & Sword Aviation, Pen & Sword Maritime, Pen & Sword Military,
Wharncliffe Local History, Pen & Sword Select, Pen & Sword Military
Classics, Leo Cooper, Remember When, Seaforth Publishing and Frontline
Publishing.

For a complete list of Pen & Sword titles please contact:
Pen & Sword Books Limited
47 Church Street, Barnsley, South Yorkshire, S70 2AS, England
E-mail: enquiries@pen-and-sword.co.uk
Website: www.pen-and-sword.co.uk

Contents

CHAPTER 1

In the Beginning

The prehistory of Japan is shrouded in myth and legend, but once upon a time when the world was still young the Sun Goddess Amaterasu was appointed ruler of the heavens. Her brother Susanowo was very violent, so the gods banished him to the earth. Amaterasu was frightened and hid in a cave along the coast. All the earth was plunged into darkness and the raging seas lashed upon the rocky coastline. The bewildered gods endeavoured to persuade Amaterasu out of her cave but were unable to do so. So they decided upon trickery. Placing a mirror and necklace on a nearby tree, they hoped the sun goddess would be enticed out. Amaterasu from within her cave saw the reflection from the mirror. Going to the entrance and nervously stepping onto the shoreline, she walked towards the tree. Immediately the ambushing gods seized her, and once again sunlight bathed the countryside and seashore. Meanwhile her brother had killed a monster and among its eight tails found a sword, which he had presented to his sister out of affection. Therefore, it happened that the grandson of Amaterasu descended to rule Japan, taking the three sacred treasures as the Imperial regalia. Eventually the great-grandson of Amaterasu, the Emperor Jimmu Tenno, ascended the Imperial throne in the year 660 BC. Thus, the line of Emperors was of divine origin, as were those related families of his personage, and with his peoples he sustained a special relationship.

Through civil war, saved from the Mongolian invasion by the kamikaze, or Divine Wind – a typhoon in 1381 that scattered the Barbarian fleet of ships – Japan had survived and prospered. Japanese trading ships sailed from the southern ports of the Home Islands to the South Seas, to India and China. International trade prospered, but meanwhile the Tokugawa Shogun, the ruler, had gained complete mastery of the four Home Islands, making the Emperor an impotent puppet confined to the Imperial Palace in the city of Kyoto, and strictly controlled by the Tokugawa family. The shogunate had developed into an office that combined the authority of a prime minister with that of a commander-in-chief of the armed forces. Thus, all the power was in the hands of this one family for approximately the next two hundred years.

Contact with European influences and peoples came when a Portuguese ship, lately out of Macao, the Portuguese settlement on the China coast, was wrecked on the south coast of Japan. The ship, having been blown off course for days, was hurled across the sea amid a raging storm of white-capped breakers set against a darkened sky in which the thunderclouds stormed away over the heavens. Through the torrential rain, the ship had staggered from wave crest to deepening trough, the crew desperately clinging to the crumbling timbers with all shrouds blown away long ago. Suddenly, when all seemed lost, the survivors and what remained of the ship were violently hurled onto the shores of the sub-tropical coast of southern Japan, somewhere along the shores of Kyushu. This was a fortunate accident, indeed, for the survivors, who were warmly welcomed. The Japanese thereby acquired the new European muskets, which their gunsmiths promptly reproduced with much artistry. In the following year of 1549, a Japanese privateer landed in southern Japan, bringing the Jesuit St Francis Xavier to the Home Islands. Immediately the Jesuits set about converting the people. Soon the Roman Catholics could claim over a quarter of a million Japanese converts. The Shogun saw the conversion to Christianity as a political advantage when the Spanish friars landed and a Spanish seaman was heard to say that Spanish conversion to Christianity had been but a prelude to Spanish empire building in the Americas. The Japanese political establishment took a contrary view most seriously. But eventually the Shogun looked upon Christianity as a dangerous, disruptive, political menace combining revolutionary ideas with dangerous practices. Repressive measures followed, leading to the crucifixion of six Franciscan priests in 1597. By 1614, there was a general policy of purging Christianity in an endeavour to persuade those converts to renounce their religion – the religion of the Barbarians! The height of the persecution came in 1637, when 37,000 Japanese Christians fled to Shimoshima, where the Shogun besieged them and a Dutch ship bombarded them. Of course, the end was inevitable. All 37,000 souls were slaughtered, a fate from which they were unable to escape. But by the grace of God Christianity in Japan was not to die out; it lived on, and to this day, a memorial may be visited commemorating this terrible event in that part of Japan. International trade brought not only St Francis Xavier and the Portuguese but also the Dutch in 1609 and British merchants in 1613. Meanwhile the Tokugawa shogunate determined upon drastic measures to rid the Home Islands of the Western Barbarians. Japan was to become self-sufficient, and international trade would have to go. As a result, the British merchants withdrew completely in 1623; the Dutch fared better, being banished in 1668 to Decimo, a small island in Nagasaki Bay. Here they might do business with Japanese traders once a year, and no more. The Spanish merchants had previously left in 1624 and no one was to return for more than two hundred years.

The result of the reactionary policy of the shogunate was to prohibit the travelling abroad of any Japanese citizen for whatever reason. Likewise, foreign citizens were not allowed to land on Japanese soil, and foreign sailors, if shipwrecked, could be executed. Thus, the lucrative trade with the Spice Islands from the sophisticated cities of China and the teeming land of India was cut off. Japanese merchants who could remember the good old days of high living, the overseas riches, the new and exciting foreign ways, had to start all over again. It was forbidden to construct large ocean-going ships. The Daimyo of Satsuma, the overlord of Kyusho and the Ryukyu Islands could build only small fishing boats for coastal trips and inshore fishing. Thus, Japan was totally isolated from the outside world. In the realms of religion, a compromise was established by the usage of classical Buddhism and chauvinistic Shintoism. The net result was that the Barbarians and Barbarian ways were effectively blotted out from the Japanese way of life. For 250 years, Japan was to enjoy an isolated peace, a Golden Age in which literature and the arts were to flourish side by side with the simple life. The classical Noh dramas of Chinese origin to Bunraku puppet theatre with plays written by Chikamatsu Monzaemon, the Japanese Shakespeare, and Kabuki plays concerning historical, domestic and dance dramas were brilliantly developed. Wood-block printing, called Ukiyo-e, was brought to a high standard, producing prints for humble people beyond the confines of the Imperial court. In the field of literature, novels by Saikaku became renowned, with a resurgence of the art of the poem, such as the haiku style, with its references to one of the seasons and no more than ten words. The most famous haijin was a poet known as Basho, who lived in the late seventeenth century. However, just before the descent of the 'Purple Curtain', Will Adams, a former crewman of a Dutch ship, had been wrecked on the southern coast. Making friends with Ieyasu Tokugawa the Shogun, Will Adams had been persuaded to settle in Japan, remarry and teach the Shogun all he knew of navigation. This Adams had done, being known as Miura Anjin and becoming the pilot major of a fleet of five sailing ships. Thus the first Japanese Navy had been formed sometime in the 1640s. Upon the death of Adams a shrine was erected – the Kurofune (Black Ship) Shrine to his memory on the site of the house in which he had lived.

During the seventeenth century, a rapid expansion of trade took place all over the Far East. American traders had relations with China, and Russian settlements were appearing along the eastern coast of Siberia. Then in 1839 Commodore Beddle USN attempted to negotiate a trade treaty with the Shogun. Unhappily in Japanese eyes, the Commodore had lost face in failing to discipline a Japanese seaman who had jostled the American commander. This resulted in complete failure. The Americans attempted a further expedition, announced in 1852, and in July of that year Commodore Perry USN, commander-in-chief and ambassador, arrived off the Japanese coast. America was not the only country interested in opening-up Japan to international trade, for also in 1852 Admiral Putyatin of the Russian Navy sailed into Nagasaki, arriving in October of that year to open negotiations, but without achieving any lasting success. Naturally, at the Shogun's capital a political bombshell had exploded in that the Daimyo (the Lords) were asked to consider the American offer to negotiate a treaty. The aristocracy were split into two opposing factions: some were for ousting the foreigners, while others were for trading, and the unresolved question was still under debate when in February 1854 Commodore Perry returned with seven warships and 2,000 marines. Under such a display of naval power the Treaty of Kanagawa of March 1854 was but a natural outcome. The ports of Hakodate and Shimoda were opened for American trade and supplies.

Much of the opposition to the negotiations of the Treaty of Kanagawa was centred on the Satsuma clan, which was governed by a branch of the Imperial family named Shimazu, after an island port. These people were a seagoing principality in sub-tropical south Kyushu, the land of the Sun Goddess. They had borne the brunt of the Mongol invasion in 1274 when the kamikaze wind had scattered the enemy fleet. Ships had been provided by the clan in support of the 1592 invasion of Korea by the Shogun Hideyoshi. The Satsuma clan had conquered Okinawa and the Ryukyu Islands in 1609, and more ominously they were to lead the Strike South Faction in the 1930s, which culminated in the 1940 Pacific air and sea campaigns. Across the Straits of Shimonoseki lived the Shoshu clan in south-west Honshu, who had supported the Satsuma during the Korean invasion and controlled 11,000 hereditary lords; together with their neighbours, the Satsuma could muster 38,000 hereditary warrior lords. Both had opposed the Tokugawa's policy of replacing external aggression with internal consolidation. As a consequence both had lost much land to the Shogun during the seventeenth century as a price for their views. During a much later act in the play, both clans would lead decisive roles in the story. For the time being, both clans opposed the Shogun's policy, strongly objecting to the presence of the foreigners on sacred soil. The powerful clans of Satsuma and Mito organised violent, lawless groups of men to oust the foreigners by firing on the legations and pursuing a policy of assassination of Western diplomats and merchants. These violent killers were called Ronin, or 'wave men' – that is, they were tossed from venture to venture because all their clan loyalties had been lost. To the Westerners

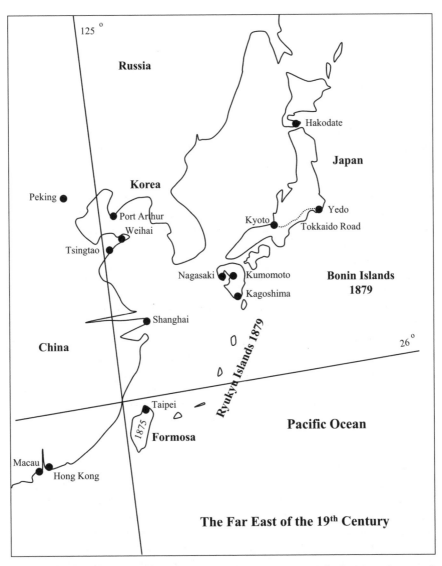

The Far East of the 19th Century

these Ronin were swaggering, reckless, two-sworded assassins and former samurai. At this time throughout Japan the cry 'sonno-joi' was heard, meaning 'Revere the Emperor! Expel the Barbarians!' The national cry was heard from island to island. Meanwhile the Emperor Komei had become obsessed with the idea of driving out the Barbarian foreigners, and this ideal was supported by the Lord Shimazu. Sometime previously Prince Asahiko had been in trouble with the authorities, and the Lord Shimazu had been able to obtain the release of the Prince. This was important, for Prince Asahiko had direct access to the Emperor. Furthermore, the Prince believed the old order was good enough and modernisation was undesirable. Unhappily the Prince wished to rid the Imperial court of the Choshu leadership – the very neighbours of Lord Shimazu. However, they were able to persuade the Emperor to cancel some of the powers of the Tokugawa Shogun, and a policy of intimidation of foreign colonists was officially initiated. During 1860 the first Japanese diplomatic delegation departed from the Home Islands to take up posts abroad in the United States of America. The American government had placed the US Navy cruiser USS *Powhatan* at the disposal of the Imperial government. For the first time the Hinomaru was flown from the mast-head of the American warship. The flag was a red sun disc on a plain white background, which became the national flag flown on all Japanese warships and land bases. As the American cruiser left the sheltered waters of the inland sea and the summit of Fujiyama disappeared below the horizon, the fanatical Mito clan murdered the Gotairo, or Prince Regent, sometime during the course of 1860. This far-sighted man had negotiated the original Treaty of Kanagawa with the Americans and had advocated the opening-up of Japan to foreign traders. The Mito clan had attacked the legations and killed Henry Heusken, the US Consul. The acts of the Mito clan and the Barbarian Expelling Party had consistently been supported by Prince Shimazu Saburo of Satsuma since the first visit by Commodore Perry in 1852. Shimazu was a political moderate who was proud, courageous and authoritarian, but nevertheless believed in a reconciliation of the Shogun and the Mikado. He came to be of the opinion that Western technology possessed too much potential to be easily dismissed out of hand.

Conflict on the Tokaido

It had been the policy of the shogunate to ensure the wives and families of the Daimyo should live at court while their husbands attended to business on their estates. This had been enacted as an insurance policy against would-be usurpers of the Imperial state, and in accordance with custom Prince Shimazu Saburo of Satsuma rode out onto the Tokaido (East Sea Road) from Edo (Tokyo) to Kyoto. The prince was escorted by servants and samurai guards on his way to the Emperor's capital at Kyoto. In magnificence the procession rode down the road. The grooms on foot tended the horses of the Daimyo and his servants. Before and aft of the important personages were the mounted samurai equipped with two swords. As they rode, the sun twinkled on the personal armour and sword scabbards of the guards. The jackets of the servants and grooms were emblazoned with crosses within circles on sleeves, jackets and scabbards.

As the pageant progressed through village and town, all citizens made way for the Daimyo and his entourage, as was required by the law of the day. Upon nearing the village of Namamugi, the escort commander noticed two European gentlemen and a lady riding on horseback in the opposite direction to the cavalcade. Naturally he had no reason to suppose the law would be flouted by the European Barbarians, for the legislation was quite clear. The horses taking Prince Shimazu of Satsuma gaily trotted forward. The houses of the village were now in sight, and the road commenced to narrow as it went through the village. The European riders foolishly continued, to the consternation of the servants, who gesticulated to the foreigners to turn back or at least to stand aside. Now the European party consisted of three English men and an English lady out riding enjoying the countryside. There was Mr Charles L. Richardson, who proposed later to return to England on leave as a merchant from his business house at Shanghai, and had come over to Japan to meet some old friends. With Mr Richardson rode Mrs Borrodaile, a lady on holiday from Hong Kong and the sister-in-law of Mr William Marshall, an English merchant from the city of Yokohama, who had been accompanied by his colleague Mr Woodthorpe Clarke. Both these latter gentlemen had been instrumental in establishing the Yokohama Municipal Council sometime in 1862. The area around Yokohama had originally been a marshy, desolate region given over to the Europeans for the purpose of organising a trading post. Gradually a town had grown up almost along Western lines, and the more republic-spirited members of the citizenry, mindful of the dangers of immorality and the surrounding lawlessness, had determined upon the enforcement of law and order in each field of municipal activity. Thus Yokohama never did recognise the Emperor's law, and local organisation was run exclusively along European standards. Meanwhile the three English merchants and their lady guest had been at dinner. Everyone had been entertained by Mr Richardson's views and comparisons of trading conditions experienced in Japan compared with those existing in the city of Shanghai, China. After coffee and the completion of dinner, the party had quite naturally determined upon a breath of fresh air and a little light exercise. Horses were saddled and riding habit donned, and leaving the comfortable sanctuary of the guest house, the lady, accompanied by the three gentlemen, rode away to view the scenery with the prospect of an enjoyable afternoon's gallop.

The four riders came on down the narrow Tokaido and presented a problem to the guard commander of the Daimyo's entourage travelling the opposite way. He realised that the Europeans had broken the Emperor's law and was determining what to do. An argument had commenced regarding dismounting and the right of way. Shouting broke out as the grooms violently gesticulated, the escorting horsemen scowled, and suddenly, like a flash of lightning, a samurai reached with his sword and slashed Mr Richardson across the side. The victim screamed in burning pain, his terrified horse panicked and turned about, Mrs Borrodaile, amid the uproar and equally terrified, strove to escape. In her attempt a samurai slashed at her long hair to obtain a souvenir, but the lady and horse sped away screaming in fright to Yokohama, leaving a bloodstained scene. Meanwhile, both Mr Marshall and Mr Clarke had been attacked and were lucky to escape with nothing more than sword slashes on the shoulders and about the sides. The Prince of Satsuma attempted to restore order amid the disorganised mêlée, and once again the entourage was able to proceed in an orderly fashion to the Emperor's capital. In the meantime the dying body of Charles Richardson ended up under a tree at the roadside. A charitable peasant had taken the body into care, and commenced to nurse the broken and shattered man. But shortly afterwards a samurai warrior returned to the little house occupied by the charitable peasant and the dying Richardson. Raising his sword, the samurai struck a death blow that put Richardson out of his agony. The remains of the dead man were buried somewhere along the roadside.

Eventually Mrs Borrodaile rode into Yokohama, bloodstained and hysterical, to break the horrible news. The news spread like wild fire through every European section of the town. Every able-bodied Westerner, military and civilian alike, took out his weapons; revolvers were reloaded, swords were made ready in their scabbards. Saddling up, everyone was determined to ride off in hot pursuit. Lieutenant-Colonel Pyse of the British Consulate, Kanagawa, rode out with the Legation escort. Also present in the contingent were French, United States, Prussian and Dutch marines with two doctors. The armed party advanced down the road from Kanagawa to the village of Namamugi with care and in combat order in case of an ambuscade. Naturally the body of Richardson was not found, but the partly mangled remains were discovered near a grove of trees on the roadside. On investigation they learned that Richardson had painfully crawled to a nearby tea shop and pleaded for water, after which he met his end as described above. At about this time, Marshall and Clarke, dreadfully wounded and covered in blood, arrived at the United States Legation at Kanagawa. By now the Prince of Satsuma's entourage had arrived at the court of the Emperor at Kyoto. Prince Shimazu paid his respects to his Imperial Majesty, and after a brief stay left for Kagoshima, his clan capital, suspecting a swift British reprisal.

Evening had descended upon the warehouses, offices, clubs and houses of Yokohama. The posse had returned from the village of Namamugi in great excitement, with revenge in mind. After dinner, when tempers had cooled, the British Chargé d'Affaires sent round orders for disbanding the British contingent, and instructed all to go home. It was made quite clear that only diplomatic channels would be used to solve this international incident. Nevertheless, the municipal authorities were not so complacent as the British Chargé d'Affaires. The marines were ordered to patrol the streets of Yokohama during the night in case further activities should take place. By the grace of God the evening was spent peacefully. However, a few months later after the report from the British Chargé d'Affaires at Yokohama had been received by the Foreign Office in London, Lord John Russell had issued an official demand on the Shogun's government for the payment of £100,000 in reparation for the murder of Richardson. The government in Edo (Tokyo) was left in no doubt as to the feelings of Her Majesty's Government, in that severe measures would be taken if the reparation was not paid within twenty days; also, that the assassins were to be arrested, tried and finally executed. The Prince of Satsuma would be required to pay a further £25,000 to the injured parties and the relatives of Mr Richardson. On receipt of the demand, the Tokugawa Shogun felt forced to pay: even if for no other reason, the overwhelming collection of heavy firearms would have persuaded the shogunate otherwise.

Retribution from the sea

Sometime in January 1863, Prince Asahiko celebrated his 39th birthday and invited Lord Shimazu to a party, with Konoye Tadahiro as guest of honour. During the course of the festivities Prince Asahiko and Lord Shimazu, in the presence of Konoye Tadahiro, had pledged eternal friendship. A bond had been sealed between the Imperial house of Fushimi and the aristocratic house of Satsuma. Touchingly enough, this event took place in the gardens of Prince Asahiko's villa, where the bond was sealed under the sacred maple tree. Thus two of the most important houses of Japan were united. By June 1863 the Shogun, having received the London demands from Lord John Russell, was highly perturbed on being informed that a Royal Navy squadron was gradually steaming up the coast to the city of Edo. Panic spread and rose to fever pitch as the warships crept ever nearer to the Shogun's capital. Orders were issued and Edo was evacuated. The Shogun proposed to pay the demand. On 24 June 1863 the Royal Navy warships anchored in Yokohama Bay. At the same time heavy boxes containing Mexican dollars were brought from the Shogun's treasury to the offices of the British Legation. Here the currency was tested by Chinese shroffs to ascertain whether the currency was genuine coinage. When they had satisfied themselves as to the authenticity of the money, it was crated and put aboard the British warships riding on anchor chains in the roads of Yokohama harbour. At last the Shogun had been brought to heel. Nevertheless, the Prince of Satsuma refused to pay the additional £25,000, and Lieutenant-Colonel Neale of the British Legation was determined to go with the Royal Navy squadron to Kagoshima in southern Japan, the clan capital of the Prince of Satsuma. Here the Colonel hoped to present in person Her Majesty's Government's ultimatum. Eventually the British ships slid out of Yokohama, watched by the spying Japanese. Heading south-west along the coast, the Pacific squadron consisted of the flagship, HMS *Euryalus*; two corvettes, HMS *Perseus* and HMS *Pearl*; a paddle-sloop, HMS *Argus*; a dispatch vessel; and two gunboats, HMS *Racehorse* and HMS *Havoc*. This was the China Station squadron under the command of Admiral Kuper, who arrived off Kagoshima on 11 August 1863. The following day a letter was sent ashore delivering the ultimatum. Everyone patiently awaited the reply from the Prince of Satsuma, but on 12 August a communication was received that proved unsatisfactory to the senior British officers. About noon, boarding-parties left the British fleet to seize three Japanese steamers, which were plundered and later burnt. Shortly afterwards the Japanese shore batteries high up on the sub-tropical Kyushu shore line opened fire. The British ships were surrounded by tall plumes of water spouts, but HMS *Euryalus* delayed returning fire for one or two hours. The money crates had been piled on deck two months previously next to the warship's magazine doors, and the gun crews were unable to bring out ammunition to load the guns. Before action could commence, the ship's company, rather undignified, had to move each crate to a safer location. Meanwhile a typhoon-force wind had blown up, which made the Japanese gunners' task more difficult. Eventually action commenced as the British warships opened fire. The Japanese gunners were very accurate; a 10 in. shell exploded on the flagship's main deck, and another on the upper deck, and Captain Josling and Commander Wilmot were killed on the bridge of HMS *Euryalus*. Meanwhile HMS *Racehorse* went aground in the gale and had to be towed off at the same time as HMS *Pearl* was considerably damaged in the engagement. Having bombarded the land batteries, the China squadron also bombarded the city of Kagoshima with rockets, which roared high into the sky, cascading down and causing death and destruction. Among the casualties were the famous Satsuma porcelain factories, smashed into smithereens. On the following day, 13 August, a second bombardment was undertaken by Admiral Kuper. This time the remaining land batteries were pulverised and the city was given a further pyrotechnic display, with more death and destruction. A few remaining ineffectual defensive guns replied, killing sixty-three British sailors, but the fact remained that the Satsuma ransom was still unpaid, and Admiral Kuper sailed away to Yokohama empty-handed but for the indemnity paid by the Shogun. However, the Daimyo of Satsuma had been very impressed by the artillery expertise of the Royal Navy. He initiated investigations into the manufacture of Western-style weapons and acquired a healthy respect for the British Navy. Meanwhile Lord Shimazu had unexpectedly arrived in the Imperial capital and had informed His Imperial Majesty the Emperor Komei of the outrageous action of the British Navy: great fires had swept the city, killing many innocent people and causing the destruction of much property.

Investigating Barbarian artillery
However destructive the British action may have been, the fact remains that the Daimyo of Satsuma was very impressed with the artillery capabilities of the British warships. Now six years previously, in April 1857, the Royal Navy had sent some 24-pounder guns to a Mr Whitworth's engineering works in Manchester, England. The guns had been specially bored out and rifled to fire a nine-pounder shot. By the use of Paixham shells the guns could be used to fire 24, 32 and 48 lb shot by increasing the length of the cylindrical shell to take the extra weight of the propellant. During October 1859 gun-firing tests had been conducted by Captain Hewlett RN of HMS *Excellent*, firing at a target from HMS *Alfred*. The target had been protected by four-inch iron plates and seven-inch oak sides, which had been successfully pierced using a 68 lb iron shot. Eventually the demonstration gun had burst due to excessive gas pressure. These Whitworth guns formed the main armament of many Royal Navy gunboats until they fell into disfavour because of the large number of accidents caused by bursting.

The standard armament since Nelson's day had been the 32-pounder gun. Recently improved, it now possessed a crude but efficient sighting apparatus and a new flintlock firing mechanism. The 32-pounder gun was mounted on a new truck carriage known as the Marshal Slide Carriage. A new system of gun nomenclature was introduced whereby sizes were distinguished by size of calibre and not weight of shot. All these developments had been undertaken as a result of the experimental work of a Mr Armstrong of Elswick, Lancashire, England, during 1858. The original work on gun design had been conducted by a Mr Longridge, using a revolutionary method of construction. In this method the barrels of large artillery pieces were manufactured by shrinking a series of wrought iron tubes around wire onto an inner barrel. This type of construction ensured that the gun was free from the effect of 'warping' strain when fired. The inner barrel was rifled and the gun was breech loaded for firing Paixham-type shells. Previously roundshot had been fired, but the new shells were elongated, with a pointed head. Studs were located on the base of the shell that engaged with the inner barrel rifling on being fired. Together with the breech loading, these guns were capable of a very high rate of fire at great accuracy. On being tested in 1859, some of the Armstrong guns had proved a failure, probably due to faults in manufacturing technique rather than to a faulty design. Nevertheless the Royal Navy had been sufficiently impressed with the test results to place orders with the gun manufacturers for artillery pieces firing 6, 12, 20, 40 and 110 lb shells. It followed that the Kagoshima bombardment had been an opportunity to use the Armstrong breech-loading guns in action for the first time. During the course of this engagement the Royal Navy squadron had fired 365 rounds. Unfortunately twenty-eight accidents had taken place to twenty-one of the guns due to premature bursting of the breech mechanisms. Unhappily for the Royal Navy, the fate of the breech-loader was now sealed, for the fleet reverted to the use of the muzzle-loader for the next fifteen years. Notwithstanding the bursting of some of the artillery pieces, the Japanese shore batteries had caused considerable damage to Admiral Kuper's ships before they had retired to Yokohama. However, in the following year a further action was undertaken by the Royal Navy, but accompanied by French, Dutch and United States warships, in forcing the Straits of Shimonoseki. HMS *Conqueror* with six other gunboats bombarded shore batteries and successfully landed marines. After a short engagement the marines blew up shore installations and fortifications. This was the territory of the Prince of Nagato, an anti-foreigner who wished to disrupt the trade of the West. But the power of the foreign Barbarians was all too powerful, and soon normal trade resumed. British-built warships were purchased by the Prince of Satsuma, and a policy of pursuing Western know-how and technology was instigated. The defeat of the Barbarian foreigners seemed to be an impossibility. The anti-foreign party now endeavoured to win over the foreigners, endeavouring to acquire barbaric foreign knowledge. The Prince of Satsuma and others were convinced that such a policy could keep the Barbarians at bay and keep the homeland free from corruptive foreign influences. It was recognised that a need to build the armed forces was necessary.

In 1867 the first British Naval Mission was sent to Japan. The British government loaned the services of Commander Tracey RN, with other naval officers and ratings. The mission lasted six months. Meanwhile a revolution broke out to restore the Emperor to full temporal and religious power and break the control of the government of the Tokugawa shogunate. The Emperor was backed by the great aristocratic houses, which saw an opportunity to build up the nation against the Barbarian foreigners. It was during this period that one of the great princes engaged Lieutenant A.G.S. Howe, Royal Marine Light Infantry, as a gunnery instructor for his own personal navy. This Royal Marine officer possessed great organising ability and was acclaimed by some in 1867 as the father of the Japanese Navy. The pursuit of foreign knowledge, equipment and experts increased in intensity. In April 1868 His Imperial Majesty the Emperor Meiji swore an Imperial Oath of Five Articles. Article number five stated, 'Knowledge shall be sought throughout the world so that foundations of the Empire may be strengthened.' The idea was now official government policy, and was pursued with great thoroughness. During the following year, 1869, various political adjustments took place in which the Princes issued the famous 'Memorial of the Daimyo of the West'. This statement set out ten basic agreements with the Emperor:

1. The Princes surrender their feudal rights and territories to the Emperor.
2. They state that they believe Heaven and Earth belong to the Emperor.
3. Everyman is the Emperor's retainer.
4. This agreement constitutes the Great Body.
5. The Emperor governs the people by confirming rank and property.
6. The Emperor gives and he takes.
7. No one can hold a foot of land.
8. This constitutes the great strength.
9. Imperial orders should be issued for remodelling the clans.
10. Civil and penal codes, military laws and detailed engines of war, as well as all affairs of the Empire, proceed from the Emperor.

As a result, the civil war was now over and the Imperial government firmly established. Meanwhile, in 1870, the national flag was adopted, called the Hinomaru, which means 'Round the Sun', as the national emblem for shipping. It was during the same years that the Japanese Navy was placed under the guidance of British officers. For it was considered that Britain had the best organised Navy, so that Japanese flag officers were trained in Britain and much naval building was undertaken in British yards for the Japanese government. By 1873 the second British Naval Mission had arrived in Japan under the command of Commander Douglas, consisting of thirty officers and men, and it remained in the country from 1873 to 1879. The mission established a naval college at Etajima, built as a replica of the Royal Naval College, Dartmouth, including the use of bricks taken from Dartmouth. A lock of hair from Lord Nelson was enshrined in the Memorial Hall, and to retain tradition a Western-style meal was served once a day aboard ships of the Navy, complete with silver knives, forks and spoons and an English bill of fare. The Japanese combined the traditional Royal Navy spirit of offensive warfare and the ideal of the captain going down with his ship. Like the British, the Japanese were to have an aversion for war against commerce, particularly during the Second World War. By 1877 Japanese Imperial forces were besieging Kumamoto fortresses in Kyushu, with little success. The government ordered two naval engineers, Shinpachi Baba and Buhei Aso, to construct two balloons. On 21 May 1877 engineer Baba in a balloon of 14,000 cubic feet capacity attained an altitude of 360 feet. This was the first manned balloon flight in Japan. In June 1878 2nd Lieutenant Ishimoto Shinroku reached a height of 300 feet in a balloon with a capacity of 10,900 cubic feet. Although these balloons were used during military campaigns to suppress a rebellion, as soon as victory was achieved all interest in flying was lost.

Three years previously Japan had invaded Formosa and claimed the Bonin Islands. The year previous to this invasion an expedition had been undertaken against two Formosan tribes who had killed some Ryukyuans now under Japanese protection. By 1879 the Ryukyu Islands were incorporated into the empire, but without the well-organised Navy the territorial extensions could never have taken place. In 1894 Japan possessed twenty-eight modern naval vessels, with full maintenance and dockyard facilities. The following year the Navy had six battleships and six cruisers, all originally constructed in British yards to the most modern designs available. Eventually, in 1901, Japan held the naval balance of power in the Far East and could operate naval forces more cheaply than a squadron of the European powers, and on 31 January 1902, because of this situation, Britain and Japan entered a treaty agreement that was to last until 1923. This treaty was only broken at the behest of the United States, which by 1923 was exerting a strong political pressure to isolate Japan and weaken the British position in the Far East. But for twenty-one years a strong political link connected Britain and Japan, maintaining peace in the Far East. Both parties affirmed they were open-door powers, and stated:

> If either Great Britain or Japan in defence of their respective interests should become involved in war with another power the other high contracting power will maintain a strict neutrality and use its efforts to prevent the other powers from joining in hostilities against its ally. Furthermore if any outside power should join in the hostilities the other party will come to its assistance and will conduct the war in common.

The treaty further provided that:

In peacetime docking and coaling facilities for both powers should be organised.
Both Great Britain and Japan would work together to concentrate a force larger than any third power.
Japan was given practical assistance with the use of Royal Naval equipment and cooperation in an emergency.
Great Britain had the support of a first-class Japanese fleet in the Far East to protect British economic interests.

That the United States of America should have worked against such a treaty at that time is surprising. But the political manoeuvring that forced Great Britain to retract in 1923 should have been seen by British politicians as indicating the road along which the United States was slowly plodding: to eventual war and the destruction of an empire that had much to offer the West. It was to be replaced by an international communist movement and a world military power immensely powerful but with a politically immature population incapable of defending the West, let alone Japan. Indeed, a sleeping giant was awakening!

In 1903 an event of great scientific and political importance took place at Kitty Hawk, North Carolina, USA among sand dunes more than a hundred feet high and in very strong winds. This was the first powered, manned flight of the Wright biplane. Hitherto only glider flights had been undertaken by the Wright brothers, but on 17 December 1903, equipped with a 13 hp engine producing 450 revolutions per minute, the machine successfully rose nine feet and landed 120 feet from its starting point. The propellers were driven by bicycle chains and the aircraft rested on a trolley. But during 1904 Japan was involved with Russia in a Far Eastern war. Port Arthur had been invaded by Japanese Army and Navy units. The Army had ordered the construction of two small captive balloons from a Mr Isaburo Yamada, while a further balloon was purchased from an Englishman, a Mr Charles Spencer. These balloons were used for reconnaissance and artillery-spotting duties at the front. Meanwhile the Russians had been defeated. The Japanese had captured Port Arthur, Dairen and all the railways built by the Russians in south Manchuria. Since the Tokyo government wished to be recognised as a member of the Imperial European Community Japan poured $1 billion into bandit-infested territory. This action brought law and order into the country, as well as thousands of Chinese, Japanese and Korean merchants who quickly increased the gross national product. In the following year, 1905, the Wright brothers conducted further experimental flights.

This was the year in which a Treaty of Alliance was signed by Great Britain and Japan which was to secure peace in the Far East. The Japanese Army Staff were now convinced of the importance of heavier aircraft, after

15

Wilbur Wright had actually flown for a duration of thirty-eight minutes.

With these rapid technical developments taking place and the constant swell of political reorder, the prospect of increasing commercial intercourse since the signing of the friendly trading treaty of 1844 with China caused the United States to look westwards. It was President Theodore Roosevelt who said at the time, 'Our future history will be more determined by our position in the Pacific facing China than in our position in the Atlantic facing Europe.' These were profound words, which could subsequently spark off a policy leading to confrontation at a later historical period.

Into the air
In 1907 Japan could justly claim to be the leading naval power in the western Pacific. At this time Lieutenant-Commander Eisuke Yamamoto, a naval general staff member, expressed to his chief a desire to see Japan participate in aeronautical research and development. So a meeting was arranged with the Navy Minister, Admiral Minoru Saito, and the Army Minister, General Masatake Terauchi. Subsequently an agreed aeronautical policy was established. The next year the Wright brothers were paying a visit to Europe to demonstrate their new biplane. This machine was a special two-seater, originally developed for the US Army, which achieved an endurance record of two hours twenty minutes and a height record of 361 feet. Unfortunately this machine crashed, killing the observer and injuring Orville Wright. During March 1909 Army Lieutenant Hino and naval engineer Baron Narahara each designed an aircraft. The naval machine first flew on 5 May 1909 and was the first flight of an aircraft designed and constructed by the Japanese. Unhappily subsequent designs were not successful. It was decided, therefore, to build aircraft of foreign designs until such time as Japanese aeronautical designers had gained sufficient experience to develop more successful models of their own. At this time the Japanese Army and Navy decided to embark on separate design projects, building to their own departmental specifications. On 25 July 1909 a most spectacular flight took place to win a prize of £1,000 offered by the London *Daily Mail* newspaper. This was between Les Baraques, just north of Calais, in France, and the cliffs of Dover in south-east England. The successful machine was flown by Louis Blériot, a Frenchman, in thirty-seven minutes over a distance of thirty-one miles. Taking off from France at dawn, Blériot set out and was engulfed in fog over mid-channel. The engine quickly overheated, but pouring rain soon dispelled the fog and lowered the engine temperature. Making a landfall at Margate, England, Blériot was soon able to find his way to Dover. The importance of such an aeronautical achievement was not lost in Japan, for

the significant over-channel flights had increased the important views held by pro-aviation-minded officers. By now Japan possessed eleven battleships, thirteen armoured cruisers and seventeen protected cruisers, and was the most powerful naval force in the western Pacific. Already the Australians found it necessary to maintain a naval squadron in the south Pacific.

Cementing the foundations

In July 1909 the only organisation concerned with military aviation in Japan was the Rinji Gunyo Kikyu Kenkyu Kai, or Provisional Committee for Military Balloon Research, established under the chairmanship of Major-General Gaishi Nagaoka. The terms of the committee were to study the new powered aviation, and represented the combined efforts of the Army, Navy and Education Ministries. The Navy was represented by Engineer Lieutenant Masahiko Obama. The Army representatives consisted of Captain Tanin Yamaya, Lieutenant Shiro Aihara, Engineer Colonel Jinro Inoue, Engineer Major Kumao Tokunaga, Infantry Captain Kumazo Hino and Artillery Captain Kikutaro Sasamoto. From Tokyo Imperial University were Dr Scientist Aikitsu Tanakadate, Dr Eng Seinen Yokota, Dr Ariya Inokuchi, Dr Scientist Kiyo Nakamura, chief of the Central Meteorological Observatory, and Baron Engineer Sanji Narahara. Up to the year 1920 this committee was instrumental in laying down a highly successful aviation policy that prompted Japanese initiative in the field of aeronautics and contributed greatly in the development of military aviation. All aircraft flown in Japan during this year were imported from abroad. These included a Henri Farman Box Kite biplane powered by a Gnome Rotary engine of 50 hp, with a maximum speed of 38 mph. From Germany two Grade aircraft were imported, being modified box kites from the French Gabriel Voisin organisation. Finally, in December 1909, Captain Yoshitoshi Tokugawa flew a French Henri Farman aircraft, with Captain Kumazo Hino flying a German Grade from the Yoyogi military parade ground. These were among the first flights in Japan with Japanese pilots. Meanwhile the committee had farsightedly constructed the first wind tunnel at the Army Telegraph Corps ground at Nakano, from where it was later transferred to Tokorozawa. Japan could now match the Barbarians in aviation matters, with interesting results to come. However, the sleeping giant from across the Pacific was casting envious eyes at the Oriental prospect with notions of commercial aggrandisement.

CHAPTER 2

The Spreading Wings

In the year 1910 the Japanese commenced importing aircraft and equipment from America and Europe. This period was to last until 1915. In the meantime both army and naval aviators were to attain flying proficiency and acquire expertise in ground maintenance, as well as handling. For 1910 the Tokyo government had been able to persuade a reluctant Japanese Diet to vote a budget of ¥480,000 for the military procurement of aviation equipment and services. A Curtis seaplane was purchased from the United States. This machine was a single-float seaplane in which the two-man crew sat side by side with backs to the engine. It was a biplane powered by a single push-engine of 50 hp. Meanwhile the central Provincial Committee for Military Balloon Research had dispatched two officers abroad to qualify as pilots. Captain Yoshitoshi Tokugawa was sent to France, later to return with a Henry Farman biplane. At the same time Captain Kumazo Hino had arrived in Germany, and upon completing his pilot training he returned to Japan with a newly built Hans Grade monoplane. At the end of the year, on 19 December, amid great public jubilation, a demonstration took place at Yoyogi airfield outside Tokyo. Two Army pilots rose into the blue sky from the green airfield, recording the first-ever powered flights of aircraft in Japan. Powered aviation had arrived!

In the following year a budget of ¥670,000 was voted, and a policy decision was made to divide the flying service into two organisations, to provide aerial support for the Army and for the Navy quite separately. At about this time the first naval airfield was constructed at Tokorozawa. During 1912 a naval training airfield was established at Oppama, not far from Yokosuka, in Kanagawa prefecture, as further personnel were sent to Europe to qualify as pilots. Of the officers dispatched, three naval lieutenants went to France for flying instruction, and two further lieutenants arrived at the Glen Curtis Flying School in the United States. Lieutenant Chikuhei Nakajima was to become the founder and president of the Nakajima Aviation Company Ltd upon resigning his naval commission. At the sumptuous naval headquarters in Tokyo the naval general staff had established a Naval Aeronautics Research Committee, otherwise known as the Kaigun Kokujutsu Kenkyu Kai, under the direction of Captain Yamaji Kazuyoshi IJN. This committee was part of the Naval Technical Bureau, and commenced technical investigation on 26 June 1912. Towards the end of the year Lieutenant Yozo Kaneko returned from France, while Lieutenants Sankichi Kono, Chuji Yamada and Chikuhei Nakajima boarded a cruise liner bound for Yokohama from America. Among the trunks were the two French Maurice Farman Longhorn seaplane trainers and two recently purchased Curtis training aircraft. On the forthcoming 2 November 1912, Lieutenants Kono and Kaneko made the first-ever naval flights from Oppama air station in Kanagawa prefecture. The Maurice Farman Longhorn seaplane was a pusher biplane with a front elevator, seating two and propelled by a 75 hp Rolls-Royce Hawk engine. This machine was intended for reconnaissance or training missions.

At the same time, a Deperdussin monoplane was procured, equipped with an engine of 60–100 hp. The Naval Air Service during 1912 had cost the Japanese taxpayer some ¥300,000 and had seen the first flight of a naval aviation formation. For the year 1913 a budget of ¥300,000 had been allotted to the Naval Air Arm, which provided funds for the conversion of the transport vessel *Wakamiya Maru* into a seaplane transport. This vessel had previously been launched some three years previously, but was now re-equipped to carry two assembled and two knocked-down seaplanes, plus auxiliary equipment. Immediately experiments were commenced to provide aerial support to the fleet while offshore and out at sea. The Imperial Navy was now determined to build its own Maurice Farman seaplanes in the old torpedo plants at the Yokosuka naval arsenal. At the same time detailed designs were drawn up for a prototype biplane designed by the Navy's own aviation engineers. However, the procurement of foreign aircraft continued. A Sopwith Tabloid Scout was bought from Britain, a type later to win the 1914 Schneider Trophy. This machine had an 80 hp Gnome Rotary engine that gave the plane a top speed of 92 mph. A Short Type-184 seaplane was also obtained for torpedo-bombing and reconnaissance duties. This aircraft was a twin-float biplane using a 75 or 100 hp engine. Later an Avro 504 biplane was sent to Japan, having a top speed of 62–82 mph.

Operations in the First World War
Meanwhile, with the war in Europe, the Far Eastern political situation had become more fluid, for Japan had attained parity in battleship numbers with the United States, with ten warships apiece. As a result, His Imperial Japanese Majesty's Navy had assisted in the pursuit of the German warships *Scharnhorst* and *Gneisenau* out from Qingdao, China. At the

battle of the Cocos Islands in December 1914 the Japanese Navy had assisted in the sinking of the German cruiser *Emden*. During the summer, while in alliance with Great Britain, the Imperial Navy had attacked the German-Austrian Navy in Jiaozhou Bay and assaulted the fortress at Qingdao. In September 1914 HIJMS *Wakamiya Maru* commenced operations from Jiaozhou Bay, China, attacking German installations with four improved Farman seaplanes. A hangar had been built on the back of the aft deck of the seaplane tender. These seaplanes were three-seaters, maximum speed of 60 mph., altitude ceiling of 10,000 ft and an endurance of four-and-a-half hours. Using these machines, an unsuccessful engagement was fought with a German Rumpler Taube aircraft over Qingdao, and an unsuccessful bombing attack was made on the cruiser SMS *Kaiserin Elizabeth*, dropping naval shells made into bombs by fitting a streamlined tail with fins. Despite these failures, a German minelayer was sunk. In the course of these aerial attacks forty-nine sorties were undertaken, dropping 199 bombs, flown by a total force of seven pilots. Unhappily *Wakamiya Maru* was to strike a mine, which necessitated flying the aircraft from shore establishments. The seaplane tender was withdrawn to Japan in November for refitting and repairs. During this campaign the Naval Air Force had consisted of three Curtisses, two Deperdussins and five Farman aircraft, with a total of twenty pilots all fully qualified. Meanwhile, in Japan during the course of the year, the Imperial Aviation Association of Japan was incorporated with an active committee and a membership of 5,000. The patron was His Highness Prince Kuni, president Marquis Okuma and vice-president Baron Sakatani. The association bought two monoplanes from Germany, later to be used by the Imperial Army over Qingdao, and trained approximately thirty civilian aviators. It was noted in international circles that the association did not join the International Aerial Convention. With a naval aviation budget of ¥400,000, aircraft production and procurement would expand. Lieutenant Nakajima resigned his naval commission and went into business as the president of the Nakajima Aircraft Factory. The head office was in Ohta-Machi, Gumma prefecture, with the Tokyo office at the Yuruku Building, Marunouchi, and production facilities located in Nishi Ogikuko, near Tokyo, with an aerodrome at Ojima, Ohta-Machi. The entire enterprise was launched with ¥6,500,000 capital.

In the following year the Japanese Imperial Navy escorted Australian and New Zealand troop ships to Europe and extended their patrol activities into the south Pacific and Indian Oceans with prior agreement of the British government. At the same time the famous 'Twenty-one Demands on China' were presented by the Imperial government, offering China a Japanese protectorate for the direction of military, political and economic life on the mainland. This was the first violation of the US 'open-door' policy to China to which all major European powers had adhered. But for the Japanese, access to the rich mineral resources of the Chinese Yangtze Valley and the Dutch East Indies was of paramount importance, which the Western powers had hitherto successfully denied her in the Far Eastern scramble. From the year 1915 until as late as 1932, a Japanese aviation copying period was embarked upon. Foreign manufacturing licences for the production of aircraft and engines were purchased while foreign aeronautical engineers were engaged to design on behalf of Japanese companies. Japanese aeronautical engineers were enabled to gain valuable practical experience in the art of aviation design. Thus the first naval aircraft was the Type RO Model A produced at the Yokosuka naval dockyard, designed by Lieutenant Kishichi Mageoshi and Engineer Lieutenant Chikuhei Nakajima IJN (retired). This machine was a seaplane powered by either a Salmson engine of 140 or 200 hp or a Gnome Rhône of 160 hp rating. Meanwhile the Naval Aeronautical Research Committee had been incorporated into the Temporary Submarine and Aircraft Investigation Committee. The Naval Technical Bureau was disbanded and the Second Section of the bureau assumed administration of technical matters relating to aircraft development. A national aviation society was created in December 1915 under the presidency of Lieutenant-General Nagaoka, the reputed Japanese Zeppelin enthusiast, aided by forty public men and scientists, whose intention was to issue an appeal for public funds for aviation development. The society had offices at Yobancho, Kojimachi, Tokyo, and was incorporated into the Imperial Aviation Association in August four years later. All this aerial activity had been accomplished on a budget of ¥500,000, and Nakajima had built three tractor biplanes during the year. During February 1916 the British government, as a result of increased sinking of Allied shipping due to rising U-boat activities in the north Atlantic, found it necessary to ask for increased naval assistance from the Japanese. In reply, the Japanese required an improved attitude towards Japanese nationals in Australia and in British colonies. Assurances were duly given by the British government, and Japanese warships were based at Singapore and elsewhere. British destroyers were withdrawn to the conflict in the more northerly climes. Meanwhile a chair of aviation was created at Tokyo University on a budget of ¥60,000. Four weeks later the Imperial Navy established Naval Air Corps ordinances on 17 March 1916 to concentrate on organising maritime military aviation. The first Kokutai (Naval Air Corps) was established at Yokosuka, formerly the Naval Aeronautics Research Committee, under the command of Captain Shiro Yamanouchi HIJMN. At this air station the first flight training of cadet pilots took place.

While pilot training was in progress the Imperial Navy placed procurement orders for aircraft designed and constructed abroad. These machines included Deperdussin seaplanes from France, as well as Sopwith Tabloid fighter seaplanes and Short 184 reconnaissance seaplanes from Britain. Some Japanese aircraft designs were manufactured by the Ordnance Division of the Yokosuka naval dockyard. The machines constructed were the Yokosho twin-engine biplane, four Hogo-Otsu biplanes and a Farman seaplane. For these activities a budget of ¥760,000 had been voted. At the time the Committee of Aeronautical Research established at Imperial University, Tokyo, planned to build an aeronautical institute, and very shortly afterwards building commenced. For future equipment and construction a fund of ¥630,000 was established to be spent over a period of five years.

In European waters the German U-boat campaign had reached its height by 1917, when the British government asked for further aid from Japan. In return Japan promised additional naval assistance to the already hard-pressed Allied escort fleet. In a diplomatic communication of February 13 1917, the government of Great Britain agreed to the terms, and as a result three Japanese naval squadrons were dispatched to the Indian Ocean, to Australian waters and to the Mediterranean Sea. The ships comprised light cruisers and destroyer escort. The price paid by the Allied governments was to support the Japanese claim to the Shandong province of China and the claim to the German Islands in the Pacific Ocean at the forthcoming peace conference at the end of the war. In the following May the heavy cruisers *Izumo* and *Nisshin*, together with the four destroyer escorts, joined the Imperial fleet. Japan was by now a major naval power whose forces extended beyond the confines of the Pacific Ocean. Meanwhile the Imperial naval staff established an aeronautical section of the Naval Affairs Bureau with the sole responsibility of administering the Aviation Branch of the Navy. At the same time the 2nd Kokutai (Naval Air Corps) was originated at Sasebo, and an all-Japanese-designed and - constructed scout seaplane was successfully launched. This machine was the Rogo-Ko biplane powered by a 200 hp engine. Naval purchases from Britain comprised the Sopwith Schneider seaplane and the Short 320 seaplane. These transactions had been accomplished on a budget of ¥800,000 when Baron Shibusawa and Baron Kondo of the Imperial Aviation Association initiated a collection during May among industrialists which realised an additional ¥1,000,000 for the purchase of land for the laying-out of aviation stations for the Navy. At this time the Nakajima Aircraft Company was being financed by the Kawanishi Company, with the undercover name of Nikon Aircraft Company Ltd, with financial funds supplied from the Army and Navy Budgets. With the success of the Yokosho-type reconnaissance seaplane designed by Lieutenant Mageoshi HIJMN, equipped with a 200 hp Hispano-Suiza engine, and the Rogo-Ko biplane, Japanese designs had become established realities. In England Lieutenant-Commander Kaneko HIJMN had arrived with the intention of studying the new British technique of landing on the deck of aircraft carriers. Posted to HMS *Furious*, he participated in flying activities and aerial research, learning the method of landing and taking-off from this new class of ship. Eventually he returned to Japan, and in the course of writing his report proposed the construction of both aircraft and aircraft carriers. The Imperial naval staff in Tokyo were very impressed with the organisation and administration of the British Royal Naval Air Service, and in particular with the findings enumerated in the reports of Lieutenant-Commander Kaneko, who had returned home in 1918. As a consequence orders were placed with the Sopwith Aircraft Company of Kingston-upon-Thames, Surrey, for the delivery of six Sopwith TI Cuckoo single-seat torpedo-bomber biplanes. These machines were equipped with one 200 hp Wolseley Viper engine, and eventually they became the model for all subsequent Japanese torpedo-bombers. The Navy also purchased the Terrier T3 flying-boat for evaluation. These highly successful acquisitions had been procured on an aviation service budget of ¥3,500,000. However, a Mr Yamashita, described as a 'ship millionaire', contributed ¥1,000,000 towards Army and Navy funds for the purchase of aircraft from the United States and Europe. During March 1918 the Navy established a Naval Aircraft Testing Station at Tsukiji, Tokyo, where the first large wind tunnel was built. In the following month the Imperial Aviation Association purchased land at Ichijima, Tokyo, and commenced the construction of an aerodrome. A member of the Aviation Association – Mr Masao Goto, established a long-distance flight record by flying from Tokorozawa to Osaka, a total distance of 300 miles, in six hours and twenty minutes. On the river Arakawa, a Dr Kishi opened an aviation school in the Tokyo prefecture. Covering an area of 50,000 *tsubo*, or 197,500 square yards, the school opened under the direction of Lieutenant Inouye HIJMN as chief instructor. However, on 11 November 1918 the Armistice Agreement was signed, and hostilities between German and the Allies ceased.

Reparations and acquisitions

After November 1918, as a result of German reparations payments, Japan acquired the manufacturing licences for the large-scale production of duralumin sheet metal and components. At the same time, sample products of German Heinkel airframes, together with BMW and Daimler-Benz engines, were obtained, with detailed drawings and specifications. On detailed consideration by Dr Aikitsu Tanakadate of Tokyo Imperial

University and Rear-Admiral Shiro Yamanouchi in charge of aeronautical engineering administration of the Imperial Naval Technical Bureau, a naval research programme was initiated to exploit the manufacturing potential of the all-metal plane and advanced aviation engineering technology recently acquired. Unhappily, during the course of international negotiations, the United States and China refused to ratify the Treaty of Versailles, and to keep naval parity with the USA Japan proposed to build further warships. The Tokyo government announced a naval building programme that projected the construction of eight battleships and eight cruisers. However, some Japanese Deputies in the Diet felt this programme was beyond the country's industrial capacity. During the course of 1919 the Japanese Naval Air Services were flying various different types of aircraft. These comprised fifteen miscellaneous types of machine of one sort or another, ten Curtiss flying-boats and forty Sopwith biplanes of the one- or two-seat variety equipped with a Gnome engine of 100 or 160 hp, and some fifty Short biplanes, two-seaters with Sunbeam engines of 150 and 225 hp, as well as Farman biplanes utilising a 70 hp Renault engine, and 150 Curtiss aircraft using either Salmson 120 hp or Benz 100 hp engines. The total aerial fleet comprised 265 aircraft. The increase in the size of the Naval Air Service had necessitated the creation of two further Naval Aviation Corps at Sasebo and Kure, with approximately fifty flying-officers on establishment. At the same time the Naval Air Service had acquired sufficient technical expertise to enable its personnel to build the Farman biplanes and install the Renault engines. But with a Naval Air Service budget of ¥12,000,000 for the year 1919, the procurement of machines from abroad continued unabated. From France fifty Nieuport Nightjars or Sparrowhawks were exported to Japan. These single-seat fighter aircraft were the carrier versions of the Nighthawk. From the British Royal Naval Air Service supply came twelve of 156 aircraft exported to Japan, comprising the Parnall Panther with folded fuselages to facilitate carrier storage. At about the same time the Japanese Naval Air Service shared five Sopwith Pup fighters with the Army Air Service after importation from England. Once again a record flight was attempted with the Yokosho-type reconnaissance seaplane, involving a course from Oppama to Kure, and thence to Chinhae, Korea. On the return flight the machine covered the Sasebo–Oppama section in eleven hours at an average speed of 70 mph. Japanese naval officers in Germany had learned of the construction of an aircraft – the Zeppelin R.IV all-metal bomber, designed by Dr Ing. Adolf Rohrbach. Upon careful consideration, the Navy entered into a contract with Dr Rohrbach for the design of six all-metal flying-boats based upon the plans of the land-based bomber. It was planned for these six machines to be manufactured by the Mitsubishi Shoji

Trading Company. At this time the companies commencing airframe and engine manufacture comprised Mitsubishi Heavy Industries, which previously constructed internal combustion engines, as well as Kawasaki Aircraft Ltd, an offshoot from the shipbuilding industry. Both these aircraft companies had aviation divisions centred on the city of Kobe – the Hitachi Aircraft Company, a branch of the Gas and Electric Company, and the Nikon Aircraft Company, now re-established as the Nakajima Aircraft Company – while the Navy built a new naval arsenal at Hiro for aeroplane manufacture, as well as organising aircraft manufacturing facilities at the Imperial naval dockyard, Yokosuka. A further aircraft-manufacturing plant was located at Ootamachi under the title of the Japanese Aircraft Factory. Aviation centres were established at Sasebo and Yokosuka, which possessed a naval flying school. The total personnel establishment at this time consisted of some 331 officers and men. Officers named as controlling the Naval Aviation Services were as follows:

> Rear Admiral J. Matsumura IJN
> Captain S. Yamanouchi IJN
> Captain S. Harada IJN

The flying-officers consisted of:

Commander Y. Kaneko	
Commander S. Fukuoka	
Commander Umekita	
Lieutenants	17
Sub-Lieutenants	27
Plus officers	65
Men	50
Under-training officers	100
Men	69
Superintendent Captain K. Yamagi IJN	
HQ Oihama, near Yokohama	

The French Aviation Mission

Meanwhile the Army Air Service had requested the French government to send out a French Aviation Mission. Very shortly Colonel Faure landed in Japan with sixty-two other French aviation personnel, and among the baggage were Nieuport, SPAD, Caudron, Farman and Breguet trainer aircraft. This mission not only increased the efficiency of the Army Air Service, in which the Breguet trainer was adopted as the Army Type-KO training plane, contributing to the expertise of pilots and air crew as well as ground personnel, but it advised the Naval Air Service on the construction and piloting of large flying-boats.

At this time the Navy commenced constructing the first land-based airfield at Kasumigaura, some fifty miles north-east of Tokyo, sometimes known as the Misty Lagoon. Other naval flying-grounds built included those located at Oihama near Yokosuka, Kure, Kagamigahara, Mikatagahara, Yokkaichi, Fukuoka, Octa, Inage, Haneda and Naruo near Osaka.

In August 1919 the Imperial Aero Club was established in offices located at 1 Chome, Yurakucho, Marunouchi, Tokyo. His Imperial Majesty contributed ¥500,000, and other sums were given by many people. The organisation was to encourage aviation contests and competitions with the sole object of promoting aviation development. The Honorary President was Prince Kuni, the member of the Imperial family who oversaw aviation progress. The first and second vice-presidents were Baron Sakatani and retired General Nagaoki. At the same time the Aeronautical Institute commenced work under the control of the Principal of Tokyo University, having directors who were professors and assistant professors appointed and funded by the Ministry of Education. The research work was conducted by twelve Sections – physics, chemistry, metallurgy, materials, aerodynamics, engines, aircraft, instruments, psychology, etc. The research work undertaken by the Institute was very similar to that conducted by the British National Physical Laboratory. Four months later, on 19 December 1919, the aircraft carrier HIJMS *Hosho* (Flying Phoenix) was laid down, of 12,500 tons displacement and one of a class of two as envisaged in the 1918 Fleet Construction Programme. The sister ship was not to be built, however, despite a grant from the Army and Navy Air Service Budgets of ¥12,000,000 for the financial year 1919.

In 1920 the Navy had undertaken the construction of some naval airships of the non-rigid type. These were similar to the British SS class of airship. They were designed as a submarine scout, with a crew of five and a capacity of 2.83 m^3 of hydrogen gas. With an overall length of 52 m, a width of 11 m and a height of 15.2 m, these ships were able to achieve a maximum speed of 24.6 knots with two Rolls-Royce 90 hp engines. Their endurance was some fifteen hours. For fighting machines the Navy possessed Sopwith single-seater biplanes powered by 110 hp French Le Rhône engines, as well as SPAD single-seater biplanes. Naval reconnaissance was undertaken by three types of aircraft – the Maurice Farman pusher biplane with two seats and a 110 hp Benz engine, the Short Tractor two-seater biplane with 230 hp Sunbeam engine and the Kaigunshiki aircraft designed by the aviators' branch of the Yokosuka Dockyard propelled by a 140 hp Salmson motor. Offensive bombing operations were conducted by such types as the Short Tractor two-seater biplane with a 320 hp Sunbeam engine for torpedo-bombing, the

Kaigunshiki two-seater biplane equipped with either a 200 or 260 hp Hispano-Suiza engine and the Tellier three-seater flying-boat. At the time the Fairey III F Seal Mk VI aircraft was imported from England and had been developed from the Fairey III biplane of 1917, as used by the British Royal Naval Air Service in various versions while in active service. By far the most important development was the importing of the British F5 Felixstowe flying-boat. This machine was to form the basis of the Navy's traditional flying-boat squadrons and to be the archetype for further developments. This machine was equipped with two engines and could carry four 230 lb bombs for an endurance of ten hours. Initially some six or fifteen were imported into Japan, but Japanese manufacture accounted for four constructed by Yokosuka Naval Arsenal, ten by the Hiro Naval Arsenal and forty built by the Aichi Aircraft Company, after licence production had been arranged between the British and Japanese companies involved.

At about this time the Japanese Naval Staff decided to approach foreign governments with the request that aviation missions should be dispatched to the Home Islands for the purpose of training the Imperial Naval Air Service. A French Army Air Service mission had already arrived in Japan. Although it was expressly concerned with the development and reorganisation of the Army Air Service, valuable help had been extended to the Naval Air Service. Meanwhile the Naval Staff had come to a unanimous decision to model the Japanese Naval Air Force along the lines of the British Royal Naval Air Service. Some twenty years later, historical events were to prove themselves correct when Japan was facing the overwhelming numerical superiority of the Barbarian West, and traditional RNAS ideas and organisation enabled the Japanese Navy to hold out as long as in fact was the case. Ironically, these were ideas and organisations that the British had foolishly thrown away, and this was to cost them dear in numerous ships sunk, thousands of lives lost and an Empire forfeited to the influential political whims of a friendly Allied power bent on economic aggrandisement.

The British Aviation Mission and Aviation Engineers
Admiral Baron Kato IJN, the late chief delegate of the Japanese mission to the Washington Conference, had approached the British government through Rear Admiral Kobayashi, the Japanese Naval Attaché in London, for an aviation mission made up of Royal Air Force officers to go to Japan. The government in London delayed a decision for some time as hasty consultations were made among various Departments of State in Whitehall. Subsequently it declined the request owing to a shortage of personnel! Eventually, a further request was made, this time for a private

mission from England. The British government was asked to nominate suitable officers, and selected the distinguished Colonel the Master of Semphill to command the mission. At the same time the Navy requested the British company Short Bros of Rochester, Kent, to send a technical group to Japan. This unit comprised twenty engineers and was led by a Mr Dodds. It was ultimately responsible for assisting Japanese companies to manufacture the Short F5 Felixstowe flying-boat. At last the thirty pilots and engineers under Colonel Semphill assembled for the voyage to Japan, but at the time the Handley Page Aircraft Company of Cricklewood had become involved in financial difficulties. Located in the semi-rural home counties of England, this company had been responsible for designing the Handley Page 0/400 and 0/200 series heavy bombers for the purpose of conducting deep-penetration raids into Germany during 1918. Unfortunately the Armistice had been concluded before the independent force of heavy bombers could be organised and aerial attacks could commence. In the meantime, a multi-slotted-winged passenger monoplane had been projected under the direction of Mr George Volkert as a principal engineer, who had been with Handley Page during the previous nine years of aviation development work. At the time it is alleged that Handley Page Ltd had a bank overdraft of £400,000. An associate company, the Aircraft Disposal Company, was owed £40,000 from aircraft sales, and Handley Page owed shareholders a sum of £336,000. A tremendous row ensued upon the public revelation of the financial position. The directors of the Aircraft Disposal Company refrained from liquidating Handley Page in return for the right to nominate the management of the main and associated transport company. Handley Page agreed to these conditions and waived the rights to a sum of £176,000 owing to him from the company, while his brother Theodore was forced to resign. The Bank of Scotland had been the main creditor, so that four new directors were appointed and Lieutenant-Colonel J. Barrett-Lennard was appointed general manager. At this point George Volkert resigned, at about the time that Major Brackley, a former senior manager, called in at the office to say goodbye. Major Brackley was leaving with the pilots and engineers for Japan with Colonel Semphill's mission. He expected to be away for several years, and since the mission was short of technicians and engineers he persuaded George Volkert to join the team. Since Volkert knew Colonel Semphill as a pilot and had flown the first US-built 0/400 bomber, he was promptly made chief of design and inspection. As a special gift, Handley Page gave Volkert all the bound volumes of the Aviation Advisory Committee's reports, since a cash token could not be given. The Advisory Committee's reports were worth their weight in gold, containing valuable technical information for aviation development.

Having been picked from students of the Northampton Institute College from before the First World War, and trained by Handley Page, George Volkert's departure was a disastrous loss to British aviation. At this time Herbert Smith, with the entire design team from Sopwith Aviation Ltd of Kingston-upon-Thames, Surrey, left for Japan to join Mitsubishi, as did aeronautical engineers from Junkers and Rohrbach in Germany, including a Dr Alexander Baumann. Dornier aviation experts from Germany went to the Kawasaki Aircraft Company, as did Dr Richard Voigt from Blohm and Voss GmbH, Hamburg. From Breguet and Nieuport French engineers went to Nakajima, and Prof. Lachmann from Germany joined the Ishikawajima Aircraft Company. Thus Japan had been able to invite European aviation experts to join her expanding aeronautical industry simply because they were out of work or members of now bankrupt aeroplane companies. The ignorant, uninformed and short-sighted policies pursued by British politicians had resulted in wasted and financially starved companies. Unemployed technicians and the loss of the original scientific aviation research work was the result.

In 1920 the Japanese aviation industry was a rapidly expanding series of small business firms, enterprising and located in inappropriate factories within city limits or to be found in fields surrounded by pleasant countryside. The multiplicity of firms may be gauged from the following list. The Akabane Aeroplane Works, Akabane, Tokyo, manufactured Kishi engines and aeroplanes. Also in Tokyo was the Haneda Aeroplane Works at Haneda, while M. Ichimeri made aeroplanes at Tengachaya, near Osaka. The Aichi Aeroplane Works was situated at Tsudanuma, in the Chiba Prefecture, and the Japan Aeroplane Works Company at Ootamachi in Gumma Prefecture manufactured Nakajima aircraft. The Japan Automobile Company, at Tameika, Akasaka, Tokyo, made aeroplanes and engines, and the Aviation Division of the Kawasaki Dockyard Company, Higashi Kawasaki-Machi Kobe, manufactured machines under its own name. Balloons were made at Osaki-Machi near Tokyo by Kiku Seisakusho. Perhaps the most important manufacturer, the Mitsubishi Dockyard Company, had organised an aviation department at Wadasakicho, Kobe, but was later to expand elsewhere. R.K. Oguri was making aeroplanes at 2 Naka-Sarugakucho, Kanda-ku, Tokyo, and the Tokyo Gas and Electric Engineering Company Ltd of Ootamachi, Kojimachiku, Tokyo, has diversified its activities and was now actively making aeroplanes and engines.

In the year 1920 the formation took place of the Mitsubishi Kokuki Kabushiki Kaisha, or Mitsubishi Aircraft Company Ltd. This company already possessed an aircraft division at Wadasakicho, Kobe, but such facilities proved inadequate, and so a separate associated company was created for the express purpose of designing and constructing aeroplanes. The head office was in Marunouchi, Tokyo, with manufacturing plants situated at Shibaura and Nagoya, the latter being the largest factory in Japan for the construction of aircraft. The chairman of this new company was Vice-Admiral K. Funakoshi HIJMN (retired), with Managing Director Y. Shibuya operating a complex with a paid-up share capital of ¥5,000,000. Flying and testing was concentrated at Nagoya, with a large airfield laid out alongside the extensive factory. This new company had been evolved from the Kobe Internal Combustion Company of Mitsubishi Shipping and Engineering Ltd, sometime manufacturers of heavy-oil engines and other heavy engineering products. At about this time the company of Sale and Frazer Ltd had been established in offices in Marunouchi, Tokyo, as the largest import agent of engineering items, including aeroplanes, from England. Among the representative agencies held was one for the importation of Handley Page aircraft from England, initially for civilian usage only. Other companies now entering the aviation field included the Kawanishi Aircraft Company, the Aichi Aircraft Company and the Tachikawa Aircraft Company. All had been previously in some other engineering field: Kawanishi had manufactured machinery, Aichi produced clock and electrical equipment and Tachikawa had been in the shipbuilding industry. Meanwhile Tokyo University had set up a Chair of Aviation, and at the same time had purchased land in Etchujima, Fukagawa, Tokyo, for the building of college facilities. The university had purchased from the United States one 400 hp Liberty engine; and from England a 350 hp Rolls-Royce engine and one de Havilland DH4a aeroplane also had been procured.

During 1920 1st Lieutenant Magashi HIJMN designed a biplane seaplane trainer to replace the now ageing Maurice Farman aircraft. The new design proved so successful that seventy of the new machines were built by the Yokosuka Naval Arsenal. A Sopwith Pup machine was successfully flown from a ramp on the forward deck of the seaplane carrier *Wakamiya Maru*, which at the time was the first take-off by any naval personnel from the deck of a ship. In July 1920 the Naval Technical Bureau was reorganised, and the 6th Section became responsible for aircraft affairs. During the following month the new Hiro branch of the Kure Dockyard was created to investigate aeronautical research and the normal production of aircraft. The new aerodrome being established on the lakeside of Kasumigaura was to have an area of 1,200,000 tsubo, one tsubo being equivalent to 3.305785 m^3 (3.95383 yd^3). Eventually the equipment for the Sasebo and Kure Naval Air Corps was completed, and the Navy now possessed three squadrons of planes with a further five projected squadrons, later increased to nine to be completed by 1922.

Within one year this number was to be raised to seventeen, and for the year 1923 it would consist of 284 machines – six squadrons of land-planes, nine squadrons of seaplanes and two squadrons of training planes.

The Navy Flying School was located at Oppama Airbase near Yokosuka. Preliminary flying training was conducted using Maurice Farman pusher seaplanes, advanced training utilised the Kaigunshiki plane, the Short, Sopwith or other tractor biplanes available, and at times even the flying-boats. The officer commanding the flying-boat training was Vice-Admiral Torao Kuwabara, under whose command F5 Felixstowe flying-boats flew at Yokosuka and Sasebo. This officer later became a director of Shin Meiwa Industries Ltd in 1967, when development took place of the only known successful flying-boat design of the period to go into squadron service outside the Soviet Union. At the time the personnel organisation included the following officers:

> Rear Admiral S Yamanouchi Chief CO
> Rear Admiral S Yoshida
> Captain Umekita
> Commander   S. Fukuoka
>   K. Hanashima
>   H. Hayashi
>   K. Hirabayashi
>   Y. Kaneko
>   S. Kono
>   Y. Yashima

Aviators were:	
Lieutenant-Commanders	4
Lieutenants	49
Sub-Lieutenants	30
Total Number of Aviators	360
(Training	120)
(Flyers	240)

The unified Imperial Air Force concept

With the gradual build-up of two separate and independent air forces, thoughts among many leading officers were directed towards the advantages of an air force solely responsible for all flying activities. In 1920 a joint Army and Navy Co-ordination Committee was selected to investigate the problem. Regrettably, both the Army and the Navy tended to view the committee's findings with contempt, and as a result Army/Navy co-operation was seriously threatened. Eventually the committee broke up with no agreement having been reached. However, all was not lost, for some senior officers of the Army and Navy unofficially continued discussions with a view to the eventual reuniting of the two separate flying organisations. In the main the talks were led by the Army, for the Navy was very suspicious of losing its Naval Air Service. A compromise suggestion was reached whereby an Air Force would be organised that would not be completely independent but would be subject to some overall direction by the Army and by the Navy. Subsequently the talks broke down, and nothing further came of the affair until the late 1940s.

The Navy had organised two Naval Air Corps, each equipped with two combat squadrons of six aircraft each and one training squadron of eighteen aircraft. The 1st Corps was located at Oppama and the 2nd Corps at Kure. The Fleet Flying Squadron was attached to the 1st Battleship Fleet normally based at Hiroshima. Meanwhile the Naval Air Service was studying day and night bombing and anti-submarine operations, as well as introducing the use of airborne radio communications. Beside possessing airbases at Oppama and Kure, landing-grounds had been established at Sasebo, Ise, Oihama, with a projected airbase to be constructed at Kasumigaura, north-east of Tokyo.

During 1920 a great deal of money was received from Japanese business men abroad, mainly those residing in the USA, for the promotion of the aims of the Imperial Aerial Society or Society of Naval Aviators. Meanwhile an anonymous American living in New York donated the sum of ¥180,000 in cash to the Imperial Aerial Society. In the course of correspondence he had indicated that he considered the Japanese as the finest aviators in the world, due to their being small and quick. He requested that his donation be used for competition purposes and the development of training schemes. The letter forwarding the donation was considered to have come from a wealthy aviation enthusiast of American origin. Also from America came a Curtiss Oriole aircraft equipped with a 400 hp engine, as well as a Vought machine with a 150 hp Hispano-Suiza engine, both aeroplanes being purchased by a Mr T.K. Oguri. The Tokyo Gas and Electrical Engineering Company of Marunouchi had obtained special permission from the British government to import one Rolls-Royce aero-engine. The Navy had built 200 aircraft of the Maurice Farman type with modifications for the training purposes – a two-seater machine with either a 70 hp Renault or 90 hp Curtiss engine. Facilities for this programme had been established in an Aircraft Division at the Yokosuka Naval Arsenal by 1921.

CHAPTER 3

To the Misty Lagoon

In 1921 the organisation and administration of the Japanese Naval Air Service was spread out over various departments of the Navy Ministry, with no centralised air department for the Navy as a whole. As a consequence the Minister of the Navy had to co-ordinate all aviation matters exercised through the Naval General Staff. Thus the group concerned with aircraft materials was the 6th Section of the Bureau of Supply and Research, while aviation policy was formulated by the 3rd Section of the Bureau of Military Affairs. Likewise, training was the concern of the Air Section of the Bureau of Education. At the same time the Army and Navy was once again considering the question of a unified Air Force, very much at the instigation of the Army. The naval opposition was led by two very conservative senior naval officers; Admiral Baron Kato was Prime Minister and Minister of the Navy, having been Chief of Staff to Admiral Count Togo at the Battle of Tsushima, when the Russian fleet had been destroyed. Likewise Admiral Yamashita, the current Chief of the Naval Staff, had been Director of Operations at the time of the historic battle. But if conservatism hung like a cloud at the top, enlightenment shone from below, for in the following year a group of naval officers were sent to Germany under the command of Lieutenant Misao Wad, who had recently completed a three-year study course at Tokyo Imperial University. Accompanying the naval party were two engineers, sixteen mechanics, two army engineers and three Mitsubishi engineers. Later a further group was to arrive in Germany to study metallurgy, led by Engineer Captain Tokyi Ishikawa HIJMN, with six Army and six naval engineers, as well as four Sumitomo Company engineers. The purpose of this visit was to study the manufacture of Duralumin as used in the production of aircraft.

Meanwhile Mitsubishi had changed the name of part of its organisation to Mitsubishi Internal Combustion Engineering Company. On 15 February 1921, five British aeronautical experts accompanied by others had arrived in Japan for the express purpose of erecting an aviation department at the invitation of the Mitsubishi organisation. This group was led by Herbert Smith, the former chief designer of the Sopwith Aeroplane Company of Kingston-upon-Thames, in Surrey. This company had been merged with a motor car group owing to the lack of funds and orders from the British government. Subsequently all the engineers and technicians had been made unemployed, and realignments to utilise their talents had not been forthcoming. The chief designer had been accompanied by John Brewsher and Ernest Comfort, two leading engineers who had been responsible for initiating new designs. An embryonic works staff also went out to Japan under Jack Hyland, the chief production engineer, with R.A. Lippscombe and B.W. Walters of Tinsmiths. Also in the party was J.A. Salter, a fitter, and A.E. Venn, a rigger, as well as Captain W.R. Jordan DSO DFC, an ex-Royal Air Force test pilot. These gentlemen were to stay in Nagoya, Japan, for some three years for the purpose of introducing basic designs, initiating construction principles and supervising the building of a large aircraft factory outside the city of Nagoya for the Mitsubishi Company.

The fair-haired, steel-blue-eyed Herbert Smith was born at Bradley, near Skipton in Yorkshire, on 1 May 1889. His father was the chief accounts clerk with the London and North-Eastern Railway Company, and Herbert had been educated at Bradley and Keighley Grammar School. He then attended Bradford Technical College, studying mechanical engineering, obtained a diploma and joined the company of Dean, Smith and Grace Ltd, of Keighley, manufacturers of machine tools. Subsequently, he obtained design- and drawing-office experience with the firm of Smith, Major and Stephens of Northampton, makers of lifts and associated equipment. Then he joined the British Aviation Company under Captain Frank Barnwell and M. Henri Coanda, and eventually accepted the position of leading draughtsman with the Sopwith Aviation Company at Kingston-upon-Thames in 1914. Rising rapidly to assume the position of chief designer, he was responsible for the design of the Sopwith Pup and Camel fighters, the 1½ Strutter, the Sopwith triplane, the Dolphin, Snail, Snipe and Salamander aircraft. By September 1920 the Sopwith Aviation Company went into voluntary liquidation due to lack of work and lack of government interest. In November the Hawker Engineering Company had taken over the plant and production facilities, believing that the field of aviation in the United Kingdom was finished. The whole idea for the employment of the Sopwith engineers and technicians had been conceived by the Japanese Navy in a masterly stroke. The Aviation Department of the Mitsubishi Dockyard Company had been established at Wadasaki-cho, Kobe, but the new aircraft factory was located at Nagoya on an estate of over 700 acres. This established the Mitsubishi Aircraft Division as the largest producer of aircraft, with the largest production facilities, in the Orient. As well as these facilities and plants, Mitsubishi had established a satellite factory complex at Shibaura, near Tokyo. Meanwhile the Vickers Armstrong Company had opened offices in

Nagoya under the management of Major Winder, a former British Army officer whose task was the promotion of armament sales in Japan. This office was to channel weapons and armament designs for the building-up of the new aeronautical departments of the Nipponese Army and Navy.

Immediately the new aircraft design team and the Vickers Sales Department personnel commenced work in earnest. The French Hispano-Suiza aircraft engine was imported and a licence obtained for the production of this motor in Japanese factories. At the same time four basic aircraft types were designed and built, and a flying development programme initiated for the Japanese Navy. The Mitsubishi Type 1 aircraft was similar in design to the Martynside F4 machine, but with a Sopwith outlined tail. Equipped with a Mitsubishi licence-built 300 hp Hispano-Suiza, the fighter plane achieved a maximum speed of 145 mph. Armament comprised two Vickers machine-guns firing through the swirling air-screw and controlled by interrupter gear operated from the gun button by the pilot. The Mitsubishi Type 2 aircraft was similar to the Type 1, but was a two-bay biplane accommodating a pilot and observer/gunner. Power was obtained from a similar engine installation as that fitted to the Type 1 aircraft. The third machine was the Mitsubishi Type 3, a triplane of wide wingspan and divided undercarriage legs to facilitate torpedo dropping. This machine was powered by a 450 hp British Napier Lion engine considered to be the leading liquid-cooled aero-engine of the day. The final design, the Mitsubishi Type 4, was a two-seater biplane also powered by the 450 hp British Napier Lion engine, and designed for torpedo carrying by means of a wide split undercarriage. All model types were test flown by Captain William Jordan, formerly of the Royal Flying Corps, and now working for Mitsubishi. Later he was to land a Type 10 fighter machine on the deck of the aircraft carrier HIJMS *Hosho* in February two years later. In March 1921, 1st Lieutenant Tochiichi Kira carried out similar experiments successfully. By December 1923 the Naval General Staff had been convinced of the feasibility of training operational carrier pilots to fly machines capable of working with the fleet. The Navy failed to adopt the Type 1 because it was too heavy to be practical, and as a result only twenty examples were built. The Navy ordered Mitsubishi to redesign this aircraft, and this was undertaken by Herbert Smith; it was successfully flown sometime in 1923. The Type 2 MT became the Type 13 carrier attacker, of which 442 examples had been built by 1933. The Type 3 MT was a three-seater constructed to a total of 402 machines, of which forty were manufactured by the Hiro Naval Arsenal. In 1924 John Brewsher and Ernest Comfort resigned from the Mitsubishi design team to return to England, where they joined Vickers Aviation Company as senior designers. While on the staff of Vickers, John Brewsher was responsible for designing and developing the new all-metal wing. He was the chief designer inspiring the design of the Vickers Jockey, the COW Gun Fighter and the Vickers Venom fighter plane. This latter machine possibly inspired the design of the German Focke-Wulf Fw 190 fighter and the Japanese Mitsubishi Zero of the 1940s. With the year 1924, Herbert Smith's contract with Mitsubishi came to an end, and he returned to the United Kingdom to retire from the aviation industry for a well-earned rest.

The 1920 Semphill mission, consisting of thirty officers and men, operated from Kasumigaura and Yokosuka. It was instrumental in the formation of the Temporary Naval Flight Training Troop under the command of Rear Admiral Tadaji Tajiri HIJMS. The British aircraft detailed to be used by the Group included the Avro 504 land-planes and seaplanes, Sparrow Hawk fighters and Panther reconnaissance planes. Included in the aircraft inventory were Swift and Cuckoo torpedo-planes, and Felixstowe, Viking and Seal flying-boats on which most of Japan's naval pilots received instruction. By the end of January 1921 Colonel Semphill had come to an agreement with Rear Admiral Tadaji Tajiri in London concerning the organisation of the mission. At the age of 23 or 24, Colonel Semphill was to embark upon a most exciting and interesting operational career.

By agreement with the Japanese Naval General Staff in February 1921, the eighteen officers and twelve warrant officers of the former Royal Naval Air Service were granted acting ranks in the Imperial Japanese Navy equivalent to those held in the Royal Navy. Originally the contract was to cover one year, but eventually the arrangement was to last three years, and in some instances five years.

The command of the mission was under the direction of Colonel the Master of Semphill, with Commander C.H. Meares as second-in-command and Squadron Lieutenant R.M. Brutnell as chief executive officer, assisted by Warrant Officer 2nd Class J. Hunter. Activities were conducted by five separate sections – flying, technical, armament, photography and medical.

Flying was sub-divided into four groups, with Lieutenant-Commander F.B. Fowler in charge of flying training. Flying-boats and seaplanes were under the control of Lieutenant-Commander H.G. Brackley and Lieutenant M.J.M. Bryan, while scouts were under the command of Lieutenant R. Vaughan Fowler and preliminary flying training was in the hands of Lieutenant A.G. Loton. The ship- and torpedo-planes, as well as fleet co-operation aircraft, were directed by Lieutenant-Commander C.H.C. Smith.

A Technical Group had been organised under the command of Lieutenant-Commander F.C. Atkinson, comprising four sections, with Engineering directed by Squadron Lieutenant A.S. Sheret and five warrant officers; an Aircraft Technical Section was commanded by Squadron Lieutenant E.C. Landamore, assisted by four warrant officers; a Design and Inspection Section was controlled by Squadron Lieutenant G.R. Volkert, assisted by Warrant Officer 2nd Class F.E. Sherras; finally a Technical Instruction and Parachute Section was under Lieutenant-Commander T. Orde.

Two officers controlled the Armaments Group – Lieutenant-Commander H.J. Eldridge and Squadron Lieutenant A.W. Hatfield, while the Photographic Reconnaissance Group was commanded by Lieutenant W. Pollard, assisted by WO2 S. Manton. A Medical Group had also accompanied the Semphill Mission under Surgeon Lieutenant W.F. Jones and WO2 Writer.

In assisting the British mission, the Japanese authorities placed no major restriction on the use of equipment or training in promoting the efficiency of the new air service. Equipment for the mission had been collected from a wide selection of British aviation manufacturers. The Avro 504K and 504L seaplanes were made by Messrs. A.V. Roe, while the F5 flying-boats had been constructed by Short Bros. The fifty Gloster Sparrow Hawks had been redesigned from the Gloster Nighthawk by Henry Folland, chief designer to Sir Samuel Waring, Nieuport and General Aircraft Company, while the Supermarine MK II Channel flying-boats, powered by Puma engines, had been constructed on the Solent at Southampton, on the south coast of England.

On 14 March 1921, a Japanese naval delegation visited the Supermarine Aircraft Works at Southampton for a demonstration flight by the company's Channel flying-boat. At the time a full gale was blowing and a strong tide running with waves four to five feet high. The flying-boat was launched into rough seas with the Japanese Air Attaché, as well as the Chief of Inspection, Naval Air Service, aboad. The demonstration flight was under the control of Captain Henri Biard, Supermarine's chief test pilot, who ascended into the air from the rough water in five seconds. Flying around the Isle of Wight, the machine re-alighted in the heavy seas of Southampton Water to taxi one and a half miles to demonstrate sea manoeuvrability, subsequently disembarking the passengers. The result of this flight was a Japanese naval order for three Channel MK II flying-boats equipped with 230 hp Siddeley Puma engines.

Two days previously an advanced party of the British air mission had left England for Japan, taking the quickest route, via the USA. It was led by the second-in-command, with three flying officers and two engineering warrant officers. The equipment, on being purchased, was inspected and shipped out in sections, and by the middle of March the bulk of the personnel were on their way. In April much of the scheduled equipment had been purchased and was available in Japan. It is interesting to examine the list of British companies that had supplied the mission with its requirements. The suppliers included:

A.V. Roe Ltd
Armstrong Whitworth Aircraft Company
Blackburn Aeroplane Company
Butchers Ltd
Disposal Company
Geo. Parnall & Company
Gloster Aircraft Company
Metal Air Screws Ltd
Monarch Eng. Ltd
Napier & Sons
Nobel Industries Ltd
Rolls-Royce Ltd
Short Bros Ltd
Supermarine Aviation Company
Vickers Ltd
Yorkshire Steel Company

During April 1921 the advanced party of the training mission was able to inspect the site of the naval air station located at Kasumigaura – translated as 'the lagoon in the mist' – on the shores of Lagoon Kasumigaura some forty miles north-east of Tokyo.

The aerodrome, on one mile of lagoon shoreline, was located on a raised platform of land approximately 700 acres in area, with available spare land for future expansion up to a total of 1,400 acres if necessary.

The seaplane station was built on the lagoon shore, a non-tidal area of sixty square miles. It was some twenty-five to thirty miles long and three-quarters of a mile wide, with unique training facilities for seaplane and flying-boat operations. A light railway connected the airbase with the local railway station, road and water facilities. The quarters, specially built for the mission, were excellent, with recreation facilities that included a billiard room and tennis courts.

Machines and equipment were now arriving in Japan. The aircraft that were to be used under service conditions included the Blackburn Swift torpedo carrier, the Supermarine Seal amphibian spotter and the Vickers Viking fleet spotter. Overall some 200 aircraft were originally ordered from manufacturers, and the aero-motors included in the order were the Lynx and Jaguar engines. Aircraft spares and engine replacement components were also sent. Other spare parts included floats, propellers, Hulks Ford Starters, electrical equipment, instruments, fabrics, dopes, tools, bolts, nuts, oil, tubing, spark plugs, armaments, photographic and medical equipment.

The original concept of the training plan had concluded that preliminary training would take place with all pilots using the Avro 504K with the 110 hp Le Rhône engine, and seaplane pilots training on the Avro 504L with the BR1 engine. Scout training pilots would use the Sparrow Hawk aircraft powered by the BR2 engine. Originally, Nighthawks had been selected, with Dragonfly radial engines, but these machines proved unsuccessful. Preliminary training for flying-boat pilots would be conducted on the Supermarine Channel type, using Siddeley Puma engines, and also F5 Felixstowe flying-boats with Eagle VIII engines. For fleet co-operation work, Sparrow Hawk machines would be used, and for torpedo-dropping flights the Cuckoo aircraft with Viper engines.

During July 1921 all mission personnel were now present in Japan except one officer who had remained behind in England to conclude purchasing, inspecting and dispatch of equipment to the mission in Japan. Meanwhile the mission busied itself with planning the building layout, at the same time as aerodrome preparations commenced, clearing, levelling and draining some fifty acres of land. This work was undertaken by the Works Department of the Japanese Admiralty. At the end of July 1921 a Shinto ceremony took place on the completed airfield. This was watched by 30,000 spectators, many of whom had walked many miles to be present at this religious ceremony in which the aerodrome was consecrated to the use of the Imperial Forces, and as a symbolic sign a propeller was used on the Shinto altar. Preliminary training could now commence, with picked Japanese pilots assisting the instructors flying Avro 504 trainers, utilising the Gosport System of preliminary flying training. At about this time the Aichi Electric Machinery and Watch Works opened an aviation department for the manufacture of seaplanes. The manufacturing facilities were under the control of Captain Umitani HIJMN, a former naval air officer, and now the director and chief engineer of the Aviation Department of the company. Also in July, the USA had suggested to the Great Powers that a disarmament conference should be convened with the express purpose of limiting the competing naval building programmes. In consequence the Imperial Government dispatched Admiral Kato Tomosaburo, the Navy Minister, to Washington, DC.

In September a new course of flying instruction had commenced at Kasumigawa, with twelve flying officers, one pilot from the Army Air Service and concurrent courses operating engineer, armament and photographic classes for officers. At the same time courses were run for warrant officers, petty officers and ratings.

The Washington Naval Conference was opened by President Harding in November 1921. The Japanese delegation was led by Prince Tokugawa, assisted by Baron Shidehara, the Japanese Ambassador, and Admiral Kato Tomosaburo, the Navy Minister. For the United States the US Secretary of State Charles Evans Hughes led the conference, proposing reductions in capital ship tonnage by 40% by scrapping existing ships as well as accepting a ten-year stoppage to naval building. This would mean that the USA and Britain would possess 500,000 tons of shipping and Japan would accept 300,000 tons.

The Japanese were furious at this discrimination on the part of the United States, especially since the chief delegate from Britain, Mr A.J. Balfour, conceded to objections of the Americans to the 1902 Treaty between Britain and Japan, and proposed in its place a new treaty between the USA, Britain and Japan. But even to this diplomatic arrangement the Americans were hostile, and ultimately a new treaty, much weaker than the 1902 settlement, was settled between Japan, Britain, the USA and France. This agreement merely pledged the four powers to respect each other's insular rights in the Pacific and to consult in the event of a conflict involving any of the signatories. It was to come to pass that no consultation ever took place on the part of the USA, and it is doubtful if it was ever intended by the Americans. From the outset, Charles Evans Hughes had been blunt, dictatorial and ruthless in his pursuit of American domination over Japan and Britain. His diplomatic assault was to upset the balance of power enjoyed hitherto and pave the way for a conflict in the future.

The Japanese delegation insisted on a halt to all fortification construction being undertaken and naval base building in the Philippines, on Guam and in Hong Kong. Thus no naval bases were to be built within operational distances of Japan, no US bases west of Hawaii and no British bases north of Singapore. In return Japan had to release possessions conquered on the Chinese mainland in Shantung province and in East Siberia. It was accepted that capital ship building was to be 5:5:3 in favour of the USA and Britain.

There can be no doubt that Balfour and the British delegation betrayed Japan and the 1902 Treaty to align themselves with the USA. This left Japan isolated and threatened, and in the eyes of many Japanese officers war was unofficially declared at this time, though not verbally. As regards the British, they would be allied to a slumbering superpower, which would lead not only to the destruction of the Nipponese Empire, but eventually to the disintegration of the British Empire, with calamitous results for all of Britain's overseas friends. Britain would spend years in repenting in poverty, powerlessness and political impotence. For the Japanese the Barbarians were closing in.

CHAPTER 4

The Heavenly Castle

The Japanese, and especially the Nipponese naval expansionists, were entirely hostile to the Naval Treaty of 1921, since the ship ratio of 5:5:3 was inferior to the ratio of 10:10:7 as proposed by themselves. Unfortunately the Washington Naval Treaty and the amended Treaty of Friendship had been agreed without consultation with Japanese senior naval officers, one of the delegates claiming that war commenced the day the treaty was ratified.

Though the Naval Treaty acknowledged Japan's naval inferiority, the agreement did not include provision for submarines or torpedo-boats and other classes of warships – only the battleships. This permitted funds to be spent on aircraft carriers, cruisers and destroyers. Thus the naval development plan was able to commence in earnest. This top-secret naval plan was worked out at the Emperor's lodging-house by the Naval General Staff and others late in 1922, subsequently being agreed upon by His Imperial Majesty, the Emperor Hirohito. National resentment had been generated over the refusal by Britain to renew the Anglo-Japanese Alliance of the previous year, and the ruling of the US Supreme Court in 1922 in making Japanese ineligible for American naturalisation. Resentment reached a high peak when two years later the US Congress enacted legislation excluding Orientals from US immigration schemes.

Meanwhile, take-off trials had been conducted by the Imperial Naval Air Service with aircraft from a ramp over a gun turret on a battleship, and later from a cruiser using a Type 10 fighter, a Parnall Panther and a Gloster Sparrow Hawk. Since no provision could be made for landing these aircraft, once taken-off, the experiments were soon discontinued. Such tests could only herald the building of the first aircraft carrier, which was completed in December 1922, HIJMS *Hosho*. This ship displaced 7,470 tons and was 542 feet long; a small bridge was located on the starboard side, with three small funnels abaft the navigation bridge. The carrier attained a maximum speed of 25 knots while on trials, and housed seven fighters, ten attack-bombers and four reconnaissance aircraft in a hangar below decks, with access to the flight deck by two hangar lifts. During flying operations the three funnels could be swung down, and the pilots were assisted by the use of mirrors and lights during landing operations. Early flying tests were conducted by Major H.G. Brackley and Captain Jordan RN, who flew the Supermarine Seal and Vickers Viking from the deck of the carrier. This was the first purpose-built ship for aircraft operations in the world. Later tests were undertaken using the Blackburn Swift MK II in torpedo dropping and landing on *Hosho*. In low-flying exercises over the Pacific Ocean, it was found necessary to reduce the compression ratio of the Lion engines to 5:1.

During 1922 the Muroran Seiko Sho Iron and Steel Works Ltd commenced the manufacture of aeroplane engines at Muroran, Hokkaido. At about this time the Tokyo Gas and Electric Company was concerned with general engineering and motor cars, having obtained the licence to build Le Rhône engines in France. The Japan Aeroplane Works & Company of Ota-machi, Gumma Prefecture, opened a factory for the production of fifteen aeroplanes per month under the technical direction of Captain C. Nakajima HIJMN – a company later known as the Nakajima Aircraft Company Ltd. The Kawasaki Dockyard Company Ltd of Higashi-Kawasaki, Machi, Kobe, had opened an aviation department, building Salmson biplanes under licence and manufacturing aeroplane engines. The engineer in charge of the Design and Technical Department was a Mr G. More, an ex-lieutenant from the French Aviation Mission. Late in 1922 a most important event took place, with the launching of the aircraft carrier HIJMS *Amagi*, or *Heavenly Castle*. Having previously been launched in July 1920, with the intention of being constructed as a battleship, she was never finished, due to the provision of the Naval Treaty, and she was finally launched as an aircraft carrier.

On 11 November 1922 the Eleven Club was formed in Tokyo by Marquis Koichi Kido. The club was so named after a meeting that took place on the eleventh day of the eleventh month in the eleventh year of the reign of the Emperor Taisho. Marquis Kido, who served as Lord Keeper of the Privy Seal from 1940 to 1945, was the son of Emperor Hirohito's childhood foster-father. The club commenced with three members, but was increased to twenty under Marquis Kido's tactical direction, and met regularly in one of Tokyo's more exclusive restaurants until the night of 11 January 1945. The membership consisted of very influential persons who organised the Emperor's policy requirements at government level and subsequently took greater responsibility for the direction of the country's affairs. It was in fact the co-ordinating centre of the so-called 'big brotherhood', which were the brains of Hirohito's cabal. This cabal was the nerve centre of the Anti-Barbarian Expansionist Movement, and its operations were initiated prior to Prince Hirohito's ascending the Imperial Throne.

During 1922 the Naval Air Service had flown 4,815 hours in 11,162 flights, though unfortunately 111 machines had been wrecked in accidents. Due to poor maintenance, ninety-eight aircraft had suffered engine failure, resulting in a total of four flyers being killed, while eleven had sustained injuries.

In 1923 HIJMS *Hosho* entered service as the world's first purpose-built aircraft carrier, but the planned sister ship, HIJMS *Shokaku*, was never built. Meanwhile *Hosho* underwent a refit, the island bridge was removed and the ship became a flush-deck carrier. Later she was removed from front-line duties and became a training carrier. During the course of developing and working-up HIJMS *Hosho*, Major Brackley had been most prominent, so that by February 1923 he was appointed Air Adviser to the Emperor's Navy.

Meanwhile the Nipponese island empire was being penetrated and surveyed by American spies. During 1921 Lieutenant-Colonel Peter Ellis, US Marines, had been assigned to reconnoitre the island chain defences, and as a result of covert operations had submitted a prediction report to the US Department of Defense. In the document he had predicted that the Japanese would attack the Hawaiian Islands sometime in the future. In 1923, with the continuation of US intelligence operations, Lieutenant-Colonel Ellis USM had been sent across the Pacific with a sailing-ship disguised as a German trader to enter the restricted areas of the Caroline and Marshall Islands of the western Pacific illegally. On this particular mission he had been caught, was interrogated by the Japanese Military Police, the Kempeitai, and had mysteriously died, believed to have been by poison, while in their hands. Later in 1923, General 'Billy' Mitchell, US Army Air Corps, had made a series of illegal flights by aircraft around the Mandated Nipponese Islands, reconnoitring bases and anchorages of the Imperial Navy. As a result of his undercover flights of Micronesia, he reported in his conclusions that the islands were being prepared as forward bases for a future attack on the US naval base at Pearl Harbor, forecasting the likely time of attack as being on a Sunday morning,

Back in Japan the naval air training programme had continued, and on 10 July Major Brackley, now Lieutenant-Commander Brackley HIJMN, was able to report to Colonel Semphill, stating that the Blackburn Swift aircraft was no longer used for torpedo-dropping attacks. At about this time the Kaibogikai Foundation was erected for the express purpose of funding from civilian sources the research and development of new aircraft. During the next six years the foundation acquired or financed two Junkers reconnaissance seaplanes from Germany, two six-seater passenger KD6 aircraft, and two types of all-metal flying-boats. Further aircraft were obtained under this civilian funding, including two transport aircraft, two reconnaissance planes, a Gizu flying-boat designed from the original German Dornier Wal design, but equipped with wing floats in place of the Dornier sponsons. Thus the Navy was able to acquire original designs quite outside any financial restrictions that the naval budget might have imposed, due to privately allocated funds. It was now that the Navy held its first design competition for a new reconnaissance seaplane. Mitsubishi proposed the 2MT4 (Ohtori) a seaplane version of the Type 13 torpedo-bomber; Nakajima offered the Type 19 A2B, a version of a French Breguet machine; and Kawasaki proposed a version of the Dornier Do D. The winner of the competition was a design put forward by Yokosuka Naval Aircraft, designated Type 14 sea reconnaissance Navy Type E1Y13. This machine was a three-seater aircraft powered by a 450 hp engine. A total of 320 machines of this type were built – twenty-three by Yokosuka, forty-seven by Nakajima and 250 by Aichi.

On 1 September 1923 the entire Kanto District was awoken by a terrible noise and unnatural movements: a heavy earthquake had struck! At first it was not possible to gauge the extent of the devastation, but eventually it became clear that the destruction had been thorough and very extensive. Yokohama and Tokyo were heavily damaged, and port facilities at the harbours had been crushed, resulting in the oil tanks bursting and pouring ship fuel oil into the sea, which was now effected by a tidal wave. The result was a monumental raging fire that destroyed all the shipping at the quays and lying in the roads. Among the ships that had escaped the inferno was the aircraft carrier HIJMS *Akagi*, which had been so severely damaged that the ship would be scrapped the following year. Meanwhile the wind tunnel at the Naval Air Service Laboratory, Tokyo, was destroyed, along with the entire establishment. One of the William Froude tanks had been smashed up and a bad fire had resulted. At this time Major (or Lieutenant-Commander) Brackley organised a fund-raising campaign to assist the victims of the disaster, but the British Mission was recalled home until times could be more normal. Rescue and rehabilitation of those suffering was undertaken by the Japanese Army at the government's direction. The naval facilities that had been destroyed in Tokyo were part of the Technical Research Institute. This organisation replaced the Tsukiji Arsenal due to the influence of the Washington Disarmament Conference, with an amalgamation and reorganisation of the Warship Model Testing Station, the Aircraft Testing Station and the establishment of an Aeronautical Research Section. This was the institute that suffered so much in the earthquake, with heavy damage to buildings and equipment. The establishment possessed an 8.2 ft dia. wind tunnel and other research facilities built under the guidance of Dr Weiselberger of Gottingen University, which exceeded in size and power the 6.6 ft wind tunnel of the

Eiffel type completed in 1919 at Tokyo Imperial University. Later in the 1920s the Navy was to add further research facilities at Kasumigaura Airbase to supplement the Weiselberger tunnel. Much later, during 1930, a 9.8 ft wind tunnel was to be built at the Aeronautical Research Institute of Tokyo Imperial University at Komaba.

Despite the relative success of the 1920 expansion programme, three obstacles remained to impede its successful outcome. Within the senior ranks of the Navy an air power versus big-ship conflict had broken out, splitting much of the strategic thinking. Secondly, the Washington Disarmament Pact of 1922 resulted in severe reduction of naval equipment supply. Thirdly, during the devastating earthquake of 1923 the entire eastern part of Japan had been destroyed, by which time only ten operational squadrons of aircraft had been raised. At this stage the importance of land-based planes had been realised, resulting in the emergence of the aircraft carrier construction programme. The British Mission under Colonel Semphill had suggested the creation of flying units operating from shore bases as a means of increasing the effectiveness of the Naval Air Service.

CHAPTER 5

Training and Development

In 1924 the Aeronautical Research Section of the Naval Technical Research Institute authorised Seito Yokota and a team of young engineers to design a twin-engine flying-boat. This machine was financed from the Kaibogikai-donated funds for aeroplane research and development. It was designated the KB model. This flying-boat had a gross weight of under three tons and was powered by two BMW 185 hp water-cooled six-cylinder in-line engines manufactured by Kawasaki under licence. The machine was completed at the Imperial Army's Tokyo Arsenal and made a maiden flight in December of the following year. Flight developments continued successfully, to the delight of the young group of aeronautical engineers. Unhappily during the course of a flight in March 1926 the machine mysteriously nose-dived into the sea and became a total loss. Despite the disappointment at the loss of this very promising flying-boat, it was nevertheless the first all-metal aircraft of Japanese design manufactured from domestic materials.

The Kawasaki Dockyard Company, which had manufactured the BMW engines for the KB model, had become interested in the construction of all-metal flying-boat designs. Contracts were entered into with the German Dornier Company of Friedrichshafen for the acquisition of engineering data and licences to build specific designs. One of the leading Dornier engineers was invited to Japan. This was Richard Voigt, who was employed by the Kawasaki concern until 1933. Under Dr Voigt's guidance Kawasaki produced one Dornier Do D Falke seaplane, as well as the Kawasaki Type 87 twin-engine heavy bomber. Meanwhile part of the Rohrbach flying-boat had been manufactured at Berlin Staaken. These components were exported to Copenhagen, and eventually to Japan with financial assistance from the Imperial Japanese Navy. The flying-boat was completed and designated as the Type R flying-boat. It was powered by two 375 hp Rolls-Royce engines and carried twenty-four passengers.

At about this time the political feeling towards the Americans became very embittered. This was due to the American Immigration Acts, which virtually excluded the Japanese from entering the United States. The Americans required an equal quota figure for entry into Japan, and this naturally resulted in the exclusion from the USA of the Japanese. Offended national pride in the course of time built up a bitter hatred. This was seen in Japan as a discredited liberal policy of co-operation with all Western nations regardless.

Imported aircraft continued to arrive in Japan besides the previously mentioned Dornier Do D Falke, the first seaplane designed by Claude Dornier, and subsequently the Dornier Wal was also acquired. This was a high-wing flying-boat monoplane with twin engines mounted in tandem. The Kawanishi Machinery Company of Hyogo, Kobe, became a newcomer in the aviation field, having originally built single-engine mail and passenger planes for the Japan Aviation Company. At the same time the Ishikawajima Shipbuilding Company opened an aviation division with capital of ¥2,000,000, with its head office located at 1 Eirakucho, Marunouchi, Tokyo, and manufacturing plants at Tsukishima and Tachikawa under the company presidency of T. Shibusawa.

The year 1924 was the year of the second design competition for a reconnaissance seaplane, which had to be smaller than the Type 14. Design proposals were submitted to the Navy technical department. Nakajima proposed a two-seater sesquiplane powered by a 300 hp engine. Yokosuka suggested the Tatou-Go all-metal low-wing monoplane with a 320 hp engine. Aichi planned an all-wood low-wing monoplane known as the Mi-Go, powered by a 300 hp engine. Indeed, the latter two design proposals were similar in concept to the German Hansa Brandenburg seaplane. Eventually the Nakajima concept was accepted and became known as the Type 15 seaplane sea reconnaissance, Navy Type E2N12. Production commenced three years later, with forty-seven built by Nakajima and a further thirty manufactured by Kawanishi. At the time the Imperial Navy considered that only seaplanes were suitable for naval operations, as the concept of the land-based, long-range, over-ocean aircraft had not been conceived. Resulting from this belief, the Rogo-Ko biplane seaplane had reached a total production of 218 aircraft by 1924.

The work of the British Aviation Mission under the command of Colonel Semphill was drawing to a close. Originally the mission had considered utilising 70% of its time in an executive capacity, with 30% in purely advisory capacities. The idea was that in the final advisory period Japanese personnel would run the service, with only a small percentage of instructors in the background.

The facilities organised included a laboratory for testing metals of machine tools and all materials used by the Engineering Section. The engine repair shop and machine shop were located in the central flight hangar in the middle of the aerodrome, possessing a fully equipped machine shop. Adjoining was an engine repair shop with a foundry capable of manufacturing castings up to 560 lb, as well as a stores and a welding shop. The repair shop was capable of rebuilding six machines a week and had the facilities of a woodwork shop on hand.

A School of Technical Training was being built, and engineering officers and ratings under instruction were accommodated in workshops. By 1923 the workshops were fully self-contained, with a capacity to overhaul sixteen rotary and eight stationary engines, or two Eagle engines and one Lion engine, per week.

The Technical Section was composed of five officers, each of whom was responsible for an individual aspect of technical training, and it included engines, aircraft, design, technical training and parachute training. Included on the staff were four warrant officer engineers, four riggers and one draughtsman, each of the warrant officers being assigned to a flight or workshop to instruct and supervise the work. The Japanese engineer officers were extremely keen, though theoreticians only, tending to leave practical matters to the naval warrant officers and petty officers, who were capable of working very long hours. Later the need for a very thorough practical training was recognised, and the Japanese eventually made good machinists, mechanics and blacksmiths.

Among the equipment was the Sparrow Hawk aircraft, considered very good but greatly improved if fitted with a radial engine – either a Lynx or a Jaguar motor. The F5 flying-boats were moored out permanently and not placed nightly in sheds, owing to improvements made in the weathering qualities of the machines. Originally made of wood, the F5 flying-boat was subsequently manufactured in metal because it was found that the sea-water seeped into the wood and reduced the aerodynamic qualities and performance of the machines. Engine failures were few except in the case of the Le Rhône motors, which proved to be sensitive on fine adjustment with single ignition, but generally speaking running times were up on average before overhauls proved necessary. The compression ratio of the Lion engines was altered from 8:1 to 5:1 because the machines, operating at low levels, did not possess the required power prior to adjustment. In winter spinners had to be attached to the propellers of the rotary engines because the crankshafts iced up and blocked the free flow of the fuel. The castor oil used in the rotary engines was manufactured from Japanese beans, while some quantities were imported from the USA. This oil was not up to BESA standards, and despite mineral oils of Japanese origin being obtainable, no composite oil was suitable, and therefore it was necessary to import supplies from the United Kingdom. Aviation spirit was obtainable from Japanese oil wells but the quantity was not prolific. This aviation spirit had a low aromatic content of about 4% equivalent to US spirit, but the best Aoi Komori averaged Baume gravity of 72 degrees. The initial boiling point was 35°C, final boiling point 124°C and distillation at 100°C. This spirit was light and free from constituents of the high boiling point. Supplies were low, but aviation spirit passing BESA specification had to be imported from Sumatra and Borneo by the Asiatic Petroleum Company Ltd. Benzol was obtainable, but was insufficiently free of sulphur content to be of any use. Dope used to varnish the fabric surfaces of an aircraft to waterproof and make taut the covering was particularly trying under the Japanese climatic conditions of great dampness during the long, sultry, summer months and strong sunlight. The Royal Aircraft Establishment at Farnborough had developed an aluminium nitro covering termed V84 in addition to the use of pigmented nitro-cellulose dope PDN12 in the proportion of three coats of dope plus two coats of nitro, providing a suitable surface that could be effectively painted. A proprietary doping scheme specification was recommended for tropical countries as a result of the work of the investigators. The following results on test were obtained:

	Acetate Base	PDN12 + V84 Dope
1. Part	Aileron from Avro 504	Aileron from Sparrow Hawk aircraft
2. Exposed	New	After 40 hrs or two months' service
3. Date exposed	22/02/22	25/01/22
4. Examined	29/05/22 Alum. varnish cracking	Conditions perfect
5.	24/06/22 Varnish peeling, dope cracking	Conditions perfect
6.	20/10/22 1) Fabric exposed in parts 2) Dope flaking off	No change
7. Total time exposed	8 months	9 months

Wooden propellers of British manufacture had given considerable troubled service in Japan. The airfield at Kasumigaura was a large area in which long grass grew profusely in the late spring and early summer. As a result, student pilots in a tail-up position cut the grass with the aircraft propeller, and those made of wood broke easily. Consequently, after a morning's flying much damage was experienced, which was corrected only later when metal propellers were fitted. As early as 1921 the Leitner-Watts propeller was the only example manufactured in metal in two- or three-bladed varieties. This was advantageous in the tropics for seaplanes and transport aircraft, and with the conversion of all training machines to the use of metal propellers the problem was solved. Wood was also available for aircraft construction, since the Imperial forests of Japan possessed 290 separate species, divided into seventy genera. This compared with European standards of only seventy separate species with thirty-five genera. Hinoki was equivalent to, and a substitute for, a Grade A spruce suitable for long struts, only incurring a small increase in weight. There were two types of this wood – a white, which was the superior of the two, and a red. A Grade B spruce had many substitutes – Korean pine, Formosan pine, Manchurian pine and Katsura; kiri was suitable for non-structural numbers, while kurumi, or Japanese walnut, could be used, as well as shii, a pointed pasania or keyaki, otherwise ash, in the production of propellers. Woods that were under investigation at that time included oba shirakabaha, or white birch, and narra, but no equivalent was discovered for mahogany. Generally Japanese soft woods were superior to European or American when affected by climatic exposure, being resistant to moisture as well as naturally proofed against rotting, splitting or decay.

A three-week parachuting course was held for pupils, though not every student was obliged to jump. Two different types of parachute were used as standard – the Guardian Angel D4 and the Mears type. The Mears was useful where stowing was concerned, but was difficult to function in still air and was not suitable for dropping from kite balloons. It was made of cotton, of smaller bulk and superior glissading, but a recommendation was made for future production in silk. Courses usually commenced with dummy drops, drops from kite balloons and descents from aircraft in flight, when courage was shown by pupils in jumping. The D4 parachute was used as an introduction to folding, and all parachutes were examined every year during May and October. However, in the summertime, all parachutes were unpacked, unfolded and aired to prevent the ravages of dampness. Of the two chutes, the Mears was considered superior. Fifty drops with 150 lb dummy weights were made from 400 or 500 feet down to 200 feet, opening in one and a half seconds with one unrolling strip not attached to the aircraft. Later drops were made from 600 feet with parachutes opening at 150 to 200 feet.

When flying training was in progress, over forty machines could be observed in the air at the same time over the Central Training Station. The courses included cross-country flying and overseas flights, which were completed on the F5 flying-boats with 1,500- or 2,000-mile flights divided into suitable stages of 400 or 500 miles. Longer flights were undertaken over a period of nine hours. Night-flying was also conducted, necessitating the use of Holt flares and/or Lucas Cranwell electrical sets.

The fleet courses were held at the Yokosuka Naval Base, and an experimental flight was established, which conducted the same type of work as was done at RAF Martlesham Heath in England. A high-speed course was set up with electrically timed cameras and a camera obscura. Surprisingly enough, all courses were conducted in English, which was unified and simple to use by all concerned. The mission's flying instruction was divided into six sections, comprising preliminary training, seaplanes and preliminary flying-boats, scouts (i.e., fighters), deck landing and fleet co-operation, with an overall command section. Each section had an officer in command as instructor, with a Japanese officer as the assistant instructor, plus pupils and auxiliary personnel. Prior to the commencement of flying training, all prospective Japanese pilots had to pass the special medical examination, which was set to the same entry standards practised in the RAF. From then on, preliminary flying was carried out on the Avro 504 flights, from which pupils graduated using the established Gosport System of instructions. This was followed by one of any eight courses selected on the basis of utilising to the best advantage the pupil's natural aptitude for a particular type of flying. The courses included scouts, Panther ship aircraft, fleet training, turret flying and deck landing, torpedo dropping, seaplanes, flying-boats long course with cruises of 1,500 to 2,000 miles around the Japanese Islands and a specific fleet training course. Progress was measured by grades: thus Standard A flying training consisted of eleven tests in the air to the instructor's satisfaction. Technical Standard A for officers comprised the testing and use of all aircraft performances. Armament Standard F torpedoes consisted of examinations both oral and written, together with certain tests in aerial dropping of dummy weapons. Photographic Standard B grade included tests and written examinations to be passed by pilots. In all these examinations marks were awarded out of 100%, and anyone with 75% or less was rejected, to be re-examined at a later date. Second-class certificates were granted to those whose examination results were between 75% and 89%, and first-class certificates were obtained by these achieving 90% or more. Technical examination boards were erected to test ratings in the various trades open to the pupils. As previously stated, the preliminary medical tests were very similar to those appertaining in the United Kingdom, and were followed by aptitude tests. The average age of the candidates presenting themselves for flying training was 25, and within the first fifty trainees only one was rejected, due to poor vision. Respiratory tests and reflexes were above average, while balancing was always perfect, vision as well as colour vision was normal, but the capacity to maintain an even ocular muscular balance under strain was below UK standards. All the goggle equipment was fitted with Crookes anti-glare lenses, with three differing intensities, but no case of glare or retinal fatigue was experienced. Games were widely played in the summer season, while many games were introduced during the winter months. Professionals were recruited into the service – particularly the photographic branch, for the Japanese were good photographers. Instructors in this branch were under the charge of one officer and a warrant officer. Photographic records were kept of the progress of the students under instruction, as well as photographic mosaics of the naval air station layouts.

Two types of camera were in use at this time – the Type 18 for oblique work and the Type LB2 for vertical work. However, problems with storage were experienced, though only limited material was taken out to Japan because of the climatic problems of heat and humidity. In early summer special cooling-tanks were used to preserve the hardening agent, while plates, films and papers were kept in airtight cases. The best conditions for aerial photography in these latitudes were in the autumn, winter and springtime. The worst season was of course during the summer, when only ten per cent of the days proved suitable, and this would have been less but for the special plates and filters. Regrettably the plates and papers manufactured in Japan were not up to the standard obtained in the UK. Courses for pilots, observers and specialist officers were conducted in which a cine-camera was used to make instructional films for training purposes.

Armament training was given by one officer and an assistant, with Japanese officers and men training as pilots and observers who proved themselves adaptable and keen. Difficulty had been experienced in collecting the necessary equipment, but Vickers and Lewis machine-guns were acquired, and the use of bombs up to 500 lb weight was standard. The torpedoes were similar to those manufactured in the UK, though entirely Japanese built. Range instruction took place on the ground, in a rocking fuselage or in a fuselage with engines. An aerial bombing range had been constructed at the Central Training Station, with targets moored in the lagoon, and practice bombs were used that contained stannic chloride, but additional practice was obtained by use of the Batchelor Mirror or a seaplane tender. Aerial camera-gun and machine-gun practice was obtained while flying over the lagoon attacking an aerial-towed drogue. Live armament attacks were only practised from the Kagoshima Airbase in southern Japan.

At this time in Japan five firms were manufacturing seaplanes and flying-boats, including Mitsubishi and the Kawasaki Dockyard Company. Each firm would manufacture for either the Army or the Navy, but never for both organisations. Machines manufactured in Japan under licence at this time included the Avro 504, the F5 flying-boat, the Salmson, Nieuport and several Mitsubishi aircraft, as well as Yokosuka and Nakajima types. Limited aircraft production was undertaken by several naval dockyards. Two British firms also were directly involved in aircraft manufacture in Japan at this time: Short Bros was making the F5 flying-boat at the Naval Aircraft Factory in the Yokosuka Naval Dockyard, and the technical staff of the former Sopwith Aviation Company of Kingston-upon-Thames had been engaged to set up engineering facilities and designs at the Mitsubishi Aircraft Division at Nagoya. Engines for aviation purposes included the building of the 300 hp Hispano-Suiza motor, the Le Rhône and a few Benz engines.

Detailed plans had been worked out for the construction of a factory at the Central Naval Air Training Station for research and development of new naval aircraft. A high-speed course had already been laid out and was in full operation during 1924. Since 1914, research work had been conducted at Sendai University, located a few hundred miles north of Tokyo, where original work was accomplished in the electromagnetic properties of iron and steel alloys, high-tensile steel and light alloys for aviation construction. In Tokyo the Naval Air Service Laboratory was to be found since April 1916 researching aeronautical problems. The equipment included wind tunnels – Eiffel type and a William Froude tank, and later the installation of a National Physical Laboratory type. The main problem in all this research work was the question of language, so the researchers learned English, French or German in order to read original research papers. However, a suggestion had been made in an endeavour to simplify matters whereby Japanese would be written in Romaji (i.e. Roman) characters to write down the spoken language sounds.

Accident records were compiled in a similar manner to those kept by the RAF. The following statistics suggest that errors of judgement were the chief cause of fatalities. No fires in the air or on the ground had been recorded, nor had machines failed in the air as a result of any defective structure.

(a) Accidents at the Imperial naval air station, Kasumigaura

Training period	Flying hours	Rate killed per 1,000 hours flying
Jun to Dec 1921	983	0.00
Jan to Dec 1922	3,335	0.59
Jan to Dec 1923	7,278	0.41

(b) All Japanese Imperial naval air stations

Training period	Flying hours	Rate killed per 1,000 hours flying
Jan to Dec 1921	2,798	0.714
Jan to Dec 1922	4,987	0.802
Jan to Dec 1923	10,090	0.495

(c) Naval aviation statistics for 1924

Number of flights	48,000
Number of hours flown	19,000
Killed	9
Injured	21
Machines wrecked	160
Engine failures	200

The general conclusions of the British Aviation Mission to Japan were as follows:

1. There was little difference in the time taken to complete the Gosport course of instruction between Japanese and English pilots.
2. Originally instruction commenced with the Japanese being able to understand the English language.
3. Unhappily the instructors were not as readily understood as had been wished.
4. The Japanese pilots and crews displayed little sign of nerves.
5. The Japanese were always ready to attempt the most difficult of manoeuvres.
6. Regrettably at this period there was little interest in high mechanical efficiency.
7. The average ability of the Japanese pilots was high.
8. Due to the cult of Bushido, religious education, etc. tradition produced a racial, not an individual, type.
9. Pilots and crew were indecisive in unexpected situations.
10. They lacked the instinct sense of prompt action.
11. They possessed the virtue of high courage.
12. Both officers and men would go to any lengths to obey orders.
13. Had powers of great endurance.
14. With adequate technical training the pilots developed great interest in their machines.
15. The small stature of the Japanese crewmen was an inconvenience, requiring the adjustment of the controls.
16. The Japanese mechanics and riggers were keen to learn and hard working.

During the course of all this aviation activity the naval air station was opened one day per week to organised visits by schoolchildren, students, parents and grandparents. Every town or village was proud to send its quota, the people being orderly, with no incidence of pilfering being reported. Thus public opinion was being educated in aviation matters, for lectures on flying were given throughout the country by naval air officers.

Training and Development

On returning from his European tour, His Highness the Prince Regent was pleased to inspect the work of the British Aviation Mission, and was provided with an aerial escort by members of the mission. At about this time the Prince of Wales paid an official visit to Japan aboard the British battleship HMS *Renown*, and was escorted by the planes of the British Aviation Mission when two hours' steaming time away from the port of Yokohama. The official escort for the Prince of Wales comprised four formations of aircraft, totalling twenty machines, piloted by members of the mission. Because of the excellent work done by the mission installing the traditions and training schemes of the old British Royal Naval Air Service to the personnel of the Japanese Navy, His Imperial Majesty the Emperor was graciously pleased to decorate each individual member of the mission. Even His Royal Highness the Prince of Wales was made an honorary general in the Japanese Army!

In the meantime the lighter-than-air craft had not been neglected, for Japanese personnel had been trained in their use by the RNAS personnel. A large airship shed had been constructed adjoining the Kasumigaura Naval Air Station, though no mooring mast had been built. The Astra Torres airship was used, equipped with Sunbeam engines, but the SS Twin airships were never a success in the hands of Japanese personnel.

During 1924 conversion work commenced on the battleship HIJMS *Kaga*, to rebuild the vessel as an aircraft carrier. Originally laid down at the Kawasaki Shipyard in Kobe on 17 November 1921, she was due for scrapping under the terms of the Washington Naval Treaty. However, with the destruction of HIJMS *Amagi* during the 1923 earthquake, *Kaga* was reinstated. HIJMS *Kaga* displaced 38,200 tons and was 812$\frac{1}{3}$ feet long, and had a beam of 106$\frac{3}{4}$ feet and a height of 30 feet. She was powered by four geared turbines developing 127,400 shp, which gave her a speed of 28$\frac{1}{3}$ knots. With an internal capacity for ninety aircraft and a complement of 2,019, she was armed with ten 8-inch and sixteen 5-inch guns, as well as twenty-two 25 mm anti-aircraft guns. Enjoying some twenty-one years' active service, she was finally sunk by American Douglas Dauntless dive-bombers at the Battle of Midway in June 1942.

In 1925 the National Annual Fighter Tournament was established to encourage fighter development between the Army and the Navy. Generally speaking it was found that the naval aeroplanes and pilots were far superior in training and technical development to those of the Army. At the time there were two schools of thought concerning the introduction of monoplanes: the high-speed group favoured the planes that took advantage of weight and speed, as was found in the fighter developments of the US and German aviation industries; while the compromise group considered speed and manoeuvrability as the two most important virtues, as exemplified by fighter planes manufactured by Great Britain, Japan and Italy.

The ensuing debates that took place in Japan between these two groups were in an atmosphere in which a thriving aircraft industry had re-established a Civil Air Transport Service flying between Tokyo and Osaka on a series of regular scheduled flights. At the time, civil war had broken out in China, and Japan found it necessary to prevent the extension of hostilities by armed intervention in Manchuria. It was now necessary to expand the Japanese armed forces. Meanwhile the Kawasaki Dockyard Company had established an aviation department with a head office at Higashi Kawasaki-cho, Hyogo, Kobe, with a capital of ¥3,000,000. This company had obtained licence rights to build Dornier Metal aircraft, BMW aero-engines and Vincent Andre radiators. Later aircraft were constructed to the designs of Dr Richard Voigt, the German aeronautical engineer who worked for the Kawasaki Company. One such aircraft obtained by this company was the Dornier Komet III, a high-winged single-engine monoplane powered by a Rolls-Royce 360 hp engine, and capable of carrying six or eight passengers. At this time a further aircraft carrier was completed – HIJMS *Akagi*, a former battle cruiser of the Amagi Class originally laid down in December 1920. This vessel had been part of the provisions of Japan's 8-8 program (eight battleships and eight battle cruisers) under the provisions of the signed Washington Naval Treaty. The hull was converted and rebuilt as an aircraft carrier, which emerged with three flight decks in the constructional manner of HMS *Furious*. By now Japanese ship and aircraft designs and construction were well beyond the cradle stage in providing equipment for an active fleet.

In the following year the nation was stunned by the news that His Imperial Majesty the Emperor Taisho had died!

CHAPTER 6

Bringing the Eight Corners of the Earth Under One Roof

The new Emperor was His Imperial Majesty the Emperor Hirohito, who took the name Showa on the occasion of his accession. The word Showa means enlightened peace, and was chosen because this was to be the commencement of the Showa Restoration – the new policy adoption. This new policy would parallel the period of the Meiji Restoration of 1868, and would destroy the power of the capitalists as well as banish all Western influence within the country.

It was the time of the third design competition for an aircraft to replace the Type 10 carrier fighter. Four manufacturers prepared plans and submitted proposals to the Navy. The Aichi Aircraft Company had suggested a version of the German Heinkel HD-23; the Mitsubishi Company offered a biplane design prepared by engineer Joji Hattori known as the LMF9 Taka (Hawk), powered by a 450 hp engine.; the Kawanishi Company suggested a sleek biplane known as the K11 evolved by engineer Eiji Sekiguchi, which was to be powered by a 500 hp motor. Unhappily all these design proposals failed, and only the fourth participant was successful –the A1N1-2 monoplane. Designed by the English Gloster Aviation Company as the Gambit, it was modified for naval use by the engineer Takao Yoshida of the Nakajima Aircraft Company and powered by a 450 hp Bristol Jupiter engine. It was significant that this machine was designed by a British aeroplane company but was to be built by a very successful company organised by a former Japanese naval officer. The new Nakajima carrier fighter incorporated a new wing form developed in England by the Boulton Paul Company, which proved strong and reliable. Eventually Japanese rights were purchased for the manufacturing licence, and the new wing form was incorporated into the designs of the Nakajima Type 87 carrier fighter, the Nakajima A1N1 Type 90 carrier fighter, the Mitsubishi B2M1 Type 89 attack-bomber and the Navy Type 7 and 9-Shi bombers.

Ever since 1919 duralumin had been researched and developed from pieces of German Zeppelin airship frames, which had been handed to the Sumitomo Company with recommendations for investigations. By the year 1922 a Mr Paygan, a British engineer, was assisting in the metallurgical studies and developing manufacturing techniques. In the following year the Sumitomo Company had opened a factory at Ajikawa near Osaka after a group of engineers under the management of engineer Fujii had returned to Japan from Europe. Meanwhile Admiral Yamamoto had become very interested in the new metal for aircraft construction, and had lent official support to engineer Buntaro Otani of the Naval Bureau of Aeronautics and Mr Shinnosuke Furuta of the Sumitomo Company to conduct further research work. Eventually rapid strides were made in development work, so that by 1938 ESD duralumin was available in quantity for the production of the Japanese Zero fighter plane. Because of the original research undertaken by Sumitomo, the Japanese were able to manufacture all-metal aircraft long before the United States of America. Meanwhile the Nakajima Aircraft Company had acquired manufacturing rights for the production of Lorraine aircraft engines of 400 and 450 hp, as well as the right to build Bristol Jupiter engines of 450 hp.

At this time the aircraft of the Navy went to sea aboard the aircraft carrier HIJMS *Akagi*, which displaced 26,900 tons and accommodated fifty aircraft, as well as HIJMS *Hosho*, a carrier of 9,500 tons, driven by a 20,000 hp engine at a speed of 25 knots, which could fly thirty aircraft. Unhappily, although these two ships were attached to the Grand Fleet, no one knew how to use the new weapon. As well as the carriers, the old seaplane tender HIJMS *Wakamiya Maru* was attached to the Grand Fleet. This vessel displaced 7,600 tons was propelled by 1,600 hp at a speed of 11 knots and could fly fifteen seaplanes. On land seventeen squadrons of eight aircraft each existed with two balloon squadrons comprising two airships – the Astra and the SS-type, both acquired from England – and twenty M-type balloons. The aircraft were deployed at four naval air stations, with five squadrons of forty aircraft at Yokosuka, as well as all the balloons and the airships. At Sasebo Naval Air Station three squadrons of twenty-four machines, at Ohmura two squadrons of sixteen aircraft and at Kasumigaura Naval Air Station seven squadrons of twenty-eight aircraft were located. In total seventeen squadrons equalling 136 aircraft were in full operational status. The aircraft types comprised the following machines:

1. Fighter planes – Mitsubishi type 1921
2. Reconnaissance – Mitsubishi type 1921, 1925 seaplane and the Brandenburg seaplane
3. Long-distance reconnaissance – Dornier and Rohrbach aircraft
4. Fleet spotters – Mitsubishi 1924 type
5. Bombing and torpedo dropping – Mitsubishi type 1924
6. Patrol – F5 flying-boat
7. Amphibians – Vickers Viking Supermarine Seal

The Japanese naval air station commanders at this time were:

Commandant Yokosuka NAS; Commander D. Tachikawa HIJMN
Commandant Sasebo NAS; Commander S. Wada HIJMN
Commandant Ohmura NAS; Commander T. Yamada HIJMN
Commandant Kasumigaura NAS; Rear Admiral T. Komatsu HIJMN

In April 1927 the Japanese Navy established an aviation bureau – the Koku Honbei, organised to accelerate the naval aviation building programme. This new department was entirely independent of the Navy's Technical Bureau, and came under the direction of the 1st Chief Admiral Eisuke Yamamoto. The 2nd Chief was Vice-Admiral Masataka Ando, who led the fight for a large-scale naval air organisation for the research and development of new prototype aircraft. This date marks the rapid rise of Japanese naval air power, with Rear Admiral Isoroku Yamamoto as Bureau Chief of the Technical Division. At the time the Washington Naval Settlement was up for review at a conference to be held in Geneva. The object of the conference was to apply the limitation principles of the warship to all cruisers, destroyers and submarines. Unhappily, as described earlier, the United States' view violently alienated the British delegation, causing a serious disagreement, which resulted in the delegations failing to agree and the conference collapsing; it was decided to reconvene the conference in London in 1930 despite the fact that the United States of America was endeavouring to impose her view on both the Japanese and the British. But while the Americans might try to restrict the size of the Nipponese fleet, the Japanese were attempting to develop a naval air arm as rapidly as possible so as to defend themselves against the overwhelming superiority of the United States Navy.

In early 1927 the Navy decided to replace the Mitsubishi B1M1 Type 13 carrier attack-bomber, and accordingly specifications were issued to Aichi, Kawanishi, Mitsubishi and Nakajima. For the Mitsubishi Aircraft Division prestige was at stake, with a serious loss to the company if the new contract should go elsewhere. The new machine was to be a torpedo-bomber with a long-range reconnaissance role. Since reputation and existence were at stake, Mitsubishi requested layouts for the new aircraft from the Herbert Smith design team of the former Sopwith Company of Kingston-upon-Thames, from the Handley Page Company of Radlett, Hertfordshire, and from the Blackburn Aircraft Company of Brough, Yorkshire. One prototype was to be built in England, designated 3MR4, with a stipulated 600 hp V12 Hispano-Suiza Type 51-12LB engine, the manufacturing rights having been acquired by Mitsubishi previously. The memorandum from Major Bumpus of the Blackburn Aviation Company envisaged a large, 49 ft span biplane of about 7,500 lb weight with a Hispano-Suiza engine, similar to the two-seater Blackburn Ripon aeroplane. Further design studies concluded that a slightly larger machine would be required, known as the T.7, with a maximum speed of 139 mph. It would possess a high aspect ratio, narrow-gap mainplane and other characteristics of the Blackburn Beagle. Aircraft being imported at the time into Japan included the Boeing F2B, the Heinkel HD-23 and from Italy the N3 airship. But the N3 airship was short lived. Imported from Italy, the craft was involved in a violent explosion, which completely destroyed the airship. For the Navy a shadow was cast over future airship developments. So-called proving flights were regularly made by Type 15 flying-boats between Yokosuka and the island capital of Saipan and back, a total distance of 2,545 nautical miles. This was the capital of the Japanese Pacific Island complex.

HIJMS *Akagi* had been rebuilt. Originally laid down on 6 December 1920, she was named after a Japanese volcano (*Akagi* means Red Castle), and completed in March 1927. With an overall length of 856 feet, displacing 26,900 tons and propelled by nineteen boilers giving a maximum speed of 32½ knots with geared turbines, she possessed three flight decks in steps forward. Smoke was ejected through two funnels on the starboard side, and the carrier was armed with ten 8-inch guns and twelve 4.7-inch AA guns for self-protection.

By March 1928 the aircraft carrier HIJMS *Kaga* had been completed, after having been towed to Yokosuka partially built to be fully fitted out. Displacing 29,600 tons and propelled by geared turbines fired by twelve boilers, *Kaga* could maintain 27½ knots. The boiler effluent was exhausted via two huge trunks along both sides of the ship in the manner of motor car exhaust and silencer systems, turning outboard at the stern. Overall she was 783 feet long with a triple flight deck in the manner of HIJMS *Akagi*, with a hangar capable of housing sixty aircraft. She was armed with ten 8-inch guns and twelve 4.7-inch anti-aircraft guns for self-defence. During 1928 the aircraft carriers *Akagi* and *Kaga* participated in intensive tactical usage studies, and as a result of these exercises it was established that with the present aircraft aboard the ships had a limited combat range of a hundred miles. Subsequently the Navy would demand longer-range aircraft to be used in conjunction with high-seas fleets. On 1 May the Kobe works of Mitsubishi was transferred back to the Mitsubishi Shipbuilding and Engineering Company Ltd for the manufacture of heavy oil engines for other than aviation purposes. This reorganisation permitted the former aviation division to be renamed the Mitsubishi Aircraft Company Ltd.

On 23 May 1928 the finalised design of the Blackburn T.7B (3MR4) design was submitted by Mr G.E. Perry of the Blackburn Aircraft Company to Mitsubishi. The machine offered was to be propelled by a 625 hp Hispano-Suiza engine Type 51-12LBR and was a three-seater.

Mitsubishi readily accepted the proposed design, remembering the efficient design of the Swift and Cuckoo aircraft, which had introduced torpedo dropping to the Naval Air Service way back in 1921. A prototype order was placed immediately, and necessary licence agreements covering manufacture were quickly drawn up. The machine was, in fact, a two-bay staggered biplane with Handley Page slots on the leading wing edges and Frise ailerons on all four wings. The lower mainplanes were thirteen inches longer than the upper, and both upper and lower wings were fabric-covered, all-metal structures. The design construction followed traditional Blackburn philosophy. The fuselage was a three-piece weldless steel tube structure faired by aluminium panels aft of the cockpit. The rear fuselage was fabric covered but filled with flotation bags, while the stern bay housed ballast weights, and use was made of centre-of-gravity adjustable fixtures when changing from a two-seater to a three-seater configuration. Maximum fuel comprised 202 gallons housed in gravity top tanks in the centre section containing two 37 gal tanks. A 44 gal tank was between the pilot and engine behind a fireproof bulkhead, and an 84 gal tank was located in the floor below the pilot's position. The reconnaissance version of the machine housed a pilot, a wireless operator/bomb aimer and an observer/gunner, all of whom had ease of crew communication. The floor aperture of the centre cockpit had a mounting for a course-setting bombsight or could be used to position a fourth Lewis machine-gun mounting. The torpedo-bomber version only had a crew of two, with fuel for 100 gallons of aviation spirit. Armament comprised one Vickers machine-gun on the port side of the front fuselage, synchronised to fire through the revolving airscrew. A further two Vickers machine-guns were located on a double rotating mounting over the observer's cockpit. With a 2,000 lb torpedo released by the pilot, only 100 gallons of fuel could be carried. Alternatively, with full fuel load, two 250 lb bombs could be carried between the undercarriage legs and released from the rear cockpit, where a high-altitude bombsight could be mounted with fusing and bomb-release controls.

The Manchurian Incident
In Manchuria General Ishihara and Colonel Itagaki of the Japanese Kwantung Army Command had considered for some time previously the political and sociological consequences of Japan's partial occupation of Manchuria. Japan was in the country to guard the railways, mines, farms and businesses created by the financial power of the yen on investment. The two senior officers considered Manchuria the answer to Japanese poverty. As a source for raw materials and a market for finished goods, Japan would have to gain complete control of the country, which was

loosely governed by the Chinese warlord, old Marshal Chang Tso-lin. As a haven for all ethnic groups – Chinese, Manchu, Korean, Japanese and white Russians – General Ishihara visualised Manchuria as a haven for social democracy, which would be a buffer between the USSR and China. This plan could be effected only by the Kwantung Army Command, which obtained the agreement of the central government in Tokyo. However, His Imperial Majesty the Emperor and the War Minister were against a plan of open aggression, and so an assassination plan was laid. A Kwantung Army staff officer commanding an engineer regiment succeeded in dynamiting the train transporting Marshal Chang Tso-lin on 4 June 1928. The old Marshal was fatally wounded. Following on from this event, Prince Mikasa, the youngest brother of the Emperor, was convinced that the assassination of Marshal Chang Tso-lin was the basic cause of the war between Japan and America, by actuating the famous Manchurian Incident and being the turning-point in his brother's role as Emperor.

Meanwhile the Kawanishi Aircraft Company was set up out of the organisation of the old firm. The head office was located at Wadayama-dori, Hyogo, in Kobe, with a manufacturing plant in Hyogo. The president of the new company was R. Kawanishi, who with a capital of ¥5,000,000 had founded the company in November 1928, having taken over the aircraft-manufacturing facilities from the Kawanishi Machine Works. Licences were obtained for the manufacture of Short flying-boats and Rolls-Royce engines, for which the company were sole agents.

Planes built during 1928 included the Type 87 Model 1 primary trainer designed by the Yokosuka Naval Air Depot, equipped with an Armstrong Siddeley Mongoose engine of 130 hp. The Model 2 primary trainer had a seven-cylinder Jimpu engine of 130 hp, manufactured and designed by the Tokyo Gasu Denki Company. The Hiro Naval Arsenal produced a twin-engine flying-boat, the H2H1, with engines of 600 hp, which commenced service a year later, in 1929. The Nakajima Company designed the Type 90 A2N1-3 carrier fighter with a 580 hp Kotobuki II engine, of which 106 were built subsequently. In the previous year, 1927, the Imperial Aeronautical Society of Japan, a civil organisation, had commissioned the Kawanishi Company to build the Model 12 – the Sakura. This was a shoulder-wing monoplane with a span of 62 feet five inches and an all-up weight of 13,250 lb, equipped with a Kawanishi BMW 500 hp V12 water-cooled engine. Two examples were built, but during initial flight tests both the strength of the structure and the performance proved quite unsatisfactory. Designed with a range of some 5,000 miles, the long-planned transoceanic Pacific flight project had to be abandoned.

At the Aircraft Research Section of the Naval Technical Research

Institute at Kasumigaura, engineer Junichiro Nagahata designed a parasol-wing, carrier-based reconnaissance aircraft, which was manufactured by Kawasaki. Funding was provided by the Kaibogikai – voluntary fund raising to support the development of naval aircraft. Mitsubishi applied for a patent for the split noda flap design, but foreign aircraft companies were quick to apply the idea –Handley Page's leading-edge slots, for example. However, most aircraft of 1928 were too slow to require the use of flaps for landing purposes.

During all this aviation activity the size of the front-line squadrons had been increased at the respective naval air stations:

Yokosuka NAS	c/o Captain H. Wada	5 squadrons = 80 aircraft
Sasebo NAS	c/o Captain M. Takahara	
	HIJMN	3 squadrons = 48 aircraft
Ohmura NAS	c/o Captain J. Nakamura	
	HIJMN	2 squadrons = 32 aircraft
Kasumigaura NAS	c/o Rear Adm Y. Edahara	
	HIJMN	7 squadrons = 112 aircraft
	Total front-line aircraft	17 squadrons = 272 aircraft

The Naval Air Service headquarters was located at Kasumigaseki, Tokyo:

Director – Vice-Admiral A. Yamamoto

Three branches:
1. General Affairs – Captain K. Mayehara HIJMN
2. Technical – Rear Admiral S. Furukawa HIJMN
3. Education – Rear Admiral K. Usui HIJMN

In 1929 the Kawanishi Aircraft Company, with the full agreement of the British government, dispatched its chief engineer, Yoshio Hashiguchi, to visit Short Bros in Rochester. He inspected both designs and manufacturing methods of producing flying-boats, as well as wishing to negotiate securing a licence to build Rolls-Royce engines. Previously he had been very impressed with the Rolls-Royce engines, which equipped the British-built F5 flying-boats. Eventually Kawanishi was to acquire the rights to manufacture Rolls-Royce Buzzard engines and to build large-type flying-boats, though not before the company had moved its works from Hyogo to Naruo, on the coast between Kobe and Osaka. Meanwhile the Imperial Navy had issued a specification for a long-range flying-boat. As a result of a request from Kawanishi, Short Bros permitted demonstration flights of the Calcutta and Singapore I flying-boats. The Japanese liked the Calcutta boat very much but disliked the Bristol Jupiter engines, which were already manufactured under licence by the Nakajima Aircraft in Japan. It was agreed that Rolls-Royce Buzzard engines could be manufactured and installed, as had been done in the original Singapore I boat. It was agreed by Short Bros that the design should be a somewhat enlarged Calcutta with three engines, and accordingly the company built one prototype, C/N S753, designed under number S15 and designated the KF1. The Short Kawanishi KF1 flying-boat was assembled at Rochester alongside the Rangoon boats, from which it differed little. The hull was wide, flared at the chines, though the waist was not widened in the same ratio. The hull aft of the wings was flat, as in the Singapore MK III, with stainless-steel planing bottoms, while the central tailfin was equipped with auxiliary rudders. The pilots sat side by side in an open cockpit, with the hull containing stations for an engineer, radio operator, wardroom, galley and living accommodation. The plane was armed with one bow, two midships and one tail machine-guns mounted on Scarff gun rings. The three Rolls-Royce Buzzard engines each drove a two-bladed wooden airscrew with low-drag vertical radiators, and the wings had been thickened to take fuel tanks permitting a range of 2,000 miles.

Eventually Mitsubishi negotiated a commercial order with Supermarine to acquire a Southampton MK II all-metal flying-boat propelled by Napier Lion engines. This transaction was concluded subsequent to the highly successful flight of three Supermarine Southampton flying-boats by the Royal Air Force from Britain, around Australia to Hong Kong and thence to Singapore. This example was delivered to the Japanese at the Oppama Naval Air Depot, though the operational headquarters became the Kure Naval Arsenal, at which it was based. In the meantime the Bristol Aeroplane Company Ltd of Filton, near Bristol, had exported the airframes of two Bristol Bulldog fighters to Japan, where they were assembled and equipped with the Japanese-manufactured Jupiter engines at the works of the Nakajima Aircraft Company. Besides importing these aircraft, the Navy purchased an example of a German Heinkel HD-56 as well as a US-built Vought Corsair.

The young fledglings

Since before the First World War, Prince Kuni Asaakira, father of the Empress Nagako, had championed the cause of aviation development. His brightest pupil was a certain Captain Yamamoto HIJMN. Under Prince Kuni's sponsorship, the captain had studied at Harvard University for two years, where he took a special interest in studying the manufacture of petroleum and aviation gasoline and had undertaken a personal, private visit hitch-hiking to the Mexican oil fields. In 1921 Yamamoto had been an instructor at the Naval Staff College located at the University Lodging House in the grounds of the Imperial Palace, Tokyo. During 1923 he became responsible for air development to the Naval General Staff. In the following year he assumed total charge of the naval air development station at Kasumigaura – the Misty Lagoon – a lake the size of the US Lake George, some thirty miles north-east of Tokyo. At this time Captain Yamamoto worked with Empress Nagako's cousin Prince Yamashina, both having been trained by Commander Inoue Fumir HIJMN, senior flight instructor at Kasumigaura Naval Air Station. The Imperial personage and Captain Yamamoto had, between them, successfully developed torpedo-bombing techniques. These techniques were taught to other pilots, and included landing on wharves and barges. Later still, Captain Yamamoto had been appointed Japanese Naval Attaché to Washington, and subsequently was appointed a fleet skipper. In 1928 he joined the Naval General Staff as a planner, with a propensity for non-conventional weapons and warfare. Later he was to be offered the situation of chief technical adviser to the Japanese delegation at the 1930 London Naval Disarmament Conference.

On 26 November 1929, HIJMS *Ryujo* was laid down as an aircraft carrier of 10,000 tons displacement, conceived on lessons gained in operating the carriers *Akagi* and *Kaga*. *Ryujo*, or *Fighting Dragon*, had a single flight deck with an athwartships arresting system in the French manner, plus the use of barriers. On the request of the Naval General Staff, a second flight deck was added at a later date. Unfortunately, in 1934 the torpedo-boat HIJMS *Tomozuru* capsized owing to the design becoming top heavy. As a result of investigations, *Ryujo* was also found to be top heavy, and compensating construction had to be undertaken to right the situation. Two sets of 5-inch anti-aircraft guns were removed and a ballast keel was built onto the ship. During November His Imperial Majesty the Emperor instructed the forthcoming delegation to the London Naval Disarmament Conference to hold out for a ship ratio of 10:10:7. The delegation was to be led by the Navy Minister Takarabe and the future Prime Minister Wakatsuki Reijiro, with Captain Yamamoto as technical assistant, who undertook day-to-day discussions of his recommendations. On the way to London he stopped in Washington for a period of three weeks. It had been decided that battleship construction was too expensive and that aircraft carriers could be replaced by island runways, but the ratio of cruisers to destroyers desired by Japan was 10:7. President Hoover's Naval Aide was Captain Buchanan USN, who initiated the ideal of a 10:6 ratio, which was unacceptable to the Japanese delegation. Four months later the United States and Great Britain agreed with Japan on a cruiser-to-destroyer ratio of 10:6.9945, with no limitations on aircraft construction. When Captain Yamamoto returned to Japan on 18 May 1930 with the Navy Minister Takarabe, a diplomatic victory had been achieved. Later the United States was to agree to delaying construction of three new cruisers so as to give Japan 73% naval parity with America until a much later date at some time in 1936.

In November 1929 the Blackburn T.7B (3MR4) prototype received its final inspection, and was first flown without markings by A.M. Blake on 28 December. Tests were conducted with and without a torpedo. During the same month the Mitsubishi Company added new machines and tools to its Tokyo factory for the production of aircraft armaments and machinery. Aircraft production was concentrated at the Mitsubishi Aircraft Company in Nagoya, covering a factory area of 59 acres and employing some 2,000 executives and employees. This aircraft plant was the largest in Japan, and the most modern, and it possessed the largest machinery facility for the production of aircraft and aircraft engines in Japan and the rest of the Orient.

On 3 January 1930, 3MR4 was packed for export to Japan. The new machine was accompanied by Mr Perry and a working party from Messrs Blackburn. On arrival in Japan the Blackburn working party assembled the new machine, which was painted with red discs. Following assembly, the machine was closely inspected by the Mitsubishi engineers, who decided to commence immediate production. The early flying days of this new machine were highlighted by an unfortunate crash. It would appear that the magneto starting-handle could not be rotated unless the main oil cock was turned on first. Unhappily the Mitsubishi engineers saw fit to reposition the main oil supply cock without reference to the British designers. The result was that one Japanese pilot took off having failed to turn on the main oil supply, the engine seized and the aircraft crashed upside down into a paddy field.

In January 1930 Admiral Kato Kanji, Chief of the Naval General Staff, in association with others, opposed the ratio of 10:10:7 for heavy cruisers as compared with the United States and Great Britain. It was felt that parity was only fair, as well as an operational necessity, but the matter was debated by the civilian Cabinet, which decided upon acceptance of the 10:10:7 ratio. By March a compromise percentage had been worked out, which, though it fell short of Japanese aspirations, was compatible with their demands and subsequently was acceptable to the Navy Minister. However, a Naval Staff conference held in Tokyo was opposed to the agreement. By April the Japanese Cabinet met to discuss a reply to the proposals in the light of previously rendered advice. Nevertheless, despite deep misgivings in professional quarters, the draft treaty was considered acceptable, and on 22 April the government had signed the draft agreement. Feelings were running high, so much so that on 10 June Admiral Kanji tended his resignation, as Prime Minister Hamaguchi had refused to yield over the matter. As a result, the treaty was pushed through a stormy session of the Japanese Diet and was scrutinised by the Imperial Privy Council. On 1 October the Privy Council accepted the terms of the treaty, and as a result limitations were imposed by the Japanese government on capital ship building for five years, as well as on the construction of cruisers, destroyers and submarines. The Navy now became hostile to party politics and politicians in general, and aligned itself with the Army against any civilian government, believing the country to have been betrayed. In November, in a very busy and bustling Tokyo railway station, the Prime Minister, Mr Hamaguchi, was shot dead by a member of the Sakurakai activist group. Eventually the London Naval Treaty was scooped by both the United States and Great Britain. For the United States a certain diplomatic blow had been successfully delivered against Japanese defences, weakening her naval stand and further escalating anti-Japanese feeling. As the treaty left the Japanese Navy at a great disadvantage, the United States and Great Britain were looked upon as enemies. Consequently the Nipponese Naval General Staff initiated a policy to expand air power to maximum advantage and to offset the imbalance of naval power with aircraft. The treaty did not place limitations on aircraft carriers of less than 10,000 English long tons or 11,200 US short tons, and so HIJMS *Ryujo* was planned to displace 10,000 tons. The only effective limitation would be the naval budget, which would be imposed by a civilian administration. Nevertheless, before resigning, Admiral Kanji had advised the rapid deployment of a further twenty new squadrons of naval aircraft as necessary to offset an imbalance in the maritime defence.

During June 1930 Mr Wakatsuki, the chief civilian delegate, returned home from the London Naval Conference, as the conservative fleet faction of the Navy urged His Imperial Majesty the Emperor not to sign the new treaty. In reply the Emperor insisted upon the preparation of a co-ordinated defence plan, which would govern military expenditure during the next decade. By 23 July the War and Navy Ministers had established a Board of Field Marshals and Fleet Admirals under the control of the cousin of the Emperor. Eventually an acceptable plan was forwarded to His Imperial Majesty for his approval. The Emperor made it known that he would not sign the new treaty until the Cabinet had approved the long-range budget for the secret rearmament plan.

In March 1930 the British Bulldog flew, powered by a Jupiter VII F engine developing 440 hp and attaining a maximum speed of 196 mph. The machine had a span of 30 feet, with a length of 23 feet 6 inches and a height of 9 feet 5 inches, with a wing area of 230 square feet, and the all-up weight measured 2,850 lb. After various successful flights, the Japanese government requested the Bristol Aeroplane Company to send out a working party to design similar machines and supervise their production using local labour and available grades of steel. Accordingly L.G. Frise and H.W. Dunn left the Filton company, to travel to Tokyo. The special Japanese version differed from the Bristol Bulldog II in possessing larger fuel tanks in the upper wings, with a modified undercarriage and tail unit. The first Nakajima prototype had a Jupiter VII engine cooled by close-fitting cylinder helmets. The second Nakajima-produced prototype was fitted with wheel spats and a Townsend engine cowling ring. Both prototypes were highly successful, attaining a maximum speed of 196 mph. But the Japanese declined to place production orders with the Bristol Aeroplane Company for the new machine, and the dispirited working party returned home to England. It was very noticeable that the Bristol aeroplane features appeared in many following Nakajima products, but Bristol was never given recognition for its engineering ideas.

The Blackburn Aircraft Company had designed and built the Lincock Mk III, which aroused considerable interest in the Orient. This machine was powered by a Lynx Major engine developing 270 hp and propelling a two-bladed wooden airscrew, and having an exhaust collector ring around the engine. It made a very successful maiden flight on 6 June 1930 from the company's aerodrome at Brough, South Yorkshire. In all, five machines of this type were manufactured, with two being exported to China and the remaining three to Japan, where they were designated Type F2D. These planes were armed with two Vickers machine-guns located in troughs along the sides of the fuselage, which gave the pilot ease of access for clearing gun stoppages. The undercarriage consisted of a transverse axle with oleo-pneumatic legs, with a helical tail spring fitted externally on the sternpost. Remarkably, the all-up weight was kept down to 2,000 lb.

During August 1930 the KF1 flying-boat was nearing completion, since the erection of the wings had commenced, so that the machine would finally be built by October of the following year. In September the two Lincocks were ready for dispatch to China. Shipment took place in December 1930 aboard SS *Glengarry* from the Royal Victorian Docks, London, and were consigned to Arnhold & Company Ltd, Sassoon House, Shanghai. It is said that they were used against the Japanese in 1932. Meanwhile the London Naval Agreement was signed by Emperor Hirohito on 30 September, at the same time as long-range naval development plans had been laid by Rear Admiral Yamamoto, who believed in the future of naval air power. Naval pilots were drilled in all-weather flying techniques from the four tiny aircraft carriers at the personal direction of Admiral Yamamoto himself. Some pilots crashed and were killed, but air crews were ready for a big demonstration at the annual naval manoeuvres. His Imperial Majesty the Emperor boarded one of the battleships, which steamed out to the manoeuvres in the Inland Sea. By 26 October the naval exercises had been completed and were seen as a triumph for Admiral Yamamoto, since the carrier-borne planes of Fleet White had technically sunk the battleships of Fleet Blue.

Now the KF1 flying-boat was launched in Britain and flown for seventeen minutes by the test pilot, John Parker, who found nothing amiss. The very next day the machine was flown again with success. On 21 October taxiing trials took place to check the maximum temperature and water manoeuvrability. The following day the flying-boat was flown to Felixstowe on the east coast of England by Flight Lieutenant Weblin of the Aeronautical Experimental Establishment for handling tests, and upon successful completion the Kawanishi KF1 landed at Rochester, to be crated up as deck cargo aboard a freighter bound for Japan. A certificate of air-worthiness, No. 2899, was granted to the Kawanishi Company for the KF1 flying-boat on 2 December 1930.

That year marked the commencement of the 1st Expansion Programme to be completed by the year 1938. The programme required the building of one aircraft carrier of 9,800 tons displacement and the organising of twenty-eight land-based squadrons of naval aircraft. Unfortunately for the Naval General Staff, the government refused the building project for a 9,800-ton carrier, and requested the naval air squadrons be limited to fourteen, which would give a projected strength of thirty-one squadrons by 1938. It was planned to organise the thirty-one squadrons into nine Kokutai, or Air Corps, inclusive of three Air Corps at Ohminato, Saeki and Yokohama. The squadrons comprised four fighter, eight and a half torpedo-bomber, two carrier-based reconnaissance, six sea reconnaissance, two small flying-boat, four large flying-boat, three trainer, and one and a half experimental units.

In 1930 the naval aircraft weapons were rather small by modern standards. The fixed aircraft machine-guns were licence-built Vickers machine-guns of 7.7 mm calibre, while the free gun was a Lewis machine-gun of the same bore. Bomb development had produced missiles with explosive charges ranging from fifteen to 1,000 lb. Originally the naval torpedo was the 18-inch Type 44, but within twelve months the Navy had adopted the 18-inch Type 91, which became the Navy's main torpedo. Meanwhile minor construction work had been completed about HIJMS *Kaga*, consisting of smoke-discharge tubes with stern exhausts turned outwards, though in operational conditions this arrangement proved a disaster. The flying-boat built by Short Bros – the Kawanishi Type 90-II, or H3K2, equipped with three Rolls-Royce Buzzard engines of 825 hp and an all-up weight of 39,000 lb – was built under licence in Japan, and a total quantity of four were eventually constructed.

Between the years 1920 and 1930 companies manufactured numbers of aircraft for the Navy as follows:

1.	Yokosuka Naval Arsenal	161
2.	Hiro Naval Arsenal	234
3.	Nakajima Aircraft Company	527
4.	Sasebo Naval Arsenal	324
5.	Mitsubishi Aircraft Company	617
6.	Aichi Aircraft Company	427
7.	Kawanishi Aircraft Company	77
8.	Watanabe Aircraft Company	22
	Total	2,389

A new aircraft company now appeared: the Aichi Watch and Electric Machinery Company Ltd created an Aircraft Division with its head office at 15 Chitose, Funakatcho, Nagoya. The president of this division was K. Aoki and the director was B. Masumoto. The new company was floated with a capital of ¥5,000,000. The Mitsubishi Aircraft Company was building aircraft and equipment under licence at the same time.

The licences covered the production of Hawker trainer planes, a Blackburn reconnaissance torpedo-plane and Curtiss fighter aircraft, as well as aero-engines of the Hispano-Suiza and Armstrong Siddeley type. Production rights were held for the manufacture of Reed-Levasseur metal propellers, Claudel carburettors, Lamblin radiators and reduction gearing for aeroplane engines, Herzmark and Letcombe engine starters and Handley Page slotted wings.

The Navy, at the direction of Vice-Admiral Ando, turned its attention to the question of the design study of dive-bombers. Chief Engineer Junichiro Nagahata of the Nakajima Aircraft Company studied intimately the American Curtiss and Chance Vought bombers. He conceived an aeroplane for which the basic design work and prototype construction were undertaken by Nakajima. The new aircraft was of unorthodox configuration, with small horizontal stabilisers and a large elevator and ailerons. Unfortunately the machine proved unstable during test flights, and controllability was very inadequate. The general characteristics could not be improved and as a result the design was dropped.

Further attempts were made to build a successful home design when Lieutenant Misao Wada returned from Germany. Utilising the Rohrbach type of construction, he designed the Hiro H3H1 Type 90 Model 1 flying-boat, which was built by the Hiro Navy Arsenal. This machine had three Hispano-Suiza nine-cylinder water-cooled V12 engines, each developing 650 hp; at twelve tons all-up weight the maximum speed was 142 mph. Of unusual configuration, a production run of six was built, and a seventh was installed with two 950 hp engines. Meanwhile Mitsubishi was manufacturing the K3M1 Type 90 navigational and crew trainer equipped with engines built by Tokyo Gasu Denki, the Tempu engines giving 300 hp, and a production run of 317 aircraft was completed. Nakajima was building the A1N2 Type 90 carrier fighter equipped with a Model 2 450 hp Kotobuki nine-cylinder engine based on the Bristol Jupiter. Over a hundred of these carrier-based fighters were made.

Shanghai invasion
On 28 January 1931 Japanese troops landed on Chinese soil at Shanghai, ostensibly to protect Japanese interests, which had been under attack as part of the oriental political battle raging at the time. Two days later HIJMS *Kaga* arrived off the Chinese coast with an air group consisting of twenty-four Nakajima A1N fighters and thirty-six Mitsubishi B1M attack-bombers. By 1 February the Shanghai invasion force was reinforced by the arrival of HIJMS *Hosho* with a further ten Nakajima A1N fighters and nine Mitsubishi B1M bombers, which brought the supporting air group to a total of seventy-nine aircraft and gave the Japanese overwhelming air superiority in the clear blue skies above Shanghai. Nevertheless, for some time the Chinese Army was able to resist the Japanese invasion in the Woosung area of Shanghai. During the course of this campaign it is alleged that a difference of opinion between the senior commanders of the Army and the Navy took place, resulting in the temporary withdrawal of Navy support until such time as matters were settled. By 3 March the aerial assault resumed, and eventually the Chinese quarter of Shanghai was destroyed by the *Kaga* aircraft. After the Shanghai disturbances an uneasy truce followed which had been precipitated by a Chinese mob which had set upon five Japanese Buddhist priests. Meanwhile Japanese troops had secretly massed in Manchuria during the previous fine summer months. On 18 September bomb explosions took place near the city of Mukden. By 1030 on the same morning Japanese troops fired on the Chinese 7th Brigade. Japanese soldiers assaulted the city walls and on the following day Mukden had been captured. This was the commencement of the fifteen-year war. The railway line near the city of Shenyang in Manchuria was blown up; the Chinese were accused, and this became the excuse for the Japanese Kanto Army Command to start offensive operations. The Kanto Command, or Kanto-gun, consisted of some 10,400 men originally stationed in Manchuria to protect Japanese interests, rights

and immigrants as far back as 1861. In the following year of 1932, the Chinese Army of 330,000 men were beaten back and the southern half of Manchuria fell to Japanese arms. This policy of conquest was characterised by the slogan, 'Bringing the eight corners of the world under one roof', uttered by the Emperor Jimmu in the year 600 BC – the slogan for the national policy in 1931.

The two naval aircraft types used in the Shanghai Incident reflected British design influences. The Mitsubishi B1M attack-bomber was a three-seater aircraft capable of carrying a torpedo or two bombs of 530 lb weight each at a maximum speed of 130 mph. This was the Navy's principal attack-bomber. The Nakajima A1N fighter had a maximum speed of 150 mph and was armed with two 7.7 mm machine-guns and carried two 66 lb high-explosive bombs. During the Shanghai and Manchurian Incidents the naval aircraft had been designed by foreign engineers, and were reliable but lacked performance. In Manchuria no Chinese aircraft opposition was encountered, and over Shanghai failure of the Chinese aerial opposition was entirely due to Japanese quantity rather than quality. As a result both the Japanese Navy and Army were to provide more money for aviation developments. By 1931 a total of fifteen and a half squadrons of a seventeen-squadron air force had been completed, including one squadron of balloons and one of airships. The disposition of this naval air force was as follows:

1.	Kasumigaura Naval Air Station Commandant: Rear Admiral S. Kobayashi	1 airship squadron 7 squadrons of aircraft
2.	Yokosuka Naval Air Station Commandant: Captain G. Hara	1 balloon squadron 4½ squadrons of aircraft
3.	Tateyama Naval Air Station Commandant: Captain M. Tsutsumi	Under organisation
4.	Sasebo Naval Air Station Commandant: Captain F. Sugiyama	2 squadrons of aircraft
5.	Ohmura Naval Air Station Commandant: Captain K Fujisawa	2 squadrons of aircraft

Aircraft carriers	HIJMS *Hosho*	19 aircraft
	HIJMS *Akagi*	60 aircraft
	HIJMS *Kaga*	60 aircraft

The headquarters of this Naval Air Service was housed at Kasumigaseki in Tokyo under a director, Vice-Admiral M. Ando, and divided into three branches: A General Affairs Office under the control of Rear Admiral K. Mayehara, a Technical Affairs Office controlled by Rear Admiral K. Unsui and an Education Office directed by Captain K. Akiyame.

Analysis of the Naval Air Service by squadron and air station that housed an Air Corps, or Kokutai, 1931

Air Corps	Fighters	Land reconnaissance	Land attack	Sea reconnaissance	Flying-boats	Land training	Sea training	Research	Total
Kasumigaura	-	1.5	-	1.5	-	1.5	1.5	1.0	7.0
Yokosuka	0.5	0.5	0.5	0.5	0.5	-	-	-	2.5
Tateyama	0.5	-	1.0	1.0	1.0	-	-	-	3.5
Kure	-	-	-	0.5	-	-	-	-	0.5
Sasebo	-	-	-	1.0	0.5	-	-	-	1.5
Ohmura	1.0	-	1.0	-	-	-	-	-	2.0
Total	2.0	2.0	2.5	4.5	2.0	1.5	1.5	1.0	17.0

The Chief of the Technical Branch of the Naval Bureau of Aeronautics was Rear Admiral Isoroku Yamamoto, whose senior officer was Engineer Commander Misao Wada. As a result of the air war over China new policies would have to be instituted by NBA. These comprised:

1. The initiation of detailed programmes of design and construction of new naval prototypes.
2. The anticipation of the aerial development programme for the period 1932–34.
3. The renewal programme for aircraft development to be established.
4. Each manufacturer to be notified of any new programme and ensured of a prototype manufacture.
5. Immediate need for high-performance carrier planes.

Design work on all types continued inauspiciously until a further Shanghai Incident took place in 1932. Both Admirals Matsuyama and Yamamoto were fully conversant with overseas operations, and they ensured that the design activities were supported by a large-scale technical organisation.

As previously mentioned, on 18 September 1931 the Japanese Kanto-gun, the Kanto Army Command, had captured the city of Mukden, and by Wednesday 23 September a plot had been discovered involving the naval officers of the Kasumigaura Naval Air Station. These pilots at the Misty Lagoon Airbase were implicated in a conspiracy to circumvent the decisions of the League of Nations over matters concerning Manchuria and north China. They intended heading-off the economic sanctions to be imposed by the League. The capture of the entire region of Manchuria was considered vital, as was the silencing and control of the banker industrialists in Japan. This was necessary as the Emperor had pledged a pause for reflection to the Prime Minister, Prince Saionji, a liberal, and a banker whose political defeat was of paramount importance. These pilots of the Kasumigaura Naval Air Station were naval lieutenants and commanders considered to be crack flyers in the naval aviation arm, at least equivalent to two grades above corresponding personnel in other services, especially since they would have been seaman and petty officer ranks in more normal circumstances. They were students of Admiral Yamamoto and protégés of the Empress Nagako's father, Prince Kuni, pioneer of Japanese air power. They were also comrades of the Imperial torpedo-bomber test pilot Prince Yamashina, who was the best of five aviators in the Imperial family.

Japan was still importing aircraft from abroad and now took delivery of the Boeing 100 from America and the German Heinkel HD-62, but the development of the KF1 flying-boat was of the greatest significance. Early in January 1931 a party from Short Bros was dispatched to the Kawanishi Aircraft Company at Naruo, on the coast between Kobe and Osaka. They were under the control of Mr C.P.T. Liscomb, who was accompanied by seven other engineer specialists concerned with design, stress, construction and inspection. Their mission was to superintend the assembly and trials of the new KF1 flying-boat at Naruo, which was flown to Japan at the end of February via Canada. In March the official launching took place. This machine had a span of 101 feet 10 inches, length of 74 feet 5 inches and a wing area of 2,300 square feet. Empty, the machine weighed 22,100 lb, all-up weight being 39,000 lb at a maximum speed of 124 mph, with a maximum range of 2,000 miles. On 8 April the English test pilot John Parker flew the machine under light load for one hour. The next day the flying-boat was further tested for one and a half hours with an all-up weight of 35,000 lb. Four days later, on 13 April, speed trials were conducted at 122 mph with an all-up weight of 35,000 lb and pronounced highly successful. The following day, overloaded to 39,000 lb, the KF1 was flown under test for one hour. The directors of the Kawanishi Company, as well as the Navy pilots, were highly delighted with the test results. The aircraft was immediately ordered into full-scale production. In the meantime it was decided that Mr Liscomb and his fellow engineers would stay on at Kawanishi Aircraft with the express function of instructing and supervising the construction and future development of this new flying-boat. The English engineers did not depart from Japan until 1932. In March 1932 the KF1 Type 9-2 (H3K2) flying-boat flew, built entirely of Japanese materials, but equipped with Rolls-Royce Buzzard engines imported from England. In July of the same year a KF1 flying-boat flew non-stop from Tateyama Naval Air Station to the island of Saipan in the Mariana Islands, South Pacific, a distance of 1,400 miles, under the command of Admiral Sukemitsu Ito. This was the headquarters capital of the islands in the Pacific.

CHAPTER 7

Escalating the Shanghai Intervention

For the Japanese Naval Air Service the years 1932 to 1945 were a period of original aviation thinking and initiative in which aircraft manufacturers launched their own research and development programmes.

On 31 January 1932 the following Naval Air Service units were on active operations in support of the Japanese ground forces. The aircraft carriers HIJMS *Kaga* and HIJMS *Hosho* formed the 1st Air Wing and together operated twenty-six Type 3 fighters and forty-one Type 13 attack-bombers. Also, HIJMS *Notoro* flew eight Type 14 reconnaissance planes and Type 90 MK III aircraft, and the cruiser HIJMS *Yura* had a single Type 90 reconnaissance plane aboard. All seventy-six aircraft flew in direct support of the invasion forces. The Shanghai conflict had commenced on 29 January, having started as protests by the Chinese population against the invasion of Manchuria, which resulted in Japanese persons and property being interfered with by the local civilians. As a result Japanese marines had landed, and severe battles were fought with the Chinese Army at Wusong, Nanxiang and Jia Ding, all in the vicinity of Shanghai. The Chinese air and ground resistance was more powerful than that encountered by the Japanese forces in Manchuria. In the air the Japanese Naval Air Service generally commanded, so that on the first day of the fighting the sea reconnaissance aircraft of the seaplane carrier HIJMS *Notoro* off the coast, using eight Type 14 and Type 90 MK III aircraft, attacked and bombed Shanghai with small-calibre bombs.

The first aerial combat took place over the crowded city of Shanghai high above the business quarter, when three Type 3 and two Type 13 aircraft on patrol from HIJMS *Hosho* engaged nine Chinese fighter aircraft. Among the Chinese formation of nine fighters were Chance Vought Corsair pursuit planes previously purchased from the United States. In a wild mêlée of twisting vapour trails, the outcome of the air battle was one Chinese plane severely damaged, with an injured pilot, but the Japanese sustained no losses. During 22 February three Type 13 attack-bombers and three Type 3 fighters on patrol from HIJMS *Kaga* were attacked by a Boeing P12 fighter previously acquired from the USA. After a two-minute aerial combat the Boeing P12 was shot down in flames, with the pilot – a US citizen named Robert Short – killed, slumped over the controls. Next day the Japanese Naval Intelligence Service received reports of a build-up of Chinese Air Force squadrons preliminary to the commencement of large-scale air attacks on the Japanese positions and ships. An immediate air offensive was launched by the Japanese Navy against Chinese airbases located at Suzhou, Hangzhou and Hongqiao between 23 and 26 February, in which nine Chinese aircraft were destroyed on the ground. While the airfield at Hangzhou was attacked, a large air battle took place in the vicinity of the aerodrome, with eighteen Japanese aircraft engaged by five Chinese defending fighters. In between the weaving planes and the staccato rattle of aviation machine-guns three Chinese fighters plunged earthwards, flames and smoke pouring from each machine before they exploded in balls of fire upon impact. During this period the Japanese Naval Air Service did not lose a plane to any Chinese fighters in aerial combat, but three planes were lost to Chinese ground fire. In Manchuria events were going wrong for the 330,000-strong Chinese Army, which was being beaten back by the Japanese Kanto Command Army, and in the following March the invading Army installed a puppet government to rule South Manchuria, now called Manchukuo. The campaign continued, with penetrating thrusts from bases in Changchun, Qiqihar and Harbin. These Japanese attacks captured Hailar, Jiamusi and Heiko.

Back in Japan on 1 April 1932 at Taura in the Kanagawa Prefecture, the Naval Aircraft Establishment was created next door to the Yokosuka Experimental Air Corps grounds, which itself was absorbed into the new organisation as the Science Division under Engineer Captain Nagamasa Tada HIJMN. Other divisions of the new branch were as follows:

General Affairs Division	Captain Goro Hara HIJMN
Engine Division	Engineer Captain Koichi Hanajima HIJMN
Flight Testing Division	Captain Daijiro Ichikawa HIJMN
Aircraft Division	Captain Hideko Wada HIJMN
Ordnance Division	Captain Tozo Nakamura HIJMN

This new organisation was the most important contributor to the rapid rise of Japanese military aviation. Design programmes were immediately initiated for carrier fighters, dive-bombers, attack-bombers and reconnaissance seaplanes. Specifications were issued to civilian manufacturers on a competitive basis as well as a separate Navy project for a land-based bomber aircraft.

In February 1932 the Mitsubishi 7-Shi carrier fighter was available for evaluation and testing. It was observed that the fuselage was a clumsy shape with a body out of proportion, a semi-monocoque duralumin structure with longitudinal stringers only, with transverse frames for bracing. The centre-line of the machine-guns ran above the engine, which was a double-row radial 14-cylinder air-cooled Mitsubishi A4 rated at 710 hp. The upper fuselage was consequently cambered with a high cockpit for good visibility on carrier landings. Since the machine-guns had to be installed forward of the pilot to facilitate easy access for repairs as required, the panel fasteners protruded above the surface of the fuselage skin. The wing was a two-spar fabric-covered structure with a diagonal spar connecting members taking the torsional movement. Manufacture of the domestic duralumin wings was a production question, so wing spars and torsion members were made into a Wagner beam structure with tension field webs. This idea was originated by Captain Misao Wada and Engineer Lieutenant Commander Juniro Okamura of the Hiro Naval Arsenal. The flight tests commenced in March 1932 under the control of pilot Yoshitaka Kajima at the Kagamigahara Airfield, with the assistance of Sumitoshi Nakao, who at the time was testing the Mitsubishi 7-Shi attack-bomber. At about this time the Nakajima prototype fighter was under flight test. This machine was an adopted Army Type 91 fighter suited for naval purposes; while attaining superior manoeuvrability over the Mitsubishi entry, it was nevertheless eight mph slower. In fact both machines were performance short on the Naval Air Services specification. On the Mitsubishi plane the air resistance was higher due to surface frictional drag and large engine diameter installation. The climbing performance was quite inadequate due to propeller inefficiency and insufficient engine power. The prototype eventually crashed due to the disintegration of the tailfin while in a power dive over Kagamigahara Airfield. The pilotless plane glided, turned and gently landed on the south-eastern side of the airfield on the edge of the Kiso river, where it disintegrated due to inadequate strength of the vertical stabiliser. Fortunately the pilot, Kajima, landed quite safely nearby at the same time.

The second Mitsubishi prototype was delivered to the Naval Aircraft Establishment Flight Testing Division, where the machine was tested by Lieutenant Yoshito Kobayashi HIJMN, but in the final consideration the machine was rejected. Flight tests did continue and many technical problems were solved. Eventually the prototype was turned over to the Yokosuka Naval Air Experimental Corps, where further flying was undertaken by Lieutenant Motoharu Okamura, chief of the fighter squadron. Unhappily the aircraft was whipped into a flat spin as the pilot attempted a double roll. As it began to break up, the pilot had to bale out, and in the course of his escape the fingers of his left hand were cut off by the whirling propeller blades. Both of the Mitsubishi prototypes had been completely destroyed and one test pilot injured. The outcome was the replacing of Mitsubishi as a supplier of carrier fighter aircraft by products of the Nakajima Aircraft Company. The Mitsubishi Type 89 B2M1 and B2M2 possessed inadequate flight performance due to structural weaknesses and were plagued by engine trouble. The Naval Air Service did not accept a replacement but took into service the Aichi/Hiro B3Y1, and for Mitsubishi the failures were a severe loss of morale in its design abilities. In April 1932, at a conference held at the Naval Aircraft Establishment under the chairmanship of Vice-Admiral Yurikazu Edahara, with aircraft manufacturers and engineers present, it was accepted by all that the six Shi aircraft requirements had failed the Navy specifications and all references to these specifications would be abandoned. Admiral Edahara stressed the importance of a three-year design period, commencing with the 7-Shi aircraft requirements. For the 7-Shi carrier fighter the Navy set no specific criteria, nor was a basic design configuration laid down. Only listed requirements were quoted, in that the new fighter aircraft should have a maximum speed of 208–230 mph at an altitude of 9,842 feet, with a rate of climb of four minutes to this operational height. With a wingspan of no more than 33.8 feet to facilitate aircraft carrier stowage stipulated, the specification was concluded.

This specification gave an increase in performance over the Nakajima A1N1 and A1N2 Type 90 carrier fighter, which was popular with pilots and maintenance crews alike. It achieved a maximum speed of 136 mph at 9,842 feet, with a rate of climb to operational height of 6 minutes 10 seconds. At the time the biplane configuration was the accepted design type in America, Europe and Japan, but the NAE considered other design layouts, including a parasol-wing type. Commander Jiro Saba of the Naval Bureau of Aeronautics suggested a low-wing monoplane design for high-speed flight, and had prepared a gull-wing layout sketch. This suggestion proposed a low cantilever wing without external bracing, which at the time was not considered capable of standing high-speed stress. An elliptical wing layout was further suggested by Commander Saba to give the minimum air resistance at high angles of attack during aerial combat. These ideas would present manufacturing difficulties, though use of the M6 aerofoil section was adapted, which it was considered would eliminate torsional movement in a dive.

Meanwhile Prince Fushimi was appointed Chief of the Naval General Staff, while Prime Minister Saionji was not consulted, but only informed of the new appointment. Both the Army and the Navy were controlled by Imperial princes. At the time the officers of the Misty Lagoon were operational on missions bombing Chinese structures next to the Western-owned docks and warehouses of Shanghai. There accuracy was highly commendable, and while European installations were left intact and undamaged, no undue concern was shown despite the obliteration of Chinese property next door. Business was too good to protest! The constant bombing practice until 100% accurate results were achieved was now paying off! The demolition work was almost mathematical and precise. In the period from September 1931 to February 1932 during the Manchuria invasion and the Shanghai Incidents, the Naval Air Service had the edge on the Army in the field of aeronautical developments, though unfortunately many of the home designs were inferior. In an endeavour to overcome this technical inferiority, aircraft were imported from abroad – the Savoia S62 from Italy, the Hawker Nimrod from England and the Cierva Autogiro. In production the Mitsubishi B2M1 Type 89 Model 1 was fitted with a Hispano-Suiza Type 7 engine of Mitsubishi manufacture, developing 650 hp. With tailplane, fin and rudder of rounded shape, the elevator horn balances were reduced in size and the nose redesigned to a slimmer line. A shuttered chin compartment was deleted and a retractable radiator fitted. The design competition for the replacement of the Type 89 torpedo-bomber had produced designs by Mitsubishi for the 3MT10, the Nakajima Y38 and the Aichi A8B, none of which was considered by the Navy as acceptable. In the end the Yokosuka Arsenal B37 improved Type 13 carrier-based attacker equipped with a 600 hp Type 91 engine was accepted. A total quantity of 130 were built, but engine trouble was constantly experienced, plaguing the design. The Navy was determined, despite the disappointment of all these failures, to acquire a bombing aircraft with the potential of a large payload and long range to fulfil the role of reconnaissance and bombing operations. The Mitsubishi Company designed the Type 93 G1M, a twin-engine biplane bomber of which a total of eleven were built. Unhappily these machines suffered from stability problems and engine trouble. Consequently the Navy ordered the Hiro Naval Arsenal to develop a large all-metal monoplane with a fixed undercarriage to fill the role. However, the successful aircraft company designs then in existence and accepted included the Kawanishi E7K1 three-seater reconnaissance seaplane, the Hiro Naval Arsenal G2H1 twin-engine land-based attack-bomber, the Mitsubishi low-winged carrier fighter, which interested the Navy but had poor performance, and the Nakajima modified Army Type 91 fighter equipped for naval usage. Unfortunately the dive-bomber designs from Aichi, Nakajima and Mitsubishi had been rejected as unsatisfactory. On behalf of the Navy it must be said that all this design and engineering effort was totally unsatisfactory, failing in one respect or another the specifications laid down. Indeed, only eight examples of the Hiro Naval Arsenal's G2H1 were built for these reasons.

The London Disarmament Treaty had set specific limits on the construction of aircraft carriers and other warships. In reply the Navy had initiated the philosophy of using long-range non-carrier-based aircraft in co-operation with the fleet for offensive operations. The problems had been posed by Admiral Matsuyama, chief of the Naval Bureau of Aeronautics, by Admiral Yamamoto, chief of the Technical Division, and Commander Wada. A technical comparison was made into the design of a land-based aircraft as compared with an equivalent flying-boat design, which was submitted to Admiral Yamamoto. A long-range aircraft design was required, but flying-boats appeared unsatisfactory, possessing poor performance, a penalty on weight and air resistance encountered during landing and taking-off from water. At the time airfields for long-range bombers did not exist, as great cost and effort would be involved in constructing a chain of airbases with operational and maintenance facilities. But such land-based planes would possess an increased performance. The report to Admiral Yamamoto set out the technical merits and liabilities in the comparison of flying-boats versus land-planes. Unhappily the Navy lacked manufacturing experience, although Commander Wada was skilled in designing and the Navy possessed at least three successful examples of flying-boat design – namely the H3H1 Type 90, the H4H1 Type 91 and the Kawanishi H3H1 Type 90 Model 2 built under licence from Short Bros. It was during 1932 that the Navy decided to abolish the airship and balloon squadrons. Unhappily both the Navy and the aircraft manufacturers had failed to develop an airship for naval escort purposes. When the war subsequently broke out between the Nipponese empire and the Barbarian Western democracies, the Navy did not have suitable aircraft as anti-submarine convoy escorts. The United States Navy persevered with airship construction and successfully provided protection against enemy attack on Allied convoys.

A new expansion programme was initiated for completion by the year 1937. This new plan called for the building of eight land-based bomber squadrons, two aircraft carriers of 20,000 tons, each flying seventy-three aircraft, and three seaplane carriers of 13,000 tons, each accommodating thirty-six seaplanes.

Late one dark January evening in 1933 night-flying training was in progress off Tateyama using one of the new Kawanishi KF1 flying-boats, which was used normally on the long overseas run to the island of Saipan, 1,400 miles away. While attempting a sea landing, the KF1 stalled and crashed into the Pacific Ocean. Immediately the machine broke up, killing three of the crew of nine, including the pilot. On subsequent investigation it was determined that the accident was caused by altimeter lag and pilot error. Two modified versions of the KF1 were now built, known as the H3K2, in which modifications had been made to the bow; enclosed pilot cockpits were also fitted, as well as long exhaust pipes suppressing noise from the overhead engine installations. The chief of the Technical Division of the Navy Bureau of Aeronautics, Rear Admiral Yamamoto, was highly dissatisfied with the technical situation. He considered Mitsubishi could build a superior land-based bomber following the German Junkers design techniques fused with the ideals of the Hiro G2H1 7-Shi attack-bomber. The precise technical requirements would be left to the design team of Mitsubishi, for Admiral Yamamoto feared that the Navy was too unrealistic in its demand for developments, resulting so very often in unrealistic final products. However, the chief of the Technical Division insisted on his subordinates specifying the range required, a fuselage modelled on the G2H1 7-Shi attack-bomber design and a decision that the aircraft to be produced by Mitsubishi would be a non-competitive prototype.

The Mitsubishi Aircraft Company set up a special design team to build the G3M, with Sueo Honjo appointed as chief engineer, since he had considerable experience with Junkers-designed aircraft as well as experience designing bombers for the Japanese Army Air Force. The two flight engineers were Tomio Kubo and Nobuhiko Kusakabe. In the course of the design, Junkers ideas were incorporated with the general fuselage arrangements of the G2H1 bomber. Junkers corrugated skin was used on the rear sections of the wings, as well as Junkers standard control mechanisms, and double wing ailerons were designed into the wing structure, with miscellaneous sheet-metal parts located on the aircraft elsewhere. In April 1934 the first prototype was wheeled out of its hangar, and the maiden flight was undertaken by Yoshitaka Kajima, chief test pilot for Mitsubishi. After a series of test flights it was found that the maximum speed was 164.9 mph for a range of 2,737 miles, which could be extended by trimming the aircraft for a range of 3,760 miles. Interestingly the maximum speed equalled that of the Nakajima A1N2 Type 90 carrier fighter in current service.

At the time in 1933 that Mitsubishi set up its special design team to build the G3M, four problems remained for the naval aviation builders:

1. The poor radius of action of the single-seater fighters operating from aircraft carriers at sea.
2. The inability of reconnaissance float-planes or attack-bombers to conduct artillery spotting for the fleet.
3. The poor performance of high-level horizontal bombing against high-speed targets at sea.

4. Insufficient numbers of carrier-borne aircraft due to the limited number of aircraft carriers that could be permissibly built under the treaty regulations.

To overcome these problems it was resolved:

1. To introduce the concept of the two-seater fighter plane with a longer range and endurance.
2. The re-equipping with higher performance reconnaissance float-planes.
3. The introduction of the dive-bomber, hitherto an unknown class of plane in the Japanese Naval Air Service, for use against high-speed targets.
4. Designers would envisage the concept of large land-based bombers for over-ocean reconnaissance as well as bombing missions, which would counterbalance the lack of sufficiently large aircraft carriers.

All these ideas were embodied in the 8-Shi prototype building programme issued to the aircraft manufacturers. By the summer of 1933 it was of special importance to Admiral Yamamoto that the Pacific war could not be opened without the building of fleets of long-range bombers capable of operating from bases in the Micronesian Islands. These planes would be used for the destruction of enemy fleets before they could arrive in the vicinity of the Nipponese island empire. Thus Mitsubishi was awarded the non-competitive design contract to build the G3M series. The idea was not lost on the American Barbarians whose assistant chief of the US Army Air Corps had reiterated a tactical principle for the new forthcoming Boeing B-17 series, the Flying Fortress, in asserting that:

1. The bombing operation would have to be accomplished without fighter defence support.
2. These strategic bombers would be heavily armed.
3. The bombers would possess an ability to fly in close formations for defensive purposes.
4. The aircraft would maximise defence in the air by use of improved tactics.
5. The engines would be equipped with improved silencers in a belief that sound-detecting equipment would be unable to report the approach of a bombing formation.

At the time the US Army Air Corps was engaged in the annual air exercises in the American far western states, including exercises in advanced flying training, in bombing operations and air-to-air firing by fighter aircraft in attack formations. The intelligence of such operations would not be lost on Admiral Yamamoto, who only too readily appreciated the American intentions as a threat to Japanese independence.

In 1933 the Nipponese Navy was still importing foreign aircraft for evaluation purposes. So when a Northrop Gamma was acquired from America, a shock ran through the Naval Bureau of Aeronautics. The Northrop proved to be of an advanced design, technically superior to the outdated, slow, fabric-covered biplanes with old-fashioned equipment. As a result Nakajima was contracted to design and develop a version of the Type 90 fighter as a sesquiplane with an 800 hp Hikari radial engine. This machine was designated as the Type 95 A4N carrier fighter. The development of a dive-bomber took place in accordance with the previously mentioned policy decision. Yokosuka combined with Nakajima to enter the RZ (D2N) based upon the experiences with the 6-Shi and 7-Shi aircraft. The Navy, however, accepted the Aichi AB9 (D1A1), which was designed from the German Heinkel HD-66, which had been itself designed by Heinkel specially for the Aichi Aviation Company. Both machines were fabric-covered biplanes seating a pilot and observer. Designated the D1A1, a total of 162 were built, each powered by a 580 hp Kotobuki II engine and a bomb load of 660 lb. Dive-bomber tactics were introduced into the aircraft carrier fleet. Also, 550 warship-catapult-launched float reconnaissance planes were built from a design of Nakajima – the E8N1. This float-plane was a two-seater armed with two 7.7 mm machine-guns, and was capable of carrying 132 lb of light bombs. The Hiro Naval Arsenal developed a new flying-boat – the H4H – to replace the Type 15 and 89, which were now obsolete. This was a single-engine monoplane with a 750 hp Myojo engine, having a gross weight of 16,800 lb. A total of forty-seven were built for patrol and transportation purposes during the Sino-Japanese war. During 1933 a Hawker Nimrod had been imported from England. This machine was one of four exported –two to Denmark, one to Portugal and one to Japan. The machine had a span of 33 feet 6¾ inches, a length of 26 feet 6½ inches, a height of 9 feet 10 inches and a wing area of 301 square feet. It was powered by a Rolls-Royce Kestrel III engine, giving the aircraft a maximum speed of 193 mph with a service ceiling of 28,000 feet. The plane first flew in 1930 as the HNI, employing Hawker construction with flotation bags in the upper wings and rear fuselage, and provision was made for wheeled or float alighting gear. First demonstrated in the Argentine in March 1931, an example was sent to RAF Martlesham Heath for fighter evaluation. The first two production Nimrods were sent to Japan for demonstration purposes in October 1931 as S1577 and S1578. By June 1932 Nimrods had been cleared for deck landings aboard the aircraft carrier HMS *Eagle*, with further developments being introduced for pilots' head-rests and deck arrester gear. The thirty-sixth airframe was modified to take swept-back wings as the basis of the Nimrod II. The English Blackburn T.7B or Mitsubishi B2M1 and B2M2 were now serving as attack-bombers aboard the carriers HIJMS *Ryujo* HIJMS *Akagi* and HIJMS *Kaga*, as well as being fully operational on land with the Tateyama and Omura Air Corps. Aircraft were also being installed aboard Japanese submarines during 1933, float-planes were loaded onto the submarines I-5, 7, 8, 9, 10 and 11. From 1934 to 1941 two submarines each year were equipped to carry an aircraft. In May 1933 the conversion work on the carrier HIJMS *Ryujo* was completed, with a final tonnage of 8,000 tons and over 591$^1/_3$ feet long.

On 4 August 1933 the Chief of the Naval General Staff, Prince Fushimi, asked his colleagues in the government bureaucracy to support a major change in the Navy regulations that would have the effect of putting the Navy on a permanent war footing. Hitherto the Naval General Staff had to have clearance with all orders and regulations from the Navy Minister in the government of the day. Under Prince Fushimi's proposals, only the Emperor's clearance would become necessary. The Army and Navy Ministers protested most vigorously to the Emperor, but to no avail. By 30 August a most important Cabinet meeting took place in the presence of His Imperial Majesty. The Emperor asked penetrating questions concerning the Japanese colonists in far northern Sakhalin and in Dutch New Guinea in the south. However, despite the attacks made on the Navy positions and strong counter-arguments by representatives of the Strike North factions, who believed that the communist Soviet Union was the real enemy, His Majesty favoured striking south. On 21 September the Emperor approved the new regulations, and from henceforth all fleet movements came under his control. At the same time he requested that all matters pertaining to naval activities should be kept secret from the public. Regulations were laid down simply and sharply. On 25 September the Cabinet sanctioned the official Strike South policy, and the Army's policy of viewing the Soviet Union as the greatest enemy was not acceptable. As a result more money was provided for the Navy budget and the Foreign Minister declared that Japan would expand diplomatically, but after that recourse to the military solution would be resorted to. By 20 October the Cabinet gave the seal to the Emperor's programme to build up military strength and the Cabinet to realise the many international goals set by the Emperor. The situation presenting itself now was the exclusive victory of the Strike South faction. This group was mainly controlled by the powerful Satsuma clan, which had pioneered the cause of the Navy and provided so many of its principal officers. It believed that Japan's best interests lay in a rapid southern advance to the Dutch East Indies, Malaya and Indo-China, and to Burma where vast untapped raw materials of one sort or another were to be found. The capture and exploitation of these lands would enable Japan to become independent for raw materials and to further bar ambitions in the Pacific hemisphere to protect the homeland and island empire from the influences of the foreign Barbarians.

In January 1934 the Eleven Club, the Emperor's private cabal, met with the express purpose of deciding on a policy to free the Japanese aviation industry from dependence upon Western aircraft parts and spares. Those present included Prince Higashikuni, Kido and Secretary Spy Harada, who with the executives of the aircraft industry joined senior officers of the two military air forces. The agenda of the meeting included the policy to form a joint Army and Navy Tokyo Imperial University Aviation Institute. A few months later this institute was organised under the direction of Wada Koroku, an aeronautical engineer who was Kido's brother, and Professor Tanakadate Aikitsu, a 77-year-old Professor of Physics who had been an aviation lecturer to the Emperor. Having privately announced his intention to look south, His Imperial Majesty concerned himself with the organisation of the Civilian Spy Service in south-east Asia, which had collected intelligence information ever since 1900. The service was now called upon to intensify its activities as from 1934. The members of the Spy Service were in the main businessmen established in legitimate trade, commercial travellers moving about the business world and others. On 1 October 1934 Prince Kan'in visited the headquarters of General Count Terauchi in Taiwan. The forward plans of the spy organisation were approved by the Imperial visitor. All maps, plans and general information were to be collected from civilian sources concerning Malaya and Indo-China, Luzon and the Dutch East Indies. The analysis of the information acquired would form the operational basis for orders to be issued by Unit 82 staff officers – sometime known as the Southward Movement Society when the invasion of south-east Asia was imminent. Sometime during 1934, President Roosevelt informed Mr Henry Stimson, the Secretary of State, that Japan had a hundred-year plan of conquest, which had been originally drafted as far back as 1889. This plan listed the annexation of Manchuria, the formation of a protectorate of north China, and the acquisition of US and British possessions in the Pacific Ocean area, as well as the conquest of Hawaii, Mexico and Peru. Apparently when President Roosevelt was a graduate at Harvard University a Japanese student had informed him of this plan of conquest as far back as 1920. On 30 December 1934, influential US newspapers printed reports of the proposed programme of US fleet manoeuvres for the year 1935. These were to be the greatest game of mock naval warfare yet held in the Pacific Ocean. The area covered by the manoeuvres was in a triangle bounded by a line drawn from the Aleutian Islands to Pearl Harbor, Midway Island and back to the Aleutians. Naturally these newspaper reports dissimulated the American peace societies into action of disapproval, while the newspapers claimed that the exercises were directed against the best interest of Japan. But much more serious in Japanese eyes was the US Vinson Plan for naval expansion, which was considered as a direct threat and stimulated the Japanese Navy to expand. The Japanese Navy conceived a plan, to be completed by 1937, to add eighteen squadrons of aircraft, totalling 262 machines, to the Naval Air Service, as well as an additional 294 ship-based aircraft and the construction of two more aircraft carriers, the entire programme being expected to cost ¥44,000,000. To commence this vast aeroplane construction programme, the Navy yards at Yokosuka, Hiro and Sasebo commenced the manufacture of new machines. A further two new naval air stations were constructed at Maizuru and Chinkai. Since the Shanghai Incident the naval aviation expenditure had accumulated to ¥3,170,000 by March 1934 for sixty-one aircraft and anti-aircraft defences. There were now eight naval air stations in Japan, located at:

1. Kasumigaura housing	7 squadrons + 1 flying school	
2. Yokosuka with	5 squadrons + 1 balloon squadron and 1 aerial	
3. Tateyama with	3 squadrons (navigation school)	
4. Sasebo with	2 squadrons	
5. Ohmura with	2 squadrons	
6. Kure with	1 squadron	
7. Saeki with	1 squadron	
8. Ohminato with	1 squadron	
Total	22 squadrons of approx. 950 aircraft	

The headquarters was still located at Kasumigaseki, Tokyo, but the director was Vice-Admiral K. Shibusawa.

In March 1934 the torpedo-boat HIJMS *Tomozuru* capsized, and it was established at a subsequent naval enquiry that the ship was top heavy and insufficiently ballasted. As a result of this enquiry several ships were reconstructed or renovated, including HIJMS *Ryujo*, which underwent extensive modifications. The deck guns were removed and a ballast weight added, which gave her a displacement of 10,600 tons. She had no island structure and the exhaust gases were discharged through two uneven funnels on the starboard side. Her armament consisted of twelve 5 inch guns and she accommodated a total of thirty-six aircraft.

In the course of 1934 Aichi Aircraft Company produced the Type 94 D1A1 dive-bomber, which was a modified Heinkel He 50 from Germany.

The new plane reflected the influence of German design and was far superior to the dive-bomber produced by Nakajima. The Aichi D1A1 was the first dive-bomber to be accepted into naval service. It had a top speed of 174 mph, was armed with two 7.7 mm machine-guns fixed to fire forward operated by the pilot and one free mounted 7.7 mm machine-gun for use by the observer. This plane carried 613 lb of bombs on normal range. Unfortunately, at the same time the two-seater fighter designs of Mitsubishi and Nakajima were failures, unable to reach Navy specifications.

The original Blackburn T.7B (3MR4) or Mitsubishi B2MN1 had been developed into the B2M2 becoming the Type 89 Model 2. Armed with two Vickers guns and possessing an offensive load of 1,764 lb of bombs, or one torpedo, the machine's range was short. It had a reduced wingspan, a triangular fin and rudder, and a slimmer nose, but the aircraft was much overweight compared with the original T.7B. Production of the B2M1 and B2M2 totalled 205 aircraft, but Mitsubishi found the new all-metal constructional technique expensive compared with the original wooden construction and fabric covering. The Handley Page slots manufactured under licence were also an expensive item. The Japanese-built Hispano-Suiza engines were unreliable and several fatal accidents resulted. Development of the T.7B was as follows:

T.7B	Torpedo	Reconnaissance	B2M1	B2M2
Tare weight (lb)	3,896	3,916	4,982	4,805
Military load (lb)	3,070	1,758	2,854	3,130
All-up weight (lb)	7,966	7,500	7,936	7,936
Speed, sea level (mph)	132	135	132	140
Speed, 6,000 ft (mph)	124	128	-	-
Cruising speed (mph)	112	115	112	119
Initial climb (ft/min)	680	810	600	900
Service ceiling (ft)	11,500	14,300	-	-
Range (miles)	390	800	1,100	1,090

Meanwhile the Naval Air Service had imported a German Heinkel HD-66 for evaluation purposes. The Kawanishi Company built the E7K1 Model 94, a twin-float, fabric-covered seaplane with a crew of three, weighing 5,500 lb and powered by a 500 hp V12 Type 91 water-cooled engine manufactured by Hiro Navy Arsenal, giving a maximum speed of 160 mph and achieving an endurance of fourteen hours, which set a new world record. The Model 2 was equipped with a Mitsubishi 600 hp radial engine, which increased the maximum speed to 170 mph. Production of this machine lasted from 1934 to January 1941, 183 examples of the E7K1 being produced, and 357 machines of the E7K2 version being manufactured. Most important was the first flight of the Mitsubishi G3M twin-engine 8-Shi bomber, which took place in May 1934. This machine was designed by Kiro Honjo as a mid-winged, twin-engine aircraft with a retractable undercarriage and a maximum range of 2,880 miles over a duration of twenty-four hours. Subsequently this machine was to be continually developed by the turn of the summer of 1935. The A4N1 carrier fighter, which was a highly manoeuvrable fabric-covered biplane, was replaced by the Mitsubishi A5M1, a monocoque monoplane on which the emphasis had changed to a compromise between speed and manoeuvrability, as designed by Jiro Horikoshi. Both the A5M1 fighter and the G3M land-based bomber were the products of the Mitsubishi Aircraft Company.

During 1934 the Navy's 9-Shi specification for a new carrier fighter was developed by Mitsubishi as the A5M. This machine was to have a maximum speed of approximately 218 mph plus at 10,000 feet, a rate of climb to 16,700 feet in 6½ minutes, with a fuel tank capacity of 53.3 gallons. The plane was to be armed with two 7.7 mm machine-guns, and a radio receiver was to be installed. The approximate dimensions were span 36.1 ft and length 26.3 ft. It was proposed to use the services of the same design team that had built the 7-Shi fighter series, since the general configuration would be based upon the 7-Shi programme. It was envisaged that the wing thickness and fuselage diameter would be reduced to the lowest possible diameter. The wing had to be manufactured of aluminium alloy, with a two-spar box type of construction. Aluminium-alloy skin covering with flush rivet construction would give the minimum skin friction using German Junkers-type rivets. The landing-gear was to be fixed in streamlined cantilever legs. In the autumn of 1934 it was learned from an aviation magazine that the Heinkel He 70 four-seater mail plane of 1932 had flush-riveted fuselage and wings with countersunk bolts. The panels on the outer skin were located by flush-headed screws into dimpled recesses. Naturally such knowledge was copied and made good use of in subsequent machines. The engine used was a Nakajima Kotobuki Model 5 fitted with a reduction gearing and based on the nine-cylinder British Bristol Jupiter radial engine. This engine was much praised by the Navy.

The Japanese designers were turning from the biplane type of configuration to the monoplane layout, which was not liked at first by the pilots because the view forward and downward was obscured. The gull wing of the 7-Shi fighter was suggested by Commander Jiro Saba HIJMN of the Naval Air Establishment, but the idea was rejected by the designers as too complicated for production purposes. Though the gull wing had been used by the British on the Supermarine F7/30 machine of 1934 with success, the Japanese engineers found that instability due to stresses on the crank wing resulted in loss of control, and the idea was shelved. In fact the gull wing was not used on the second prototype, and an elliptical wing layout, as previously used on the 7-Shi series aircraft, was now successfully utilised on the 9-Shi machine. This wing had an aspect ratio of 6:1 and an aerofoil section not unlike that used on the 8-Shi twin-engine bomber. The two-spar box structure originally introduced by the Aircraft Department of the Hiro Naval Arsenal was to be used, as was the case with the 7-Shi fighter. Though the French were introducing retractable undercarriages, the increased weight and the complication of the mechanism far outweighed the increase of speed, but in any case the development of a retracting undercarriage would delay the prototype by three months.

The planes under construction at this time for the Naval Air Service were the Aichi D1A2 dive-bomber, the Nakajima B4N1 attack-bomber, the Aichi AB4B night reconnaissance flying-boat, the Mitsubishi G3M1 non-competitive land-based bomber, the Kawanishi H6K1 flying-boat and the Watanabe E9W1 submarine-based reconnaissance plane.

During 1934 a design competition took place for a sesquiplane design. Both Mitsubishi and Nakajima entered designs, but Yokosuka won the competition with the 84Y, later designated the Type 96 carrier attacker. This machine was powered by an 840 hp Hikari engine, and the plane was the last machine to be fabric covered. Over 200 examples of this aircraft were produced, for it was a trouble-free, long-range machine with an unimpaired design.

At a conference held in December 1934, Japan requested parity with the USA and Britain in warship construction, but the two Western powers rejected the proposals. Immediately Japan gave formal notice of renouncing the Naval Treaty and all its abrogations. Prior to May 1934 the Japanese Kwantung Armies had entered Jehol in western Manchuria, which forced the Nationalist Chinese Armies south of the Great Wall. The Chinese were unable to rely upon assistance from the League of Nations any further. By the armistice of Tangku, China agreed to the loss of Manchuria and Jehol, and further agreed to the formation of a buffer zone south of the Great Wall from which all Chinese forces would be withdrawn. The victory of the Kwangtung Army thus had many of the extreme senior naval officers restive for action.

CHAPTER 8

Welding of the Weapons

During the course of 1935 three attempted assassinations took place on the lives of the Finance Minister, the Lord Keeper of the Privy Seal, Admiral Saito, and General Watanabe, the Inspector-General. Also, a further naval conference took place convened late in the year at which the Japanese once again demanded ship parity with the USA and Britain. Parity was not agreed and the Japanese delegation left the conference in disgust.

In Japan a naval aviation design competition was initiated for a new low-wing monoplane torpedo-bomber. Mitsubishi produced the B5M, powered by a 1,000 hp Kinsei engine designed by Katsuji Nakamura. This plane was characterised by a fixed landing-gear and folding wing. The Nakajima Company entry was the B5N, with a retractable undercarriage and folding wings, and powered by an 840 hp Hikari engine, being the product of designer Joji Hattori. After preliminary tests the Navy liked both machines and decided to place orders accordingly. Eventually 1,250 Mitsubishi B5M Type 97-I carrier-based attackers, and 150 Nakajima B5N Type 97-II carrier attack-bombers, were manufactured. Both aircraft were used for the Sino-Japanese War until the end of the Pacific War. Meanwhile the all-important work on the Mitsubishi A5M carrier fighter continued, so much so that by January 1935, some ten months after the inception of the project, the design work was completed. The appointment of Lieutenant-Commander Kobayashi HIJMN was confirmed as chief test pilot of the 9-Shi programme, in charge of fighter pilots. The new commanding officer congratulated the design team by presenting a Haiku – a seventeen-syllable verse expressing praise – on a suitable illustrated manuscript. The flight test commenced on 4 February at Kagamigahara Airbase, some fifteen miles north of Nagoya, the home of the Mitsubishi Aircraft Company. Mitsubishi's pilot, Kajima, commenced the test flying, in which the new plane exceeded all expectations, the maximum speed being 276 mph. The Nakajima entry only achieved 247 mph, well below that of the A5M. Unfortunately the landing run of the Mitsubishi machine proved too long for use on aircraft carriers. Meanwhile the lieutenant-commander completed the flight test programme at Kagamigahara

Airbase while he resided at the Nagaragawa Hotel in Gifu City. Speed calibration tests on the A5M proved a maximum speed of 279 mph at 10,500 feet and a rate of climb to 16,400 feet in 5 minutes 54 seconds. This compared with the British Gloster Gladiator, which had a maximum speed of 253 mph at 14,500 feet, and a rate of climb to 15,000 feet in 4.8 minutes At the time two major difficulties had arisen – the experiencing of pitching oscillations, and ballooning upon landing associated with the use of split landing-flaps, as was used on the imported American Northrop machines. This US aircraft was a two-seater of very recent importation, part of the policy of importing highly successful aircraft to reveal their secrets, which would be incorporated into the next generation of Nipponese carrier attack-bombers. The next series of tests were flight-evaluation dog-fighting tests in which the squadron was under the command of Lieutenant Ryosuke Nomura and included Chief Warrant Officer Mase and 1st Flight Petty Officer Mochizuki. After each mock combat aeronautical engineers debriefed the test pilots to ascertain stress and performance details with a view to increasing the aircraft's capability. At the time Japanese pilots required the lightest aircraft and the most manoeuvrable, seeking to master every known manoeuvre of the fighter-pilot repertoire. A total of fifteen engine variations were tried on the A5M-series fighters. As previously mentioned, the A5M1 flew on 4 February 1935, powered by a 550 hp Nakajima Kotobuki 5 radial engine and having gull-shaped wings. The next variant, the A5M2, was engined with a 560 hp Kotobuki 3, and this model was followed in 1937 by the A5M2a and A5M2b. The A5M3 was to have a 690 hp Hispano-Suiza in-line engine, while the A5M4 was followed by the K version, being a special training variant. In all, some 1,000 machines of this type were eventually to be manufactured – 800 by Mitsubishi and 200 jointly by the Sasebo Navy Air Depot and Kyushu Aircraft Company. The general specification was as follows:

Engine, one	710 hp Nakajima Kotobuki 41 radial
Span	36 ft 1 in.
Length	24 ft 9¼ in.
Height	10 ft 6 in.
Weight empty	2,681 lb
Weight loaded	3,684 lb
Maximum speed	273 mph at 9,840 ft
Ceiling	-
Range	746 miles
Armament	2 × 7.7 mm machine-guns
	2 × 66 lb bombs

In the autumn of 1935 the Mitsubishi A5M fighter was flight tested by

Lieutenant-Commander Shibata HIJMN at the Yokosuka Experimental Air Corps Station, and contrary to his previous pronouncements, for which he apologised, he now claimed the new fighter an excellent machine.

The first prototype of the Mitsubishi G3M1 Type 91 land-based bomber flew for the first time in July 1935. This machine was powered by two 600 hp Type 91 engines, later modified to Kinsei Type 2 radial engines. A third variant was produced when twenty-one machines were manufactured with 840 hp Kinsei Type 3 radial engines. It was a well-known fact that these machines lacked adequate defensive armament; nor did they possess self-sealing fuel tanks. The Navy could have specified both requirements, but niceties were sacrificed for long-range performance. The G3M2 variant, of which fifty-six aircraft were manufactured, was powered by the 1,000 hp Kinsei 45 radial engine. The G3M3 version had 1,300 hp Kinsei 51 radial engines, and this variant was the type used to attack the British battleships HMS *Prince of Wales* and HMS *Repulse* on 10 December 1941. The L3Y1 was a transport conversion of the G3M1, carrying ten passengers, and was later developed into L3Y2, fitted with Kinsei 42 radial engines. The general specification was as follows:

Engine, two	1,000 hp Mitsubishi Kinsei 45 radials
Span	82 ft 0¼ in.
Length	53 ft 11¾ in.
Height	11 ft 11¾ in.
Weight empty	11,442 lb
Weight loaded	17,637 lb
Crew	7
Maximum speed	238 mph at 9,840 ft
Ceiling	29,890 ft
Armament	1 × 20 mm cannon
	2 × 7.7 mm machine-guns
	2,200 lb of bombs or
	1 × 1,760 lb torpedo

November 1935 saw defects in the Type 95 land-based attack-bomber, the Daiko, or big attack-bomber Hiro G2H1. This machine had a span of 105 feet and a gross weight of 24,250 lb, and was powered by two Hiro Type 94 Naval Arsenal water-cooled engines of 900 hp each. This gave the plane a maximum speed of 104 mph with a range of 2,300 miles carrying 4,400 lb of bombs. Only eight of this model were produced, and five were to be destroyed in a serious fire sometime during 1937. However, the plane was too large and heavy for the available power plants, and speed climb controllability suffered due to structural defects causing vibrations. Nevertheless these eight machines enabled the Navy to gain confidence in incorporating large land-based bombers. Combat baptism of fire occurred on the Shanghai front during the period 1937 to 1938.

It was observed during the era that the Blackburn T.7B or Mitsubishi B2M1 and B2M2 attack-bombers displayed the numbers 349 and 379 as tail serial numbers on machines operated by the Tateyama and Omura Air Corps respectively. The Navy was now importing the American P2Y1 and the German Heinkel He 70 and He 74 for evaluation purposes.

The Navy instituted a special research programme under the direction of Vice-Admiral Misao Wada, senior staff officer of the Technical Division of the Naval Bureau of Aeronautics, into the best weapon for use in aerial combat. The programme director recommended the introduction of the Swiss 20 mm Oerlikon cannon, a lightweight weapon firing explosive shells with a low muzzle velocity developed and manufactured by the Swiss Oerlikon Corporation. This gun was originally developed from the German Becker cannon of the First World War, which was heavy and prone to mechanical difficulties, having been developed by the German Flying Corps. Large numbers of the Swiss Oerlikon cannon could not be imported for the Nipponese Navy because the figures involved would disclose to the world the exact quantity of the new Navy fighters in production. The manufacture of the 7.7 mm machine-gun remained the responsibility of the Naval Technical Bureau. Under arrangement with the Oerlikon Corporation, the Navy was able to acquire the manufacturing rights and contract Japanese manufacturers to produce the new weapons in quantity.

The aircraft carrier HIMJS *Kaga* now underwent a refit at the Sasebo Navy Yard. The flight deck was extended to the ship's full length, with a small bridge on the starboard side. The exhausts were replaced by rectangular funnels, which discharged smoke downwards and outwards along the flight deck level. Originally the ship had three flight decks forward, but those were rebuilt. The 4.7-inch GP guns were now replaced by 5-inch armament, and the aircraft-carrying capacity was raised to between sixty and ninety aeroplanes. A new speed of 28.3 knots was achieved on speed trials after rebuilding. The smaller aircraft carrier HIJMS *Ryujo* was damaged in a heavy storm while out sculling with the 4th Fleet. The front of the bridge was rebuilt to a new design and the fo'c'sle raised one deck higher to improve seaworthiness. Useful data were obtained with the ship, which formed the basis of the specification of HIJMS *Soryu* and her class. Though restricted by the London Naval Treaty, subsequent aircraft carriers stemmed from this particular design. The bridge was on the starboard side near the bow, which proved dangerous because of the associated air currents, which reduced the space available for safe landings of aircraft. The carrier HIJMS *Soryu* formed the basis of the 2nd Carrier Division; displacing 15,900 tons, her dimensions were 746½ feet overall by 70 feet wide by 25 feet, and she was propelled by four shaft geared turbines giving 153,000 shp at a speed of 34¹/₃ knots. Armed with twelve 5-inch GP guns and twenty-eight 25 mm AA guns, she accommodated seventy-three aircraft and 1,101 crew. Built by the Kure Navy Yard on 23 December 1935, she was finally sunk at the Battle of Midway on 4 June 1942.

The largest submarine built in 1935 was the I7, developed from the Kaidai Type 3 and 4 boats and designed to act as a scouting force of submarines. Displacing 2,231 tons, her measurements were 385½ feet long by 29¾ feet by 17¼ feet and she was propelled by two shaft diesel/electric motors of 11,200 and 2,800 hp respectively, which gave 23 and 8 knots. She was armed with one 5.5-inch GP gun, two 13 mm AA guns and six 21-inch torpedo tubes with twenty torpedoes, as well as one catapult seaplane. The seaplane hangar consisted of two tubular sections, which could be retracted into the deck when the aircraft was not in use. The seaplane fuselage was stowed in one tubular section, with the wings stowed in the second section, and the plane was finally reassembled on the catapult whenever required. The submarine could be away from base for sixty days, had a crew of 100 and was capable of cruising 14,000 sea miles at 16 knots on the surface or sixty sea miles submerged at 3 knots. The I7 was built by the Kure Navy Yard on 3 July 1935, and was lost with all hands on 22 June 1943. Her sister ship, the I8, was constructed at the Kobe Yard of Kawasaki, and was eventually sunk on 31 March 1945.

The most pressing question exercising the minds of senior officers of the Naval Air Service at this time was the vexed question of aircraft carrier tactics to be used in the event of active operations. Two views existed: there were those who considered dispersal over large areas of the sea, so lessening the chance of discovery of the entire force; and those who thought that concentration of effort in one particular area, which would achieve the desired victorious end. It was to be decided subsequently by Admiral Yamamoto which view would prevail, with astounding results achieved by the Nipponese Navy.

CHAPTER 9

The Chinese Incidents

In March 1936 the Japanese government fell and a new Cabinet was chosen. Prime Minister Hirota prepared for a full-scale war in China. New policies were initiated, which prepared light industry for development into heavy industry, though producers of steel and munitions were exempt from any direct taxation but were supplied with financial loans for the expansion programmes. A five-year warship construction programme was commenced, but when presented with the budget for such a rapid military expansion the Prime Minister baulked at the expense, and was immediately replaced by General Hayashi. In June 1936 the office of prime minister was taken over by Prince Konoye, the general having resigned.

The new Mitsubishi G3M land-based bomber was regarded as one of the historic aircraft of Japan. Designed by engineer Kiro Honjo, the aircraft had a B9 aerofoil wing and planform taken from the 8-Shi bomber design, with similar construction incorporating corrugated sheeting, similar to the German Junkers, in place of the flat sheets on the wing trailing edge. On the fuselage the skin was smooth, with flush riveting, as was the practice used on the Mitsubishi A5M. Internally the fuselage was again developed from the 8-Shi bomber, but with broader fuselage dimensions. The tail design was similar to the twin-rudder format of the 8-Shi bomber but enlarged to cover longitudinal movement of the centre of gravity. The undercarriage was a simple cantilever designed to reduce air resistance at take-off. This machine embodied many of the ideas of Admiral Isoroku Yamamoto and other supporters of the naval air power concept. The first plane was tested at Kagamigahara Airfield with test pilot Kajima at the controls, assisted by Lieutenant Sada HIJMN.

Prototype Models 1, 2, 5 and 6 were powered by Hiro Naval Arsenal Type 91 water-cooled engines rated at 600 hp each, giving a maximum speed of 195.5 mph at 4,920 feet. Against expectation, this new maximum speed exceeded the performance of the then operational aircraft of the day by 23 mph. Meanwhile the second prototype was on a long-range training flight piloted by Chief Warrant Officer Chiku under the command of Lieutenant Jiro Tokui on a flight home from the Manchurian–Korean border. While at operational altitude the plane became uncontrollable and eventually crashed, killing the five-man crew. On-the-spot investigations ascertained a defective propeller and indications of aileron flutter, which apparently had been experienced on the pre-delivery flight. At once the Mitsubishi design team increased the aileron mass balance weights to reduce vibration by damping, and the problem of the flutter appeared to have been cured.

Prototype models 3, 4 and 7 were equipped with two Mitsubishi Kinsei Model 2 engines of 825 hp each. Then the G3M1 was re-equipped with a further variant of the Kinsei engines producing 840 hp, which added a further 11.5 mph to the maximum speed. The Model G3M2 had Kinsei Model 45 power units of 1,000 hp each, which developed 230 mph at 9,840 feet; by the standards of the day the increased speed was quite a remarkable achievement.

The prototype was kept under constant development by the Naval Aircraft Establishment. In the prototype the observer's seat and the astrodome were located aft of the pilot's compartment. In the first twenty-one production aircraft the observer was moved forward next to the pilot, and in later machines the position was moved back to the original position. The ventral turret with its gunner and two 7.7 mm machine-guns offered no air resistance in the up position, but under combat conditions excessive drag and server buffeting of the aircraft was experienced. Action conditions necessitated the installation of two blisters for the lateral positions, while the two 7.7 mm machine-guns in the dorsal turret were replaced with a single 20 mm cannon. At the time patents were bought from Sperry Corporation of America for its automatic pilot, as well as the radio direction homing acquired when the apparatus was still experimental, and in the forthcoming operations these were to make possible the long-range bombardment attacks on targets across the Pacific Ocean. These machines could carry an offensive load of one 1,764 lb torpedo or equivalent weight of bombs on external racks in the under-fuselage position. The production of the Mitsubishi G3M land-based bomber counted for 1,100 machines in which the Nakajima Aircraft Company constructed the air-frames and Mitsubishi assembled the entire bomber. In competition with the Mitsubishi G3M-series bomber was the Nakajima LB2, which was not accepted by the Navy. The Nakajima prototype was sold to Manchurian Airlines, where it was put into service on freight and passenger duties.

Slotted flaps were first installed on Japanese service aircraft in 1936, on the A5M Type 96 fighter, though previously flaps had been used on an aircraft back in 1927. This earlier machine was an experimental biplane of the reconnaissance carrier-borne type. The lower wings were made of duralumin with semi-monocoque watertight compartments. This aeroplane was designed by the Mitsubishi Aircraft Company under the direction of Professor Albert Baumann of Stuttgart University. The design secret lay in the full-span slotted flaps on upper and lower wings. As a result of this work, Tetsuo Noda, chief of Mitsubishi's Wind Tunnel Section, perfected a split flap for use as a high-lifting device.

However, difficulties plagued the development of many aircraft during this period. The Type 89 carrier attack-bomber B2M1 adopted in 1929 was beset by engine failures and poor stability characteristics, causing accidents and fatalities. Eventually this machine was replaced by the B1M1 Type 13 carrier attack-bomber of 1924, which proved reliable but slow, and a total of 420 were built. Later this carrier aircraft was replaced by the B3Y1 Type 92 carrier bomber, of which only ninety-two were manufactured. Until 1936 all carrier attack-bombers were biplane types, which met the Navy's specification regarding storage requirements for elevator space aboard the aircraft carriers. Designs were submitted by Mitsubishi, Nakajima and the Naval Aircraft Establishment, but the successful contestant of the design competition was Nakajima with the B4Y1 Type 96 carrier bomber. At the same time the Navy had entered into secret negotiations with the German Heinkel Aircraft Company to acquire the He 118 dive-bomber, which was designated DHXE-1. It was agreed that no one would learn of the plane's existence other than Mitsubishi, Aichi and Nakajima. The Naval Aircraft Establishment considered this machine poorly constructed, structurally weak and retaining an inferior performance. The Italian Fiat BR20 Cicogna bomber was purchased, but once again the performance was considered inferior. Eventually the Japanese aeronautical engineers considered all foreign aircraft inferior, regardless of specification! They omitted to seriously consider the benefits of simplified manufacturing techniques, excellent accessories and equipment, ease of maintenance and convenience of layout for mass production, so that the industry never seriously attempted to master foreign techniques of manufacture.

The Navy had adopted at this time the Mitsubishi Type 96 carrier fighter, the A5M Type 96 land-based bomber G3M and the Aichi Type 96 dive-bomber, the D1A2, as well as the flying-boat A1B4 for night patrol operations and the Watanabe Type 96 small reconnaissance seaplane for submarine launching, in addition to the Nakajima B4N1 carrier attack-bomber. In December 1936 the highly successful Kawanishi (Short) KF1 H3K2 flying-boat used for training missions and reconnaissance operations was finally withdrawn from operational service. The first four-engine aircraft designed by a Japanese aeronautical engineer was built by Kawanishi as a four-engine passenger flying-boat. This machine was adopted in a special militarised version by the Navy for operational usage. It was designed by Shizuo Kikuhara, initially with four Hikari radial engines rated at 710 hp each, but later these were changed to four Mitsubishi Kinsei 43 radial motors of 1,000 hp each. Designated Type 96 H6K5, this flying-boat had a gross weight of 50,706 lb at a range of 4,000 miles, with an endurance of twenty-four hours. A total of 179 planes of this type were built for military operations, with a further thirty-eight for transport use by the year 1943.

During 1936 the Mitsubishi A5M fighter Type 96 was considered as the emancipation of Japanese aviation industry, and was designed by Doctor Jiro Horikoshi. At the time Nakajima converted for naval use the Army Type Ki-11 fighter designated as the PA. It was similar to the American P26 fighter with a maximum speed of 253 mph, but the Mitsubishi low-wing fighter with a fixed undercarriage achieved a maximum speed of 280 mph with a Kotobuki V engine. This fighter plane had a rate of climb and manoeuvrability far superior to any of the biplane fighters of the time. A total of 982 Mitsubishi Type 96 A5M fighters were built, being flown during the Sino-Japanese War until eventually replaced in squadron use by the famous Zero fighter.

The naval dive-bomber, the Aichi D3A1 powered by a Kinsei engine of 1,075 hp, was designed in 1936 with quantity production in mind, commencing one year later. Between 1937 and 1942 a total of 478 machines of this type were manufactured until replaced by the D3A2 variant powered by a 1,300 hp Mitsubishi Kinsei 54 radial motor. This later machine had a production run of 816 planes, and manufacture was completed in 1944. This aircraft had been the first all-metal low-wing dive-bomber, most effective in the hands of skilled pilots, being strongly stressed to withstand structural strains and highly manoeuvrable to assume the role of a fighter aircraft after delivering its load of bombs. The general specifications were:

Engine, one	1,300 hp Mitsubishi Kinsei 54 radial
Span	47 ft 1½ in.
Length	33 ft 6¾ in.
Height	10 ft 11¼ in.
Weight empty	5,772 lb
Weight loaded	8,378 lb
Crew	2
Maximum speed	266 mph at 18,536 ft
Range	970 miles
Armament	2 fixed and 1 movable 7.7 mm machine-gun
Bomb load	816 lb

A reconnaissance biplane on floats had been designed by Mitsubishi as the F1M2, manufactured by both Mitsubishi Aircraft Company and the Sasebo Naval Arsenal for a total production run of 524 aircraft. Armed with three 7.7 mm machine-guns and two bombs of 132 lb each, a maximum speed of 229 mph was achieved at 9,840 feet with one Mitsubishi Zuisei 13 radial engine of 780 hp. With a wingspan of 36 ft 1 in., this machine mainly operated from shore bases and seaplane tenders as a fighter aircraft. Another plane, which was conceived during 1936, was the Nakajima B5N1 torpedo-bomber powered by a 770 hp Hikari engine. This aeroplane was operated as a light bomber over China, with some models converted to the training role as the B5N1-K variant. In 1939 the B5N2 was introduced into service, and over 1,200 of this type were manufactured by 1940. Powered by a 1,000 hp Nakajima Sakae 11 radial engine, this machine attained a maximum speed of 235 mph at 11,810 feet, despite its span of 50 ft 11 in. In fact forty planes of the B5N1 type flew from the aircraft carrier HIJMS *Soryu* to attack the US battleships at Pearl Harbor. Unhappily the plane was vulnerable to fighter attacks, and was subsequently withdrawn from operational service to be used on anti-submarine duties.

Once again the Navy was importing foreign aircraft, and the importation programme included the US Fairchild A942 and the German Heinkel He 118. The Navy also made a special point of importing an example of the French Dewoitine D510 fighter, which was equipped with a 20 mm motor-cannon. Later a Hispano-Suiza 12 CRS engine was acquired, also equipped with a 20 mm cannon, but though mounted for test purposes for the A5M fighter, the research programme results were inconclusive. Eventually the Navy adopted the 20 mm Oerlikon cannon, though the 12.7 mm heavy machine-gun was not introduced and the 7.7 mm light machine-gun became standard equipment. During April 1936 the manufacturing plans for the 20 mm cannon production were approved by Rear Admiral Goro Hara, Chief of the Technical Division of the Naval Bureau of Aeronautics. Meanwhile the Chief of the Naval Bureau of Aeronautics was Vice-Admiral Isoroku Yamamoto, who placed an order for one hundred Oerlikon 20 mm Type FF aircraft cannon, designated Type 99, plus 10,000 cartridges to be manufactured by the Uraga Dockyard Company, and costing ¥3,000,000. This order was followed by a contract for an annual production of sixty-five cannon and 12,000 cartridges. The Uraga Dockyard Company was controlled by Ken Terajima, company president and a former vice-admiral, who had planned the construction of separate factory facilities for mass production of new weapons. To instruct the company employed in the use of the manufacturing facilities it was planned to acquire the services of an engineering group from the Oerlikon Company in Switzerland. This was a similar arrangement to what had existed when a group had gone to Japan to assist in building a gunpowder factory at Hiratsuka, supervising tooling and the planning of manufacturing facilities. The vice-manager of the Uraga Dockyard Company and one engineer were sent to Switzerland to purchase the Oerlikon rights to manufacture seven weapons in Japan. The Swiss company agreed to supply over the following five years technical materials to improve the performance of the new weapons.

In the meantime a total fifty-eight seaplanes had been deployed on board ships of the fleet – seven battleships, nineteen cruisers, two submarines and two seaplane tenders. The aircraft capacity of HIJMS *Akagi* had been increased from sixty-one to ninety-one machines. Interestingly enough, the full-length flight deck was inclined downwards fore and aft a third of the length from bow to stern. This was to enable planes taking off to gather speed and to slow machines down when landing. The technical problem previously associated with the operation of aircraft carriers was now resolved, subsequent to the 1936 fleet manoeuvres. A decision came down on the side of dispersal rather than concentration of effort, which was considered highly questionable by some naval authorities. The Navy in its expansion programme established four new Naval Air Corps, at Maizuru, Kanoya, Kisarazu and Chin-Kai, Korea.

In January 1937 the 2nd Fleet Replenishment Law was enacted, and Japan threw off the limitations imposed by the previous treaties. She was now able to build the kind and size of fleet that was required, and a gradual change from a defensive role to an offensive posture was established.

The headquarters of the Naval Air Service was located at Kasumigaseki, Kojimachi, Tokyo, under the director, Vice-Admiral K. Oikawa, with three departments, for general affairs, technical matters and education facilities. The operational department of the Naval Air Service was directly responsible to the Imperial General Headquarters, or Daihonei, via the Chief of the Naval General Staff, or the Gunreibu Socho. The Naval Air Headquarters, or Kaigun Koku Hombu, was responsible for aircraft, engines, equipment testing, the supervision of flight training and the training of maintenance personnel. Operational control rested with the air fleet, or Koku Kantais, which was sub-divided into air flotillas, or Koku Sentais, which contained one or more naval air corps, or Kokutais, each comprising some 150 aircraft.

Aircraft aboard carriers were assigned to air divisions, or Koku Sentais, which equalled two aircraft carriers. Each air division was assigned to a naval air corps. The Naval Aircraft Battle Force was commanded at fleet level by Rear Admiral S. Sato, the 1st Fleet Air Commander, by Captain R. Horije, the 2nd Fleet Air Commander, and Rear Admiral R. Tokari, Air Commander of the 3rd Fleet. Naval air stations were located in the following places:

Chinkai, Korea
Kanaya, Kagoshima Prefecture
Kasumigaura, Ibaraki Prefecture
Kisarazu, Chiba Prefecture
Kure, Hiroshima Prefecture
Maizuru, Kyoto Prefecture
Ohminato, Aomori Prefecture
Ohmura, Nagasaki Prefecture
Rojun, Port Arthur
Saheki, Ohita Prefecture
Sasebo, Nagasaki Prefecture
Tateyama, Chiba Prefecture
Yokosuka, Kanagawa Prefecture
Yokohama, Kanagawa Prefecture

The naval aircraft complement comprised 563 aircraft with land-based squadrons, and 332 aircraft with ships and aboard aircraft carriers, equalling a total of 895 machines. The air crew, including pilots and navigators, numbered 2,711 men.

The aircraft carrier HIJMS *Hiryu* had been designed with improved seaworthiness, having been laid down after the expiration of the London Naval Treaty. The vessel had a slight increase in displacement, with a fo'c'sle deck one level higher than HIJMS *Soryu* and a bridge located on the port side amidships. General specification of the carrier was as follows:

Displacement	17,300 tons
Dimensions	746 ft × 73¼ ft × 25½ ft
Machinery	4 geared-shaft turbines of 152,000 hp, giving 34½ knots
Armament	12 AA 5 in., 37 AA 25 mm
Aircraft	73
Complement	1,101
Built	Yokosuka Navy Yard 16/11/37
Ultimate fate	Sunk, Battle of Midway 05/06/42

During this time the aircraft carrier HIJMS *Hosho* served with the 1st Fleet, using its aircraft in a close-support role in the course of land operations. Other aircraft carriers in the fleet included HIJMS *Akagi* (26,900 tons), *Kaga*

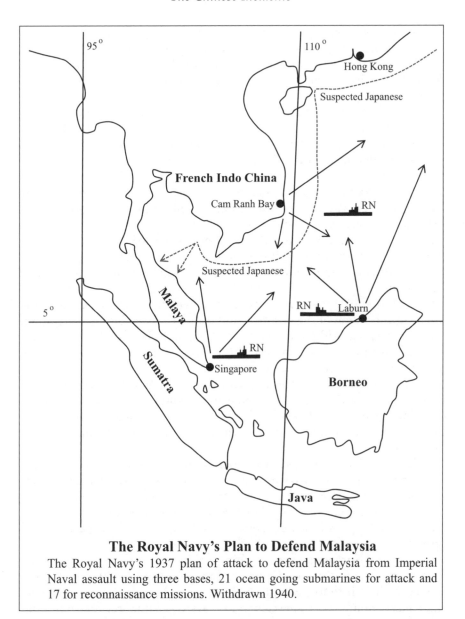

The Royal Navy's Plan to Defend Malaysia

The Royal Navy's 1937 plan of attack to defend Malaysia from Imperial Naval assault using three bases, 21 ocean going submarines for attack and 17 for reconnaissance missions. Withdrawn 1940.

(26,900 tons), *Ryujo* (7,600 tons), *Soryu* (10,000 tons) and *Hiryu* (10,000 tons); and the seaplane tenders HIJMS *Kamoi* (17,000 tons), *Notoro* (14,050 tons) and *Chitose* (9,000 tons).

During 1937 it had become known that the Japanese had a plan to attack the city of Singapore at divisional army strength if the war in China should further widen in scope. Therefore, to protect British possessions in the Far East, plans were drawn up to increase the fighting efficiency of the Royal Naval fleet on the China Station under the command of Admiral Sir Charles Little.

Two Royal Naval officers were dispatched to Hong Kong aboard the P&O liner *Rajputana* in May 1937 – Captain C.B. Barry DSO RN and Commander G.C.P. Menzies RN – , with the objective of placing the 4th Submarine Flotilla on a war footing.

Conferences were arranged with fellow officers, at which programmes, revisions, procedures and logistics were planned, and studies of the manpower situation were undertaken, with anticipated operational schemes to cover every known contingency.

The submarine depot ship at Hong Kong was HMS *Medway*, under the command of Captain Murray-Smith RN, who felt that his vessel could service fifteen submarines in peacetime and eighteen under war conditions.

On 11 June 1937 the senior submarine officer on the China Station drew up an appreciation of the numbers and duties of the submarines required for war in Japan. Although this report was placed before Their Lordships of the Admiralty in London, little was done to implement the recommendations other than in the Far East.

It was considered that in any conflict with Japan the British Crown Colony of Hong Kong was untenable and would fall immediately, so that the China Fleet would seek anchorages elsewhere, at Labuan in North Borneo and Cam Ranh Bay in French Indo-China, with the main base at Singapore.

The war plan envisaged the stationing of a second submarine depot ship with the fleet, the two ships having the capacity to service twenty-one submarines for attack operations and seventeen for reconnaissance purposes. The plan of campaign intended attacking Japanese convoys using mobile patrol tactics, sometimes known as 'wolf-pack hunter groups'.

Using high-frequency-radio ship-to-ship control, detection of the quarry by means of supersonic transmission equipment and heavily armed with 21-inch torpedo tubes, the ocean-going submarines of the Far East Fleet minimised the escalation of the war situation for some four years. Constantly training in the tactics of mass attack and continually reconnoitring an area stretching from the Bungo Channel, Japan, to the Great Barrier Reef, Australia, the Imperial Navy recognised a potential threat to the Nipponese sea lines of communication.

Two divisions of the Royal Navy's 4th Submarine Flotilla were observed on innumerable occasions by the ships of the Japanese fleet. Unhappily, in March 1940, due to the serious situation in the Mediterranean theatre of operations, all the submarines were withdrawn from the Far East Fleet to oppose the German and Italian fleets operating off the coast of north Africa. To the Imperial Navy the green light was shining in the south China seas.

On 7 July 1937 the famous China Incident took place between Japanese troops on manoeuvres and Chinese troops manning outposts on the Marco Polo Bridge near Peking. The result was to reopen hostilities between the two countries, and both powers resorted to the use of land and aerial operations. The Chinese Air Force possessed some 650–700 aircraft manned by 750 pilots organised into thirty-one squadrons. While the aircraft were first-class machines imported from Europe and America, their pilots displayed deficiencies and inadequate training. The planes were maintained by ground crews who were technically hard pressed to maintain their aircraft, and utilisation fell as low as 20% in some instances. For the Japanese to conclude successful aerial operations, an agreement had to be signed by senior representatives of the Imperial Army Air Force and the Imperial Air Service. It was arranged that the Army Air Force would provide air support to cover the offensive operations in north China. The Naval Air Service would give aerial cover over the central and southern fronts of the Chinese theatre of operations. Once the agreement had been ratified the air war over China commenced in earnest at some time during the August/September period of 1937. Off the Chinese coast was stationed the 1st Carrier Air Wing, comprising the aircraft of HIJMS *Hosho* and HIJMS *Ryujo* with the 2nd Carrier Air Wing of HIJMS *Kaga*, with a total of 264 aeroplanes for the front-line operations. Land-based aircraft of the Naval Air Service flew long-range bombing attacks from naval air stations in Kyushu, Japan, from Formosa and Manchuria.

On 14 August 1937 Japanese Mitsubishi Type 96 G3M1 attack-bombers swept in to bomb airbases, installations and dumps at Kwantoh and Hangzhou. Amid the crumbling buildings and the fires raging among the installations the world aviation leaders had been surprised by the technical efficiency of the Imperial planes and men in conducting such a long-range attack, but had it not been for American experimental Sperry direction-finding and -homing apparatus, the heavily laden bombers would not have been able to navigate to the targets and back to base so successfully. The flight across the East China Sea had covered a total distance of 1,200 nautical miles in bad weather, which had resulted in a typhoon tailwind. In all, thirty-eight bombers of the 1st Joint Air Corps, or Rengo Koku-tai, had participated and returned to Japanese bases intact. Further attacks on mainland China took place four days later, when another formation of naval attack-bombers once again swept in over the Chinese coastlines. The machines involved in these operations became known by the nickname of 'Chukos', a contraction of the words 'Chugata Kogeki', or medium bombers. During the Sino-Japanese Incident these planes flew continuous transoceanic bombardment missions from Omura Naval Air Station, Kyushu, and from Taipei Naval Air Station in Formosa, attacking targets in the Shanghai, Nanjing and Hangzhou areas, in which each flight covered a distance of some 1,150 miles or so. Over 1,100 bombers of this type were eventually constructed, with Nakajima manufacturing the airframes and Mitsubishi assembling the final product, until the introduction of the Mitsubishi replacement – the G4M bomber series. In China the naval aircraft had supported the land operations of the Japanese Army using 192 aircraft of the 1st and 2nd Air Corps, in which 560 tons of bombs were dropped on military targets in seventy aerial attacks, with Nanjing being the ultimate objective. In September 1937, airbases for the Naval Air Service were constructed in the Shanghai area, and the Type 96 Chukoh bombers were transferred to the Shanghai, Peking, region. The bomber pilots were very enthusiastic with their G3M1 bombers, and as a result the Imperial Navy called for increased production. The Navy reverted to the purchase of aircraft from abroad and acquired twenty American Seversky P35 fighters, which proved inferior in performance to the naval fighters. These imported American fighters were relegated to a reconnaissance role over the China fronts. Foreign imports also included the French Dewoitine D510 and the German Heinkel He 112 and He 118, but these machines were not considered satisfactory by Japanese pilots. Unfortunately Japanese aviation engineering groups did not examine the construction, serviceability, design, workmanship, accessories or equipment of the foreign imports. Meanwhile, during the period of 14 August to 31 September, the Chukoh bombers of the Kisarazu Air Corps and the Kanoya Air Corps flew daily raids across the East China Sea to attack military targets in Nanjing, Yangzhou, Suzhou, Hankou, Nanchang and other cities of central China. In the offensive against Nanjing the Chinese Air Force redeployed its fighter wings, and during the course of three bombing attacks on the city shot down fifty-four Mitsubishi Type 96 G3M1 attack-bombers in serial battles above the city and its surrounds. At this time, due to the slower speed and manoeuvrability of the Chinese fighters, they employed diving tactics from altitudes above the formations of Japanese bombers, which because of their unarmoured condition could prove easy victims to the guns of a fully trained professional fighter pilot. It was not due to any oversight by the Japanese aeronautical engineers that the planes did not possess armour to protect the crews and equipment or that self-sealing fuel tanks were not installed: these and other innovations were offered to the Naval Air Service, but because of the contingencies of long-distance flying the overall weight of each machine had to be kept at a minimum to increase the fuel capacity. Nevertheless, Bushido, the Way of the Warriors, would require crews to fly without such trivia until, too late, their importance was realised.

In early September the Navy's newly acquired airbase near Shanghai was the location for the 2nd Joint Air Corps, which was to co-operate with the 1st and 2nd Air Wings operating off the Chinese coast, poised against Nanjing and Chinese cities in the southern and central sectors. On 15 September the Air Corps Command changed tactics, and bombing raids were to be made at low level and not above an altitude of 1,500 feet. Flying in loose formation, the Type 96 G3M1 bombers swept into the south China areas, destroying the cities of Kwantung, Shaoguan and other centres. In this phase of the attacks some aircraft were destroyed and others damaged by Chinese ground fire. By the beginning of October 1937 the Japanese Naval Air Service possessed six airbases in the vicinity of Shanghai for offensive operations. Statistics from the Naval Air Service covering the period July to October showed that sixty-one Chinese cities had been bombed, and 181 Chinese aircraft had been destroyed in aerial battles in defence of the cities, and a further 143 during ground-attack operations against Chinese fighter stations in the course of 3,000 sorties embarked upon by the Naval Air Corps Commands. Other sources suggested that the Navy lost 21% of its aircraft in the air battles for control of the air space above the Chinese cities. The Chinese Fighter Command defence was very effective, especially against the Naval Air Service bombers, owing to their vulnerability. However, the introduction of defending fighters with the bomber formations tended to reduce the Japanese losses. Unfortunately the Chinese Air Force attacks on Japanese shipping and installations were very poor, owing to a lack of suitable planes and equipment. The planes used by the Chinese Air Forces included Curtiss Hawks, Northrop E2s and Boeing 218s, all manufactured in America. For the Japanese the Type 96 naval fighters were stationed at the Wuhu Naval Air Station south-west of Nanjing.

A dive-bomber competition held during 1937 produced the Nakajima D3N, which competed with the Aichi D3A, and upon evaluation the Aichi product was accepted by the Naval Air Service, becoming the Type 99 carrier-based dive-bomber. A total of 1,492 machines of this type were manufactured, seeing service until the end of the Great Pacific War, for the plane was reliable, stable, slow and undistinguished. The 10-Shi carrier attack-bomber competition produced two machines. The first, the Nakajima B5N1 Type 97 Model 1, possessed a retractable undercarriage similar to the Northrop BT1 attack-bomber, which had been purchased previously by the Japanese. The Nakajima plane had a Hamilton constant-speed propeller, upward-folding wings fastened with hinges, semi-integral fuel tanks and an 840 hp Hikari engine. The B5N2 machine was similar, but had a 1,000 hp Nakajima Sakai II engine, and although it was introduced into squadron service it was not reliable, subsequently being replaced by the B5N1, which saw service in the Sino-Japanese Incident until replaced in 1941. The second aircraft, the Mitsubishi B5M1 Type 97 Model 2, had an elliptical wing platform, slotted flaps and a fixed undercarriage. Though 125 machines of this type were produced, the aircraft saw limited service over the skies of China before being replaced. The Nakajima A2N fighter type replaced the A1N, but was unable to compete with the Chinese fighter planes of Soviet design, and the Nakajima machine was eventually taken out of service by the introduction of the A4N fighter type. However, even this developed type was unsatisfactory, and the A5M fighter was introduced, replacing the earlier A2N2 and A4N1 when the aircraft carrier HIJMS *Kaga* returned to Japan to embark the A5M machines. This aircraft was the first low-wing monoplane carrier fighter with a fixed undercarriage. As compared to the Soviet Polikarpov I16, the new Japanese carrier fighter possessed more manoeuvrability, could climb faster but was eight miles per hour slower than its Soviet counterpart when attaining maximum speed. As a result the A5M fighter plane attained air superiority over its opponents in the skies above China.

During 1937 the Zero design studies were commenced on 19 May. The Navy submitted design specifications for the 12-Shi carrier fighters to replace the newly acquired A5M to the Mitsubishi and Nakajima Aircraft Companies. By 15 June the Mitsubishi chief designer, Dr Jiro Horikoshi, requested the company's Wind Tunnel Department to conduct experimental studies with a new aerofoil for the 12-Shi design. Meanwhile the war in China had flared up and combat reports suggested that the conflict would be long and protracted, so that a new fighter would be required. In August 1937, Mitsubishi's chief designer held conversations with Lieutenant-Commander Goro Wada, staff member of the Naval Bureau of Aeronautics, concerning the new programme. As a result of these talks Mitsubishi considered the new fighter project of such extreme importance that the 11-Shi carrier dive-bomber project proposed by the company was cancelled. By 5 October 1937 large-scale air battles were raging over the Chinese cities of

Shanghai and Nanjing, which prompted the Navy to submit final design specifications to Mitsubishi. The recommendations and requests behind the events of the new specification were first received from the Division of Naval Operations in Tokyo, which would create a hypothetical situation to be considered. These considerations would include the identification of an enemy, likely designated combat areas, performance of likely enemy aircraft, the fire power and protective devices required, possible enemy surface vessels to be attacked, the Japanese defence strategy and the enemy's response, as well as the consideration of the international developments to be encountered at the time. On the question of armament up to the year 1937, the Navy had relied upon the 7.7 mm Type 89 machine-gun as its principal weapon, which was a version of the British Vickers machine-gun. With the introduction of the 20 mm cannon, trials were subsequently undertaken on the A6M carrier fighter armed with one 20 mm cannon mounted in each wing. Eventually, after successful experimentation and development, the A6M fighter went into action on the first operation against the Chinese city of Chongqing in central China. In massive air battles the A6M fighters swept away the Chinese defenders, and pilots filed enthusiastic combat reports. As a result Vice-Admiral Teijiro Toyota, Chief of the Naval Bureau of Aeronautics, sent letters of commendation to the Dai-Nikon Heiki Company and to Nakajima and Mitsubishi. In the United States the Boeing B-17 Flying Fortress bomber was now flying for the first time, some four years after the War Department General Staff had requested the design. This bomber could carry 2,000 lb of bombs over a distance of 5,000 miles at a maximum speed of 200 mph. Such a plane represented a serious threat to the Japanese Home Islands, as well as the island defence chain across the Pacific Ocean.

In the meantime design material had been submitted to the Combined Fleet headquarters, and results were sent to the Naval Department, Tokyo, where group conferences were arranged with the Naval General Staff, the Naval Bureau of Aeronautics, the Naval Aircraft Establishment and the Yokosuka Experimental Air Corps. Engineers, designers, pilots, tacticians and others prepared technical and operational analyses of the new fighter plane. The Navy Department considered the budgetary and industrial capabilities at the disposal of the service for the final manufacture of the new planes particularly, as the navy requirements were rarely met by the final products. The Naval Bureau of Aeronautics had drafted the initial design requirements with findings that had been evaluated by the Naval Aircraft Establishment and the Yokosuka Experimental Air Corps, as well as the operational air corps. The Navy even canvassed the ideas of the commercial companies, and eventually a joint Navy/civilian committee was selected. The draft designs were processed to attain special minimum standards of design to which a competitive commercial response could be made. Eventually a special research aircraft was constructed by the Navy for development work. The standard of design was as follows:

Purpose	Interceptor and escort fighter
Speed	Exceeding 310.5 mph at 13,123 ft
Climb	To 9,843 ft in three minutes 30 seconds
Duration	Cruising speed: 6–8 hrs
	Overload: 1.5–2 hrs
	Normal: 1.2–1.5 hrs
Take-off	Less than 229.7 mph with 30 mph headwind
Landing speed	Less than 66.7 mph
Gliding descent	690 to 787 ft per min
Manoeuvrability	Equal to A5M
Armament	2 × 20 mm cannon, 2 × 7.7 mm machine-guns
Bombs	2 × 66 lb, 1 × 132 lb overload
Radio equipment	Type 96 KU – 1 airborne radio Kruisi Type KU 3 homing goniometer
Other Equipment	Oxygen system
	Fire extinguisher
	Lighting equipment
	Instruments

Unhappily the requirements did not specify pilot armour or self-sealing tanks. Agility in aerial combat over the cities of China had amply justified the decision. Meanwhile the Mitsubishi chief designer, Dr Horikoshi, had calculated that a development programme would take about three years to bring the machine to full fruition, made up in the following manner:

- One year for the full design work
- Six months for the prototype production
- One year for flight testing and modification
- Six months for the initial production run

Thus the Japanese Spitfire was born, designed, built, tested and successfully manufactured.

The near-panic development of the successful fighter designs for the Navy had been necessitated by the successful Chinese attacks on the G3M bomber formations in aerial battles over Chinese cities. Without fighter protection the naval attack-bombers were nearly defenceless.

Originally an uneasy peace had reigned since the Manchurian and Shanghai conflicts, when the governments of China and Japan were negotiating. The Army headquarters and the Kanto Command wished to invade China, but the non-violent action had failed to bring about a crisis, so that more positive steps were necessary. These measures called for civilian disturbance, strikes, bomb incidents and assassinations. The traditional Star Festival in Japan was held on 7 July 1937, when a spark ignited a major conflagration in the suburbs of Peking: a small Japanese force, which had by agreement been stationed in Peking since 1901, was engaged in night manoeuvres when shooting broke out. There was no certainty as to who fired the first rounds, but Japanese soldiers were killed in the encounter. Immediately talks were entered into by the two governments concerned, but any chance of successful negotiations were sabotaged by the Army and the Kanto Command. Active operations commenced immediately, and Japanese armed forces occupied Peking, Tianjin and the surrounding country within five days. By 11 August further Japanese forces landed, supported by naval carriers off the coast. On 15 August four Russian fighter squadrons and two Russian bomber squadrons arrived to reinforce the Chinese Air Force. The machines were manned by 'volunteers', whose efficiency and equipment was of a high order. At the same time the American General Claire L. Chennault was engaged to organise and train Chinese fighter pilots, and was later to receive American and British fighter planes to oppose the Japanese Naval Air Service. The Chinese Air Force was under the command of Madame Chiang Kai-shek, who initiated the means to stiffen the aerial defences of the Chinese cities. As a result, within three months of combat debut, the Naval Air Service initiated a replacement programme for the G3M1 attack-bomber. Meanwhile the air war was becoming out of hand, with an escalating rate of losses. For instance, on 17 August twelve attack-bombers took off from HIJMS *Kaga* to raid the port city of Hangzhou. Unfortunately bad weather prevented a link-up with the escorting fighting force. When near the target the bombers were attacked by Chinese fighter planes, with the result that eleven Japanese bombers were shot down, with one returning to the aircraft carrier riddled with bullets. The air campaign was not going to be so easily accomplished as had been anticipated. In September 1937 the Mitsubishi Aircraft Company was officially instructed to design a land-based attack-bomber to the 12-Shi specification. This called for a twin-engine aircraft carrying an offensive load similar to the G3M type, but with a heavier armament. Performance requirements specified a maximum speed of 215 knots at 30,000 metres, with a range of 2,000 nautical miles with bombs or torpedoes, otherwise 2,600 nautical miles without an offensive load. The specification exceeded that of the G3M2 bomber then going into production. Nevertheless the bomb load was again not to exceed 800 kg or a torpedo of equivalent weight. By September 1937 the Naval Air Service comprised thirty-nine land-based squadrons in thirteen air corps, including three dive-bomber and six and a half land-based twin-engine bomber squadrons.

During October 1937 the Mitsubishi G3M2 bombers took off from the airfields of the Naval Air Service. The new bomber version of the type G3M used additional gun positions and, escorted by the new A5M2a fighter planes used as long-range escorts, had been fitted with auxiliary fuel drop tanks. The first raid by the new G3M2 bombers comprised nine bombers, escorted by twenty-seven escort fighters. The Japanese destroyed eleven of the sixteen Chinese fighters sent up to intercept the raiders. The crews were well pleased with the new equipment. At the time the new machines were raiding targets over China and the Mitsubishi design teams were reorganising for the work to commence on the new G4M bomber project.

The design team leader was Kiro Honjo, with engineers Hikeda and Kushibe. Unfortunately initial progress was slow since the design team leader shared time between design work on the new G3M2 model and the new design basic concept. Unhappily some engineers had been transferred from the 12-Shi bomber design to Dr Horikoshi's team working on the 12-Shi fighter, known as the Zero. It has been suggested that the G4M bomber design was based upon the British Vickers Wellington prototype bomber. It was known at the time that Japanese 'experts' had attended lectures given by the British inventor Dr Barnes Wallis on the geodetic system of construction for aircraft structures. But since the Japanese were unable to unlock the secrets of the new system for aircraft building, the new technology could not be utilised. Two months later, the surviving Chinese fighters were withdrawn from the Nanjing front, leaving the capital undefended.

In the course of the year the Naval Air Service purchased the North American BI9 and BI10, as well as the Kimmer Envoy, the Douglas DF and the Junkers Ju 86 for evaluation purposes. It was observed in the course of 1937 that tail unit identification numbers on aircraft from HIJMS *Kaga* were 301, those from HIJMS *Akagi* were 303 and aircraft launched from HIJMS *Ryujo* were numbered 312.

Composition of the Naval Air Service, September 1937

Air Corps	Fighters	Dive-bombers	Torpedo-bombers	Twin-engine bombers	Sea reconnaissance	Small flying-boats	Large flying-boats	Large trainers	Sea trainers	Reconnaissance	Total
Yokosuka	1.0	0.5	1.5	-	1.5	0.5	-	0.5	0.5	1.0	7.0
Kasumigaura	0.5	-	0.5	-	1.0	-	-	2.5	2.5	0.5	7.5
Tateyama	1.0	0.5	1.0	1.0	0.5	-	1.0	-	-	-	5.0
Yokohama	-	-	-	-	-	-	2.0	-	-	-	2.0
Ohminato	-	0.5	-	1.0	0.5	-	-	-	-	-	2.0
Kure	-	-	-	0.5	-	0.5	-	-	-	-	1.0
Sasebo	-	-	-	-	-	1.0	1.0	-	-	-	2.0
Ohmura	1.5	0.5	1.0	1.0	-	-	-	-	-	-	4.0
Saeki	0.5	1.0	0.5	-	-	-	1.0	-	-	-	3.0
Kanoya	1.0	-	-	1.5	-	-	-	-	-	-	2.5
Chinhe	-	-	-	-	0.5	-	-	-	-	-	0.5
Maizura	-	-	-	-	0.5	-	-	-	-	-	0.5
Kisarazu	-	-	-	2.0	-	-	-	-	-	-	2.0
Totals	5.5	3.0	4.5	7.0	4.5	2.0	5.0	3.0	3.0	1.5	39.0

CHAPTER 10

The Mysterious Islands

During 1914 Japanese warships of the Imperial Navy sailed southwards from the Home Islands and landed forces that occupied the German possessions in Micronesia. The islands in the Marianas, the Carolines and the Marshalls were seized from the German Empire. At the close of the Great War in 1919, the League of Nations mandated to Japan the territories she had occupied under Article 4 of the League's Charter. A year later, in 1920, the US Intelligence Service returned negative reports on Japan's activities in the Mandated Islands, as they had been prevented from obtaining information by strict security countermeasures.

In 1921 Lieutenant-Colonel Ellis of the US Marine Corps was smuggled into the Japanese Mandated Islands, penetrating the Marshalls and Carolines disguised as a German trader. Conducting an intelligence survey, he filed a controversial report of some 30,000 words with the US War Department concerning the area; in particular he had inspected all the military installations at Koror in the western Caroline Islands, where he was subsequently arrested on a charge of espionage. According to an Englishman, a Mr William Gibbon, the Japanese Police Commissioner ordered the death of Lieutenant-Colonel Ellis as a convicted spy, and he was duly poisoned. This death was later confirmed by Jose Telei, Chief of Native Police; apparently, it is alleged, Ellis was a great drinker, and poisoning would not have been a difficult operation to conduct. However, since the Japanese had cremated the body it was very difficult to ascertain the nature of the toxic chemicals used. A US Navy pharmacist was sent out to the western Carolines and Japan to bring back the remains, but he was drugged and had hallucinations, with little recollection of his journey. Meanwhile a second American had been caught, and he too met a similar fate by poisoning. Since Ellis had looked around everywhere on Koror, the Japanese thoroughly searched his belongings for the incriminating evidence.

In 1923 the US Navy approached the US State Department with the idea that selected American warships should make courtesy calls on particular islands. This idea was rejected by the State Department on the grounds that

114

foreign relations between the two countries would become jeopardised. To overcome the sensibilities of the Japanese Foreign Ministry, Brigadier-General 'Billy' Mitchell flew around the perimeter of the Mandated Islands, gathering information by plane. By 1925 General Mitchell's report was filed with the US War Department and remained classified material until 1961. In this report the general predicted the attack on the US fleet base at Pearl Harbor, as well as the early morning hour of the attack.

Four years later, during 1929, a little-observed event took place when the famous American woman flyer Amelia Earhart was granted the rank of honorary major in the 381st US Army Air Corps Reserve in San Francisco. At about this time on the western side of the island of Saipan, a seaplane base was constructed at Tanapag Harbor. It was to this anchorage that regular flights were made in the mid-1930s by Kawanishi flying-boats on the Japan–Saipan Mandated Islands run. However, the Mandates were not entirely closed to visitors, for during 1934 Paul Hibbert Clyde, Professor of History at the University of Kentucky, was allowed to visit the area. Dr W.C. Herre, Curator of the Zoological Museum of Stanford University, was also given permission to visit the islands. In the following year, 1935, the writer Willard Price also cruised through the Mandates with Japanese permission. The mid-1930s was the period in which the Japanese organised the Nanyo Kohatsu Kabushiki Kaisha – an islands development company, which in reality was a front organisation for the Imperial Navy and its expansionist policy. Now Palau Island was the only open island to foreigners during the era until 1937, when all islands became closed to foreigner visitors. Sometime during 1934, Amelia Earhart and her husband Fred Putman had dined one evening at the university coffee house with Dr Edward C. Elliot, president of Purdue University. In the course of a conversation, Dr Elliot had suggested that Amelia should consider joining the staff of the university, since 800 of the students were girls, and Amelia could inspire the young ladies during informal seminars. The members of the university were liberal progressives with radical tendencies, and Amelia's ideas would coincide with the ideals of the university. The following June, Dr Elliot was able to announce to the university that Amelia had agreed to accept a staff position.

In 1935 the Superintendent of Airways of the Bureau of Air Commerce, Mr William Miller, sailed aboard the warship USS *Itasca* for Jarvis Island in the south Pacific. On 25 March he landed on Jarvis, raised the US Flag and set up the town of Millersville, with the intended project of building an airfield, doing the same on the islands of Baker and Howland.

By 1936 the US Intelligence Services became extremely suspicious of the motives and machinations of the Japanese in the Pacific area. Upon evaluation of collected data, the use of the Marshall, Caroline and Mariana

Islands became suspect. Meanwhile the Japanese government had demanded naval parity with America and Great Britain, and had already withdrawn from membership of the League of Nations. Furthermore, the Washington Treaty limits had been abrogated, and the Tokyo government's military appropriations had risen from 27% of the national budget in 1930 to 46.6% during the period 1935/6.

The adventuresome ideas of Amelia Earhart had pleased Dr Elliot, especially as the university had its own airport and flying school. It was natural that Amelia would use the airport facilities, and so the Purdue Research Foundation, with the enthusiastic backing of Dr Elliot, suggested setting up the Amelia Earhart Fund. This fund was to finance a new aeroplane that would be used as a flying laboratory 'to study the effects of flying on people'. On 20 March 1936 the then most modern non-military plane was ordered from the Lockheed Aircraft Company. This machine was delivered on 24 July, the occasion of Amelia's 38th birthday. The aircraft proved to be a Lockheed 10E Electra, having a fuel capacity of 1,204 gallons and a range of 4,500 miles, at a cost of $50,000. Curiously enough, the Department of Commerce application for a licence listed the designated use as 'long-distance flights and research'. Sometime later this aircraft was fitted with military-type Wasp Senior engines capable of giving the Electra a top speed of 220 mph

Meanwhile Amelia was planning a record-breaking round-the-world flight with her navigator Fred Noonan. The main purpose was the crossing of the long over-water haul of the Pacific Ocean. At the time aircraft radar had not yet appeared, and radio navigation was still in its infancy. Therefore, to pinpoint an airfield in the middle of the Pacific Ocean, thousands of miles away, presented a formidable navigational problem, especially with the prospect of running low on fuel. So sometime during 1936 President Roosevelt requested the US Navy to arrange in-flight refuelling near Midway Island to enable Amelia to continue her flight on to the Hawaii Islands. Flight refuelling was still at an elementary stage, having been developed, so it is said, by the British in the mid-1930s, under the direction of Sir Alan Cobham, who had formed a limited aviation company to develop the idea. Tests had been made using one of the Empire flying-boats of Imperial Airways Ltd and a Handley Page Harrow transport of the Royal Air Force as a tanker.

The position of the United States in Micronesia during 1937 was such that to counteract the Japanese military strength in the islands the Americans required an advanced airbase on Howland Island. This location would place modern long-range bombers at a sufficiently short distance to attack Japanese bases in the Marshall Islands, especially the big naval base at Truk, the Japanese Gibraltar of the South Pacific. Now

Howland Island was under the jurisdiction of the US Department of the Interior, and Secretary Ernest Groening, who administered the island, came to the conclusion that his financial budget would not meet the cost of extending airfield facilities of the type required for heavy long-range bombers. As a result, President Roosevelt initiated a crash building programme at Federal expense early in 1937. Naturally the General Staff of the US Army Air Corps were delighted, for it would provide a base for reconnaissance planes and heavy bombers of the Boeing B-17 Flying Fortress type.

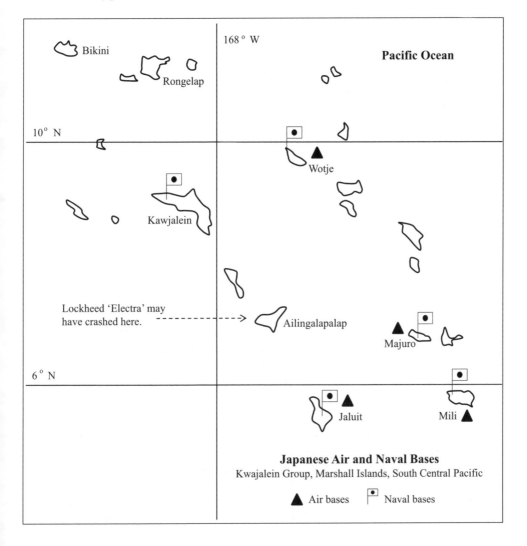

Japanese Air and Naval Bases
Kwajalein Group, Marshall Islands, South Central Pacific

▲ Air bases   ⌷ Naval bases

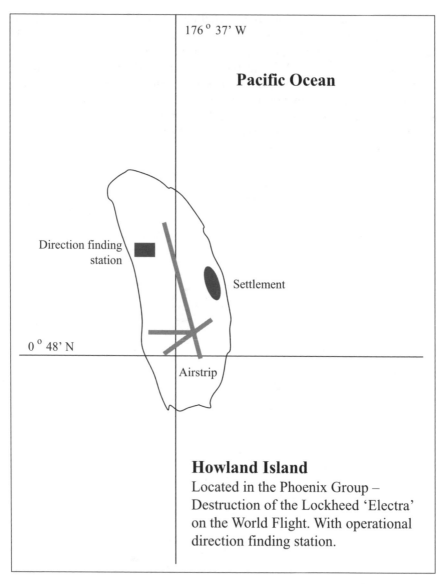

176° 37' W

Pacific Ocean

Direction finding
station

Settlement

0° 48' N

Airstrip

Howland Island
Located in the Phoenix Group –
Destruction of the Lockheed 'Electra'
on the World Flight. With operational
direction finding station.

The plans for the Earhart world flight were now well advanced, and physical tests of the two aviators by naval doctors took place at the Alameda US Naval Air Station, including the recording of dental charts of both Amelia Earhart and Fred Noonan. The Navy swore-in Fred Noonan into active service before the commencement of the flight. When Amelia and Fred returned from Hawaii, where they had been involved in a crash-landing, they resided in Berkeley, California. It was at this time that William Miller was sent to Darwin, Australia, by the US government on some secret mission well before the flight started. As previously mentioned, William Thomas Miller had been appointed Superintendent of Airways of the Bureau of Air Commerce in 1935. This was a front organisation for the colonising tasks on the Howland, Baker and Jarvis Islands in the South Pacific Ocean. Born on 8 August 1899, and educated at the University of Michigan, he had learned to fly with the US Navy and held a reserve commission as a lieutenant-commander. He was appointed the agent of the USA and assigned the task of constructing the necessary airfields. While in Berkeley, and prior to the Darwin visit, Amelia and William Miller had long meetings poring over charts of the forthcoming route. A Mrs Maetta was appointed from a local employment agency to spend time in Oakland and Berkeley helping with the secretarial work to prepare for the flight. She had met Miller at the Leamington Hotel in Oakland when he was an executive with the Civil Aeronautics Agency. Her salary was paid in cash by George Putnam, though she worked for William Miller as well.

On 8 January 1937 Miller informed President Roosevelt that an airfield on Howland would enable Amelia to land and refuel rather than taking the long direct route to Midway over the Pacific Ocean. Roosevelt was pressed to allocate funds to finance the preparations on Howland Island, and eventually he agreed to the appropriation. In March 1937 the Howland Island project was well under way, directed by the Department of Commerce, with the support of the Department of the Interior, but the Army and Navy kept a very discreet profile during the course of all this construction work. Meanwhile William Miller had been appointed Chief of Aeronautical Surveys of the South Pacific Ocean, and worked very closely with Amelia Earhart on detailing the exact route to be taken.

By 15 June 1937, the world-record-breaking flight had commenced, and Amelia Earhart with Fred Noonan had arrived at Karachi, in India, for a two-day stop, having successfully spanned the Red and Arabian Seas. While they were at Karachi Airport, mechanics from the British Imperial Airways Company and instrument specialists of the Royal Air Force had put the plane into A1 condition. Only two months previously, in April 1937, Imperial Airways had commenced using the Lockheed Electra on the London–Paris service, and therefore had operational experience in maintaining the new machines.

On 21 June the Electra landed at Bandung Airport in the Dutch East Indies, where a US engine specialist, Mr F.O. Furman, with a group of mechanics, worked for three days on the plane's engines. A test flight was undertaken on 24 June, when the machine flew to and returned from

Surabaya with engine and instrument difficulties, which necessitated a further two days' work. No known reason has come to light as to the true nature of Mr Furman's work. The aircraft had reached Port Darwin on 28 or 29 June, where parachutes were taken out of the Electra and shipped back to the USA. Delays further bedevilled the flight, which some said was due to 'irregularities in the issuing of health certificates'. However, according to Sergeant Stan Rose USMC, the Electra's radio equipment was repaired and tested. It was known that while at Port Darwin Airport an extra radio system was fitted into the Electra to be used in the event of an emergency, and became known as the 'bush telegraph'. On 2 July the logistic ships USS *Gold Star, Blackheath, Chaumont* and *Henderson* were moving into and cruising in the Marshall Islands group. But on this day the world's press broke the news that disaster had overtaken the world flight. Somewhere between Lae, New Guinea, and Howland Island the Electra had disappeared! In reality it was thought that the plane had been reconnoitring Japanese island bases – possibly Truk Island. Meanwhile the British government's Radio Direction Finding Station at Nauru in the British-controlled Gilbert Islands had picked up Amelia Earhart on 6,210 kc/s at 1030 on the day of the disappearance, saying, 'Land in sight ahead ...'. Two days later, on 4 July, the US Navy Radio Station, Diamond Head, Oahu, Hawaii, received carrier wave transmissions on 3,105 kc/s; it was a man's voice, which unfortunately was indistinguishable. Curiously, on the same day, Mr E.H. Dimitry of Oakland, California, President of International Weather Control Inc., and also a director of the Amelia Earhart Foundation, had taken a copy of messages received by three wireless operators. His organisation had been set up under the cover of inspiring the study of aeronautical navigation and science akin thereto.

According to Dr M.L. Brittain, president of Georgia Institute of Technology, North West University, University of California and Colorado, in April 1943, President Roosevelt on 4 July 1937 ordered the US Navy to send the US battleship *Colorado* to search for the Electra. The battleship was 2,500 miles closer to the area than the US aircraft carrier *Lexington*. Dr Brittain further added that many persons were known to have been sent to the Mandated Islands to discover the secrets of the illegal fortifications. But at the time there was a known feeling that Amelia Earhart was on some sort of subversive operation.

On 7 July, the day that USS *Colorado* was detailed to search for the Electra, radio direction-finding surveillance units received the transmission of a woman's voice on 3,105 kc/s. The message read, 'Earhart calling NRU 1 – NRU 1 calling from KHAQQ. On coral south-west of unknown island. Do not know how long ...' Then the carrier wave faded. Seconds later, 'KHAQQ calling KHAQQ. We are cut a little ...' Then the carrier wave faded completely. It was on this fatal day that Japanese forces commenced offensive operations in China.

By 13 July sufficient radio traffic had been collated to establish the whereabouts of the crippled Electra. The Amelia Earhart flight had come down somewhere in the area of Mili Atoll in the south-eastern Marshall Islands. A Japanese fishing boat had moved in to rescue Amelia and Fred Noonan, who had then been transferred to the seaplane tender HIJMS *Kamoi* and later transferred to the survey ship HIJMS *Koshu*. The two Americans were apparently taken to Jaluit Atoll and to Kwajalein, and finally were landed at Saipan.

The efficiency of Japanese Naval Intelligence cannot be underestimated, for the art of radio direction finding was not a monopoly of the Americans, and the possibility that Japanese Naval Intelligence or Imperial Navy radio-monitoring and surveillance units might have been party to advanced information concerning the mission equally could not be ruled out. In any event it was known that the Japanese Navy had used fishing boats for survey work over many years in all parts of the Pacific. Indeed, at one point the Imperial Navy was the only organisation that possessed accurate charts of locations and waters in the south seas, other nations having neglected a survey of the areas. Alternatively, the A5M fighters may have waylaid the Electra, which, either in its haste or possibly crippled, had flown on to come down in an isolated area of the ocean.

A former employee of the senior US military government office on Majuro Atoll, Mr John Mahan, has stated that it was known that Amelia Earhart with her navigator Fred Noonan crash-landed in the area of Jaluit, Majuro and Ailinglaplap Atoll; besides capturing the two crew members of the Electra, Imperial Naval Intelligence took the plane and some equipment. Eventually the aircraft carrier *Lexington* abandoned the search, as the survey work the ship had undertaken of the South Pacific was now complete. Meanwhile William Miller had been promoted Chief of Civilian Aeronautics Administration – International Section. However, it became known sometime in 1937 that prior to the world flight, Mr W.C. Tinus, president of Bell Telephone Laboratories, had been the engineer responsible for the design and installation of the Electra's radio equipment. This apparatus had consisted of modified three-channel Western Electric equipment as used by the airline companies. One channel operated on 500 kc/s, a second on 3,000 kc/s and a third on 6,000 kc/s. With a simple modification to the apparatus it was possible to transmit on continuous wave and medium continuous wave, with voice or with key. In evidence, one Antonio Diez, a native chauffeur to a Japanese officer, stated that the Electra did not fly to Saipan. The Japanese unloaded it from a large loader aboard the seaplane tender HIJMS *Kamoi* at Tanapag Harbor, Saipan, where it was transported to Aslito Naval Station on the southern side of the island. Aslito became fully operational as a fighter-bomber base sometime in 1938.

Upon arriving at Saipan, Amelia Earhart and Fred Noonan were separated. Deep interrogations followed, and as Fred was already injured he subsequently died. Amelia was observed by various native women, who reported her condition as extremely distraught. Amelia and Fred were either killed, executed or died under interrogation, and were buried not far from the prison to which they had been taken on the island.

The Japanese feared the airbases on Howland, Baker and Jarvis Islands, so in 1941, at the commencement of the Great Pacific War, they bombed and strafed the installations' runways and lighthouses, which at that time were manned by the Kamehameha Boys' School of Honolulu. During the course of the war, in 1943, William Miller was further promoted to Chief of the Air Carrier Division of the Civil Aeronautics Administration, but sometime during that year he died in office.

The island of Saipan in the western Pacific was controlled by Japan from 1914 until 1944. In the latter year, after a ferocious American artillery and aerial bombardment from the warships and aircraft carriers, the US marine assault craft swept in from the white-foamed breakers of the Pacific. During the course of this long-drawn-out attack on the island of Saipan, 15,525 American casualties were sustained, while the Japanese lost 29,000 of the 30,000 troops stationed on the island. This island was situated 115 miles north of Guam, a distance of forty-five minutes' flying time, and was twelve miles long and five miles wide. Three miles to the south-west was the smaller, flatter island of Tinian. In February 1944, on the island of Namur, Kwajalein, in the Marshall Islands, three US marines found a suitcase in a Japanese barracks. The suitcase was located in a room obviously furnished for a European woman. The furniture included a dresser, and upon examination the suitcase was found to contain women's clothes, newspaper clippings and a diary entitled *The Ten-Year Diary of Amelia Earhart*. All this evidence of the former presence of Amelia Earhart was handed to the US Marine Corps Intelligence Service.

A year later, sometime during 1945, Amelia Earhart's Lockheed Electra was identified by General W.M. Greene USMC in a hangar at Aslito Naval Air Station, Saipan. At the time General Greene was the G3 Staff Officer, Intelligence, with the 2nd US Marine Division on Saipan. The plane, supposedly the Electra, was taken out of the hangar, drenched with petrol and burnt to the ground.

In 1960 Lieutenant-Commander Leroy Hippe USN, Executive Officer Saipan Naval Administration, stated that twenty-two tons of Japanese

Curtiss P-40 fighters, painted with the *Flying Tigers'* shark-face emblem, China.

Douglas C-47 *Skytrain* cargo planes, with P-40 fighter escort, New Guinea.

Douglas SBD-5 *Dauntless* on anti-submarine patrol over USS *Washington* on the way to the invasion of the Gilbert Islands, with USS *Lexington* in the background, 12 November 1943.

Douglas TBD-1 *Devastator*, torpedo squadron VT-6, over USS *Enterprise*.

Grumman F2F-1 fighters, squadron VF-5, USS *Yorktown*, February 1939.

Grumman F4F-4 *Wildcat* fighters, Henderson Field, Guadalcanal, Solomon Islands, 1942.

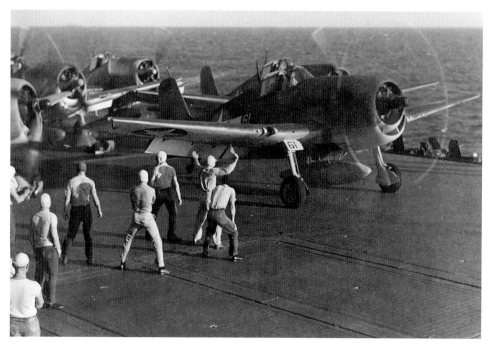

Grumman F6F-3 *Hellcat*, on USS *Saratoga*. The pilot is Commander Joseph C. Clifton, taking off for Tarawa, Gilbert Islands, November 1943.

Lockheed P-38 *Lightning*, capable of carrying 2,000 pound bomb payloads.

Chance Vought F4U-1A *Corsair*s from fighter squadron VF-17 *Jolly Rogers*, Southwest Pacific, 1944.

HIJMS *Akagi* seen from an aircraft immediately after take-off, April 1942.

HIJMS *Hiryu* (Flying Dragon), which took part in the attack on Pearl Harbor, was damaged by air attacks during the Battle of Midway and sank 5 June 1942.

HIJMS *Kirishima* Kongo class battle cruiser off Amoy, China, with Kawanishi E17K and Nakajima E8N floatplanes on the aircraft deck, scuttled after the Naval Battle of Guadalcanal, 15 November 1942.

HIJMS *Mikuma* heavy cruiser on fire and dead in the water following attacks by Douglas SBD-3 *Dauntless* planes from USS *Enterprise* and USS *Hornet*, 6 June 1942.

HIJMS *Yamato* battleship under construction at Kure Naval Base, Japan, carrier HIJMS *Hosho* on the right, and UN food supply ship *Mamiya* is in the middle, 20 September 1941.

IJN I-10, A-class Japanese submarine, Penang Port, April 1942.

IJN I-54 B-class (new-type 2) Japanese submarine, making a trial run, Tokyo Bay, February 1944, sunk east of the Philippines 28 October 1944.

records were captured on Saipan by the US Navy and sent to the United States for intelligence evaluation. Undoubtedly they contained the full history of the fate of the many American officers who had endeavoured to penetrate the secrets of the island fortifications. From 1945 until 1958 the Japanese military records were stored in the various American Federal records centres, and a total of 7,800 cubic feet of recorded material originating in the Nipponese War and Navy Ministries were very carefully examined. All this material was returned to the Japanese government in 1958, according to a Mr Darter, director of the National Archives and Records Service in Washington.

Finally, in 1960, at a Democratic Party convention, John F. Day, president of CBS News, met a technical writing teacher, one of whose pupils had been an aviation engine mechanic. He stated that after returning from Hawaii to Berkeley, California, prior to the commencement of the world flight, the engines of the Electra had been changed and were twice as powerful as before. Extra fuel tanks were fitted so that the range and speed were twice as high as originally quoted. Aerial cameras had also been fitted and were operated from positions in the cockpit. Subsequently all persons working on these Lockheed modifications were sworn to secrecy by signing an oath. Some of the people at Hamilton Army Air Force Base in northern California had been involved in this affair – as had been many others! The graves of Amelia Earhart and Fred Noonan had been located by US Marine intelligence officers soon after Saipan had been subdued, and from among the shattered wreckage and chaos of the island the last remains were secretly removed to the United States. Once back in America, reburial took place in some area best only known to the families of the two deceased. At least the myth of the disappearance during the world flight could be maintained for the benefit of the world's press headlines!

CHAPTER 11

The Zero is Conceived

In January 1938 His Imperial Majesty the Emperor had an excuse not to honour the Strike North faction, which was supported by the Army for an attack on Siberia and the invasion of the USSR. The Japanese had always held a belief that communism was the enemy of the homeland traditional way of life and its values. Indeed, as a consequence, all known members of the Japanese Communist Party had been imprisoned as a counter to their activities. The Strike South faction, on the other hand, supported the Navy and had more practical benefits to offer by the way of the possibility of exploiting the oil and minerals from the captured territories of the rich lands of the Dutch East Indies, Malaysia, Burma and Australasia. So by a secret agreement, the Strike South faction secured its ambition, and with the continuation of the war with China the Army was kept busy. The Chinese hinterland was a vast training ground where troops could live off the country and where combat readiness could be maintained until the troops were required to fight the more sophisticated armies of the West. Thus the war in China kept the Army under constant training. Nevertheless the decision to strike south would open up the possibility of exploiting one of the richest areas of the world, as well as providing raw materials for Japan's rapidly increasing industrialisation. Now for fifteen years the Navy had received the bulk of the military appropriations for weapon development. Only the Navy possessed the technology that could challenge the West. During the first eight months of the war in China it was the Navy that had provided the dive-bombers and long-range bomber squadrons to attack the inland cities in support of the general advance of the Army. The Army Air Corps was mainly good for transportation work only at this particular time. The Navy had deliberately pursued a policy to keep the Army out of ambitious adventures in Siberia, while Japan generally nibbled in south-east Asia. Unfortunately the Navy required a further four to five years to complete the perfection of its weapon development.

On 1 April 1938, one year after the opening of the Sino-Japanese Incident, the same year in which Dr Horikoshi commenced designing the Zero fighter, the government prepared plans for future eventualities known as Plan 1st, 2nd and 3rd. At the same time it passed a General Mobilisation Law to enable the various government agencies to control and utilise manpower and resources.

In May 1938 the 2nd Army under the command of Prince Higashikuni joined up with the troops advancing from Shanghai. Henceforth one of the two north–south railway lines was secured. The Prince had sought permission of the Army General Staff to open up further penetrating thrusts, but the General Staff objected with the order to face north. Meanwhile the Admirals wished to move up the Yangtze river to strike at the Anqing Airbase 150 miles further on from Nanjing. This airbase was within flying range of the British colony of Hong Kong and the converted harbours of south China. The Emperor agreed with the naval strategists, and the Army was ordered to advance on Anqing, and then on to Hankou, an additional 200 miles onward. Meanwhile in May 1938 the seaplane tender HIJMS *Mizuho* was completed, having been laid down in May 1937 under the 2nd Fleet Replenishment Law. Designed to carry midget submarines and a complement of seaplanes, the ship was built to achieve 27 knots, but on trials actually attained only 23 knots. The aircraft carrier HIJMS *Akagi* underwent a refit, in which a single flight deck was constructed with a small port-side island bridge structure. A general modernisation programme was pursued in which three elevators were constructed to facilitate the launch of her sixty aircraft.

During this period the ferocious air war over China continued unabated. On 25 February 1938 an air battle took place over the city of Nanchang. In the clear blue sky above the city eighteen Japanese fighters clashed with fifty defending Chinese fighter planes. In a running battle from operational altitude down to low level twenty-seven Chinese fighter planes were destroyed for the loss of two Japanese aircraft. In the spring active air operations were opened up in the area of the Suzhou front in which seventy aircraft of the 2nd Joint Air Corps and carrier-borne 4th Air Wing co-operated in a series of bombing attacks on targets of military importance. On 29 April, during attacks in the Hangzhou area thirty Japanese fighters clashed with eighty Chinese aircraft, and fifty-one Chinese planes were shot down in running engagements for the loss of two Japanese planes. Four Type 94 dive-bombers led by Lieutenant Junior Grade Shoichi Ogawa bombed targets in the Nanjing area on 7 July. On completion of their mission they landed on a Chinese airfield to destroy with machine-gun fire the remaining Chinese aircraft before taking-off and returning to base. By the autumn of 1938, over 167 aircraft were operational in southern China under the control of the 14th Air Corps (formerly the Kaohsiung Air Corps) in co-operation with the

1st and 2nd carrier-borne Air Wings. At the end of October 1938 Type 96 bombers had struck at Guilin, Guiyang, Kunming, Kwantung, Nanning and Liuzhou in over sixty successive attacks. In central China Type 96 bombers had made over 170 attacks against airfields, military installations and other targets located at Hangzhou, Nanchang, Luoyang, Chongqing, Changsha, Yichang and Hengyang. The Chinese Air Force was not slow to counter-attack, but lacked equipment. The Soviet Union had supplied SB2 bombers, I15 and I16 fighters, and these machines in a twelve-month period prior to October 1938 had made over 120 attacks on Japanese installations and facilities. At the same time the Chinese received some Gloster Gladiator fighters from Great Britain and a few Dewoitine 510 fighters from France. These machines were in service by December 1937. According to Imperial Navy statistics from August 1937 to October 1938, over 1,227 Chinese aircraft were destroyed: on the ground the Chinese lost 617, and a further 610 were destroyed in aerial combat. During this same period the Navy lost 111 aircraft in combat. Much of the Chinese loss was entirely due to the inadequate pilot training, not having been trained to the Japanese naval standard, which on one occasion resulted in a certain Flight Sergeant Hiroshi Kashimura landing back at base safely in a Type 96 fighter after having had half the port wing shot away.

In the United States the second US Vinson Naval Expansion Programme was announced sometime during 1938. This announcement prompted the Japanese Navy to insist on further expansion of the 1937 Programme – the 4th Expansion Programme. Construction commenced immediately on two battleships of 65,000 tons and eighty surface vessels. Among the surface ship construction plans was an aircraft carrier of 34,000 tons. Included in the programme was the construction of seventy-five land-based air squadrons and 174 ship-based aircraft. Certain reorganisations were anticipated, though the headquarters of the Naval Air Service was still located as previously in 1937. The reorganisation concerned the 1st Joint Air Corps, 2nd Joint Air Corps and the 1st, 2nd and 3rd Air Wings.

    I. The 1st Joint Air Corps comprised:
        1. The Kisarazu Air Corps – 2 squadrons of Type 96 Mitsubishi G3M2 bombers
        2. The Kanoya Air Corps – 1 squadron of Type 96 Mitsubishi G3M2 bombers
    II. The 2nd Joint Air Corps consisted of:
        1. 12th Air Corps
          • 2½ squadrons of Type 96 fighter planes

- 1 Squadron of Type 96 torpedo-bombers
- 6 Seversky 2 PA fighter reconnaissance aircraft

2. 13th Air Corps
   - 2 squadrons of Type 96 Mitsubishi G3M2 bombers
   - 1 squadron of Type 96 fighter planes

3. 14th Air Corps
   - 1 squadron of Type 96 fighter planes
   - ½ squadron of Type 96 dive-bombers
   - 1½ squadrons of Type 96 torpedo-bombers

4. 15th Air Corps
   - 1 squadron of Type 96 fighter planes
   - 1 squadron of Type 96 dive-bombers
   - ½ squadron of Type 96 torpedo-bombers

5. Kaohsiung Air Corps
   - ½ squadron of Type 96 Mitsubishi G3M2 bombers

III. 1st Air Wing – Aircraft carrier HIJMS Kaga
   12 Type 96 fighters
   12 Type 96 dive-bombers
   18 Type 96 torpedo-bombers

IV. 2nd Air Wing
   Aircraft carrier HIJMS *Soryu*
   12 Type 96 fighters
   18 Type 96 dive-bombers
   12 Type 96 torpedo-bombers
   Aircraft carrier HIJMS *Ryujo*
   12 Type 96 fighters
   6 Type 96 torpedo-bombers

V. 3rd Air Wing – Seaplane carriers

HIJMS *Kamoi*	9 Type 94/95 sea reconnaissance
HIJMS *Kaku-maru*	9 Type 94/95 sea reconnaissance
HIJMS *Notoro*	8 Type 94/95 sea reconnaissance
HIJMS *Kamikawa-maru*	6 Type 94/95 sea reconnaissance
HIJMS *Chitose*	9 Type 94/95 sea reconnaissance

On 13 April 1938 a group conference took place concerning the 12-Shi fighter programme, at which thirty representatives attended from the Naval Bureau of Aeronautics, the Naval Aircraft Establishment, the Yokosuka Experimental Air Corps and the Mitsubishi Aircraft Company. The meeting discussed the merits of foreign aircraft and the combat experience obtained in the air war over China. All foreign imported aircraft were tested by the Naval Aircraft Establishment and the Yokosuka Experimental Air Corps since at this time no Japanese designer

or aeronautical engineer was capable of flying a military aircraft. At the Naval Aircraft Establishment Lieutenant-Commander Takaeo Shibata HIJMN was the chief test pilot, assisted by Lieutenant Kiyoji Sakakihara. Now Commander Shibata had a practical approach to aircraft development, since during the Sino-Japanese Incident he had held the post of fighter group leader aboard the aircraft carrier HIJMS *Kaga*. Another opinion to the problem was held by Lieutenant Genda, who had led what was colloquially known as the Genda Circus. This unit comprised Japan's best four aerobatic flyers. The lieutenant had graduated with honours from the Naval War College as well as having been trained, it is alleged, at the RAF Officer Training College, Cranwell. The development of the fighter aircraft from the stance of a defensive weapon to that of an offensive machine had been attributed to Lieutenant Genda. When the assembled conference convened, the question of the development of the Zeke fighter was considered. The agenda questioned the most important qualities required of range, speed and manoeuvrability, even at the loss of speed and range. This view was held by the Battle Lessons Committee of the Yokosuka Experimental Air Corps, whose members were combat veterans who had studied the combat reports from the various units in China and supported Genda's viewpoint.

However, Lieutenant-Commander Shibata thought much more than manoeuvrability was required, as had been shown up by various errors in the conduct of the air war over China. He pointed out that the Type 96 Mitsubishi G3M2 twin-engine bomber had sustained critical losses at the hands of the Chinese fighter pilots, which necessitated Japanese fighter escorts. Having rejected the twin-engine escort fighter project, he suggested single-seater fighters, with speed and range, and pilot skill compensating for lack of manoeuvrability. Unfortunately there was no one of superior rank present who could give a decision on the pros and cons of the problem, which remained unsolved. Consequently the aeronautical engineers from the Mitsubishi Aircraft Company were resolved to design a fighter aeroplane, which would meet the conditions of all parties concerned. Later in the spring of 1938 the Yokosuka Experimental Air Corps advised the Naval General Staff of its views and plans on future developments. These memoranda concerned:

1. The carrier-based fighters
2. Land-based interceptor fighters
3. Twin-engine escort fighters.
4. Heavily armed escort planes, i.e. converted bombers
5. The considerations of the Battle Lessons Committee

The considerations included combat techniques and special training, new combat formations and special operations for night-fighters, techniques of high-altitude bombing by large formations of aircraft, the development of more efficient Navy formations for operational purposes, and assessment of the results of the China operations.

Controversy also ranged in the manner of the organisation which each air corps should assume. Hitherto, as previously pointed out, air corps were equipped with fighters, bombers, dive-bombers, etc. in equal numbers. Under operational conditions this arrangement had proved inadequate, and thus a decision was made that each air corps was to have either bombers or fighters, but not both. The 12th Air Corps was to be exclusively equipped with fighter planes, using the Mitsubishi A5M fighters. Other air corps were re-equipped with an entire bomber fleet. The lessons from the experimental reorganisations were collated and forwarded to the Battle Lessons Committee, and so the 12th Air Corps was able to put forward proposals concerning tactics and engineering modifications for carrier fighters, interception fighters and escort planes.

An analysis of the 12th Air Corps proposals found that both pilots and commanders favoured manoeuvrability, especially on carrier fighters, with a required flight endurance of one hour on cruising speed, and half an hour on full throttle combat. The used auxiliary fuel tanks were accepted, and full confidence was give to the capabilities of the 7.7 mm and 13 mm machine-gun. Fighter pilots came to consider the 20 mm cannon as a dead-weight. The views of the 12th Air Corps were also shared by the Yokosuka Experimental Air Corps, but were not reflected by those prevailing in the Navy elsewhere. The upshot of these reports centred on the fact that the air corps were dissatisfied with the embryonic 12-Shi fighter project, and requested the Mitsubishi designers to redesign the machine in the light of the latest thinking as a lightweight fighter. Unfortunately the design requirements would not meet the high standards of the prototype aeroplane already under construction. The Navy relented, and the original Mitsubishi design could go ahead. This represented the Navy's unbalanced view on aviation, for all over the world aeronautical engineers were working on the properties and installations of armour plating, as well as self-sealing fuel tanks. The Naval Air Service erroneously believed the increased weight would reduce range, and this mistaken view cost the service dearly in the years to come. Unhappily the China combat reports seemed to prove the Naval Air Service right, with many kills recorded for each A5M fighter lost.

All the previous views expressed were based upon the 12th Air Corps' Secret Document No. 169 of 1938, which embodied the recommended wishes of the pilots involved. Briefly, this document stated:

12th Air Corps Secret Document No. 169 of 1938

1. Manoeuvrability of fighter planes is paramount.
2. Proof of results of air raids conducted over Hangzhou & Nanchang.
3. Number of Chinese planes approximately equalled the number of Japanese machines involved.
4. Comparative performance of machines about the same.
5. Chinese pilots lacked skill, manoeuvrability and were of low morale.
6. Chinese pilot skill equalled the Japanese but victory attained by individual manoeuvrability of Japanese pilots.
7. Exceptional pilots involved in prolonged dogfights.
8. Japanese planes lost due to fuel exhaustion.
9. When outnumbered by Chinese planes Japanese pilots achieved victory by use of manoeuvrability.
10. Victories would appear to lie with fighters with most manoeuvrability.
11. Folly to disregard this element.
12. Planes required a sharp turning radius with close agility.
13. Next characteristic long range.
14. Physical time endurance on pilots, questionable fatigue factor.
15. Future aerial engagements up to 32,808 feet altitude.
16. Fighter escort missions in summer over Hankou and Nanchang indicate one-and-a-half-hour flight each way to and from target.
17. Combat altitude anticipated to heights of 22,966 feet.
18. Fighter plane escorts not always possible on long-range bombing mission.
19. 12th Air Corps suggested modified attack-bombers heavily armed.
20. Each escort attack-bomber with a ten-man crew.
21. Each escort attack-bomber will require increase in speed.
22. Faster interceptor speed required.
23. Fighter planes with above characteristics an engineering impossibility.
24. Suggest two distinct fighter plane types:

    Carrier fighters

    1. Prime flight characteristics must be manoeuvrability. Japanese fighters versus Russian-built I16B Rata (developed from American Boeing P26) in China.

Japanese fighters	Chinese I16 B
5 min 54 sec to 16,700 ft	6 min 30 sec to 16,400 ft
240 mph max speed	279 mph max speed

2. Time duration for carrier defence estimated at 1½ hours to 2 hours + 20 minutes combat + 1½ hours return to aircraft carrier. Total 3 hours 20 minutes to 3 hours 50 minutes.
   a. Long hours represent pilot fatigue and reduced efficiency.
   b. Unfortunately flights limited by inefficient, primitive navigational equipment.
3. Jettisonable fuel tanks a necessity to equal half capacity of the main tanks.
4. Total cruising endurance equal to three and a half to four hours with jettisonable tanks.
5. The maximum level speed may be 12 mph lower than enemy.
6. The combat heights achieved during the Sino-Japanese Incident ranged from 6,562 feet to 22,966 feet. New fighter planes must operate above this height using two-speed or two-stage superchargers.

Land-based fighters

1. Interceptor fighter type required for search sweeps.
2. Maximum speed a primary characteristic to exceed 46 mph, the maximum speed of the enemy bombers.
3. Plane to maintain adequate stability during diving.
4. Range secondary to speed and manoeuvrability.
5. Combat altitude must exceed 32,808 feet.
6. The Chinese bombers dropped bombs from 22,966 feet up as high at 26,247 feet. Future bombing altitude will exceed 32,808 feet.

Armament

1. The bomber armament will increase as the speed.
2. The 7.7 mm machine-gun was considered suitable for defending the bombers up to 656 feet range.
3. Future weapons will require accuracy up to 1,640 feet range.
4. An increase in the initial muzzle velocity will be required.
5. With incendiary bullets enemy may be destroyed up to 656 feet.
6. Hitherto the only way to destroy a bomber was to shoot the pilot or strike fuel tanks.
7. Effective range of incendiary bullets will increase using 10 mm or 13 mm weapons.

Special equipment requirements

1. A two-way radio telephone will be required.

2. Pilot's oxygen apparatus.
3. Electrically heated pilot's suit.
4. Bomb racks so that fighter may be used for ground strafing.
5. A radio direction-finding and homing apparatus must be installed. However, no provision appears to have been recommended against enemy fire power, i.e. armour plating or self-sealing fuel tanks.

At last, the Zero design studies commenced on 17 January 1938 with the inauguration of a joint civilian/Navy investigation committee, which met at the Naval Aircraft Establishment building at Yokosuka. The personnel present included Rear Admiral Misao Wada, chief of the Technical Division, Naval Bureau of Aeronautics, Lieutenant-Commander Goro Wada, engineering officer i/c fighters, and other officers of the Naval Bureau of Aeronautics; also Vice-Admiral Kenji Maehara, chief of the Naval Aircraft Establishment, with other officers of the NAE; Lieutenant-Commander Minoru Genda, fighter group leader; and Lieutenant Shigema Yoshimoto, a fighter squadron leader. More than twenty officers of the Naval Experimental Air Corps, Yokosuka, attended. The Mitsubishi Aircraft Company sent engineer Hattori, manager of the Experimental Section, and Dr Horikoshi, chief designer, with engineers Kato and Sono. The Nakajima Aircraft Company sent five engineering representatives.

Now Lieutenant Genda was home on five days' leave from the air battles over mid-China. He explained the combat effectiveness of the A4NI biplanes and A5M monoplane fighters and spoke of the aircraft requirements. As air staff officer of the 2nd Combined Air Flotilla, he was concerned with fighter operations over the Shanghai and Nanjing areas. After the meeting had carefully assessed the Navy's requirements, the representatives of the Nakajima Aircraft Company considered the specification impossible to achieve and walked out of the conference, to withdraw from the competition. Nevertheless, it fell to the Mitsubishi Company to accept the impossible and go ahead accordingly. As a result Mitsubishi appointed Dr Horikoshi as chief designer, assisted by Eng Mathematician Yoshitoshi Sone and Teruo Tojo. Structural engineers were Sone and Yoshio Yoshikawa, and power unit engineers appointed included Denichiro Inoue and Shotaro Tanaka. The engineer for armament and auxiliary equipment was Yoshimi Hatakenaka, while the engineers concerned with the landing-gear and related equipment were Sadihiko Kato and Takeyoshi Mori. Having organised a special design team, the Mitsubishi Company conceived design principles:

1. The engine
Provisionally selected was the Mitsubishi Zuisei 14-cylinder radial air-cooled engine rated at 875 hp, with 870 hp for take-off. The rated altitude was 11,800 feet with an rpm reduction ratio of 2,540 × 0.728 at a dry weight of 1,158 lb. The overall diameter of this motor was only 44.160 inches. The power loading was estimated to be 5.5 lb/hp, which was the Naval Air Service's original requirement for this machine.

2. The propeller
Arrangement would probably be a constant-speed Sumitomo-Hamilton two-position variable-pitch propeller, manufactured by the Sumitomo company under licence from the American Hamilton (Automatic) Company. These were the original naval specifications.

3. Weight saving
The wing was to be constructed as a single piece with small-fitting fuselage wing points. Use would be made of Sumitomo's Extra Super Duralumin developed by Dr Isamu Igarashi and engineer Buntaro Ohtami, a combination of the British E Alloy and the German Sander Alloy. Now ESD when produced as a rolled sheet or extruded tubing failed to prevent metal fatigue. The best use was for the non-clad extruded sections or the clad sheet to be covered on both sides with corrosion-resistant aluminium alloy coating. This material was mainly used in small aircraft as main spars or main Webb spars. To evaluate this situation the Mitsubishi chief designer paid a visit to the Sakurajima Works of Sumitomo Metal Industry Company Ltd at Osaka on 17 March to confer with the engineer in charge of research and production.

4. Materials
The materials were kept to the minimum owing to Japan having limited resources and a low industrial standard. Alloy steels included case-hardened steels, structural alloy fittings of a high strength, weldable structural alloy steel and nickel-chrome stainless steel for exhaust pipes. The carbon steels comprised low-carbon steel for welded parts and the use of medium-carbon steel for non-welded items. Piano wire and spring steel were also used.

5. Wing section
By 17 February 1938 wind tunnel tests were completed with the new aerofoil – the Mitsubishi Number 118. This new section had little movement of the centre of pressure, and was evolved from the B9 and NACA 23012 aerofoils. In fact the lift drag curve of the B9 was equal to that of the Mitsubishi 118, but the movement of the centre of pressure was half that of the B9.

By 6 April 1938 a full-scale mock-up of the new aircraft had been completed. On 27 April naval engineering staff arrived to inspect the first mock-up. Minor faults in power unit equipment were recognised and easily rectified. The naval engineering staff returned on 11 July for a second inspection, and at the same time decided not to amend the armament specification. On 26 December 1938, engineers from the Navy and the Mitsubishi Company reassembled at the Nagoya aircraft plant to examine an actual completed airframe.

Unhappily the Naval Bureau of Aeronautics approached Dr Jiro Horikoshi sometime in June 1938 regarding the 13-Shi design for a three-seater, twin-engine fighter plane. At the time the Mitsubishi Aircraft Company was short of design staff, and those already employed by the company were fully engaged. The company had no alternative but to report that it could not be responsible for either the design or the construction of yet a further machine. The Naval Air Service approached the Nakajima Aircraft Company with a contract to complete a prototype by October 1938.

In July 1938 the Tomioka Munitions Plant of the Uraga Dockyard Company at Tomioka, Kanagawa Prefecture, became separated from the mother company. Renamed the Dai-Nikon Heiki Company Ltd, or Japan Munitions Company, the factory commenced the mass production of the 20 mm aerial Oerlikon cannon and ammunition. In the previous March the Oerlikon Company of Switzerland had sent engineers and mechanics to Japan under a Mr Kenpon to supervise the installation of production facilities and the training of production operatives.

During August 1938 naval air officers returned to the Mitsubishi Aircraft Company at Nagoya to inspect the mock-up of the new G4M bomber design. This aircraft was cigar shaped, with a monocoque fuselage accommodating a crew of seven or nine. It has been suggested that the inspiration for this machine was to be found in the prototype Wellington bomber aircraft built by the British Vickers Armstrong Company and designed by the famous British scientist Barnes Wallis. This was the prototype Wellington bomber built with the new geodetic structure, which gave great strength to wings and fuselage. It was known that the Japanese sent representatives to the lectures given by Dr Wallis at Barrow-in-Furness, in Cumbria, at which he described the mathematics and construction of such a system. Though the Japanese took home the idea and attempted to copy the new technology, they were never able to imitate the new British method of construction. So the G4M1 bomber would be constructed in the conventional manner, similar in shape and size to the Vickers Wellington, but never attaining the weight of offensive weapons of the British aircraft. The new Japanese naval bomber was to be powered by two 14-cylinder air-cooled Mitsubishi Kinsei engines rated at 1,530 hp. The fuel tanks had a capacity of 4,900 litres but were integral in two spar wings and unprotected. The armament comprised seven 7.7 mm Type 92 machine-guns in nose, dorsal and blister positions. In the tail position was a 20 mm Type 99 cannon. The ventral bomb bay was located beneath the mid-mounted wing, where the bombs or torpedo or Baka piloted bomb were carried externally. When it was not carrying offensive weapons, a detachable panel was attached to streamline the underside of the body. After the officers of the Naval Air Service had inspected the mock-up, they departed much impressed with the new design.

It was in the month of August 1938 that successful detachable drop-tank tests were conducted by a Mitsubishi A5M carrier fighter, serial number 205, at speeds of 126.5 mph and 207 mph. Unfortunately the Type 89 and Type 92 torpedo-bombers carried conventional high-explosive bombs and possessed a low performance, which resulted in high losses over China. These machines had to be replaced, first by the Type 96 torpedo-bomber and subsequently by the Type 97. This later machine was of all-metal construction, with a retractable undercarriage. It was introduced over China by the Koshun Air Corps. During the next month, September, the minor modifications that had been requested by the officers of the Naval Air Service had been incorporated into the amended design, and the Mitsubishi Aircraft Company was authorised by the Navy to build two examples of the G4M1 bomber for flight evaluation purposes.

Next month the Mitsubishi Company had to inform the Navy that it was unable to enter the 14-Shi design competition owing to the fact that the 12-Shi fighter design was awaiting flight tests. These evaluations would take months to come, and it was a high-priority project, which had drawn upon all the resources of the company's engineers. The Navy was in disagreement regarding the interceptor design details, and requested one year before issuing final specification details. This arrangement would give Mitsubishi one year's grace, which would be put to good use.

However, Dr Horikoshi was asked to visit the Naval Bureau of Aeronautics and was requested to design an interceptor fighter plane with no restrictions on the dimensions. The order of performance characteristics would be a speed advantage of 46 mph over enemy fighter speeds, a high rate of climb, good manoeuvrability and range, a two-way radio, and an armament of at least two 7.7 mm machine-guns and two 13 mm machine-guns. The results would take time to materialise. In the previous year the Navy's standard fighter had been the Type 95 biplane, while only six production models of the Type 96F fighter had been delivered. By December 1938 the production ratio of the Type 96 to the Type 95 fighters was 2:1, but as losses of the Type 96 bomber had been rising as a consequence of the air war in China the Navy requested the Japanese aviation industry to step up production. During 1938 the Nakajima Aircraft Company designed the Nakajima 13-Shi twin-engine fighter designated J1N1. This machine was planned as a long-range escort fighter with capabilities of performing night-intruder operations and being operated as a high-speed reconnaissance aircraft. The machine made use of leading-edge wing slats and was equipped with contra-rotating air screws, but the size and weight hindered the aircraft's manoeuvrability.

In the course of 1938 the Navy had initiated various discussions concerned with the future development of operational naval aircraft. At this time the Naval Air Service had assumed responsibilities far greater than the naval air arm of any other nationality. At the time the Japanese Army Air Force was incapable not only of night operations, but also of transoceanic flights, which had a severe restricting capability on the Japanese Army Air Corps' efficiency. Strategic considerations were left to the naval flyers, much of whose policy had been decided by the Yokosuka Experimental Air Corps, the Naval Establishment and the 12th Naval Air Corps fighting an air war in China.

Meanwhile the Kawanishi H8KI flying-boat (codenamed Emily) had been designed. This machine was a formidable adversary, which utilised the British Short Bros company's method of construction and was rated by many as the best flying-boat of the Second World War. But the Navy was as busy as ever importing chosen examples of foreign-manufactured aircraft. These machines included the German Bucker Jungmann and Heinkel He 112, as well as the French Potez biplane and Caudron C600. From America was imported the Seversky 2PA-B3 and Douglas DC4 aircraft. In the meantime the war clouds were gathering over Europe and peace was beginning to gradually slip away. Within the year European hostilities would commence. The hour to strike at the Barbarians was slowly approaching.

CHAPTER 12

A Dream is Realised

During 1939 the Naval Air Service adopted a scheme called the Service Aeroplane Development Programme, in which each aircraft in the SADP scheme was allocated a code number, which was prefixed by the proposed manufacturer's company initials. For example, the M20 was the Mitsubishi J2M1 fighter project. Under this arrangement the SADP allowed personnel in charge, prior to accepting design projects, to submit to the Navy or manufacturer any problems anticipated in the forthcoming development programme.

By the spring of 1939 the air war over China had become less dramatic, though the Navy anticipated a prolonged struggle. Some demobilisation of Naval Air Service personnel took place and numbers of aircraft were reduced from 327 to 133 in the Chinese theatre of operations. The 3rd and 15th Naval Air Corps were disbanded, and the 14th Naval Air Corps was withdrawn from the 2nd Joint Naval Air Corps and eventually re-equipped with Type 96 Mitsubishi G3M2 bombers. Hitherto the 3rd Joint Naval Air Corps had comprised the 14th and 16th Naval Air Corps for operations on the South China Front, with the newly constructed seaplane carriers HIJMS *Mizuho* and HIJMS *Chiyoda* patrolling off the coast in the South China Sea. On 8 September 1939 the Japanese government took powers that endowed government agencies with conscription powers under the General Mobilisation Law. This measure ensured general conscription for all except those already in the Army, Navy, government officials, councillors, representatives of the Diet, and prefectural assemblies. However, only in 1941 was the ordnance put into effect to conscript factory workers. At this time Engineer Commander Euchi Iwaya of the Navy's technical department of the Naval Bureau of Aeronautics requested further design proposals from Mitsubishi for the 14-Shi interceptor design programme. The machine was to be the replacement for the Zero design, which was now under an intensive flight-testing schedule. But in October 1939 the Chinese Air Force made a surprise aerial attack on Hankou, killing thirty-eight senior Japanese naval officers and destroying fifty Japanese naval planes on the ground. Retaliation came in November 1939 when seventy-two Type 96 Mitsubishi G3M2 bombers of the 13th Naval Air Corps raided the Chinese airbase at Chengdu. In this spectacular operation the Japanese bombers blasted airfield installations, destroying thirty Chinese aircraft on the ground for the loss of four bombers during an engagement with forty Chinese fighters. During the course of this running aerial battle, Admiral Kikuji Okuda HIJMN, commander of the 13th Naval Air Corps, was killed as his plane crashed in flames to the guns of the Chinese fighters. In December 1939 sixty-three Type 96 bombers of the Mitsubishi G3M2 type belonging to the 1st and 2nd Joint Air Corps made three deep-penetration raids on Lanzhou. The naval bombers were joined by thirty-six Mitsubishi OB97 bombers of the Army Air Force's 60th Sentai. This was the first time that the planes of the Naval Air Service and the Army Air Force had co-operated in offensive operation. On 30 December 1939 over Lanzhou a large-scale air battle took place in which forty Chinese fighters jumped thirteen Mitsubishi A5M Type 96 naval fighters. During the course of this engagement thirteen Chinese fighter planes were shot down in flames within minutes. Naval Air Service statistics stated that 230 Chinese aircraft had been destroyed since October for the loss of twenty Japanese naval planes. These figures seemed to indicate a rising Japanese loss, which was to herald the appearance of the Zero, or Mitsubishi A6M1 fighter plane, in the skies over China at a later date.

Flight testing of the new prototype Mitsubishi A6M had continued since early 1939, while the air war was taking place in the skies over China. On 14 February the chief designer, Dr Horikoshi, visited the Naval Aircraft Establishment to co-ordinate design data with the results that the Navy had obtained from tests conducted on the plane. These tests revealed wing flutter at speeds lower than the Navy had anticipated. As a result, two Mitsubishi A6M planes had disintegrated in flight because of wing flutter, which had occurred at a lower speed than the Navy had considered feasible. On 24 and 25 February two naval engineering groups returned to examine a completed aircraft and to study armament and special equipment, while at the same time the Navy cancelled the necessity for a final examination of a complete assembly. By 16 March, eleven months after Mitsubishi had submitted the original 'Explanatory Document on the 12-Shi Carrier Fighter Design', a fully completed prototype stood ready for flight testing. The next day a final inspection by the chief designer was made, and the Flight Test Division of the Naval Air Establishment was advised of the aircraft's readiness for the commencement of factory flight tests. The first engine tests and testing of all components took place on 18 March and were passed as functioning perfectly. The next day, the prototype fighter was transferred from the Nagoya factory over the reclaimed land of southern Nagoya to the airfield at Kagamigahara, some twenty-five miles away. The local roads and runways were inadequate, so transportation was made by ox cart. Now Mitsubishi's factory at Nagoya was the largest aviation factory in the Orient, but curiously enough had no runway adjacent for flight-testing purposes. However, the Kawasaki aircraft factory at Gifu and the Tachikawa Aircraft Company at Tachikawa had their own company airfields. Once established at Kagamigahara, the new plane revealed defects in the oil cooler, which was slightly underpowered, while the top screw-caps on the landing-gear shock-absorbers suffered air leaks. By 1 April the prototype machine, equipped with a 780 hp Mitsubishi Zuisei 13 engine rated at 780 hp, made preliminary trials. An engine test run was made, followed by taxing trials with a jump flight, finally completed by a familiarisation flight. The machine was lightly loaded, at 4,380 lb, with the main and tailwheels not retracted. After each flight the shock-absorber strut leak was checked and an independent retraction test took place.

During these trials the fuel and oil capacity was only half rated. These successful flight tests had taken place on a splendid day, for the sun had risen without a cloud in the sky and a cross-wind blowing across the aerodrome at a steady seven miles per hour. The assembled party of engineers and pilots had consisted of chief designer Dr Horikoshi, Lieutenant-Commander Nishizawa HIJMN, naval superintendent in charge of flying, engineers Sadihiko Kato and Kumatero Takenaka, and company test pilots Katsuzo Shima and Harumi Aratani. Unfortunately test flying at Kagamigahara airfield could only take place during the afternoons, for in the mornings the Army Air Corps utilised the aerodrome for flying training. However, once local flying had ceased the new prototype was wheeled out of the hangar. The engine was started by engineer Takenaka, who revved the motor up and down, studying the performance, the instruments and controls, and ensuring that everything was checked. Meanwhile test pilot Katsuzo Shima, in flying suit and carrying a parachute, which he donned, climbed into the pilot's seat and checked the controls. Having satisfied himself that everything was in correct order, he waved aside the ground crew as the chocks were pulled away, and the aircraft commenced to roll at 1630 in the afternoon. Now Katsuzo Shima was a retired petty officer 3rd class, an outstanding flyer who had joined the Naval Aircraft Establishment and later transferred to the Mitsubishi Aircraft Company as a test pilot. The brakes were adjusted subsequent to taxiing tests, which were later repeated. At 1730 the aircraft took off from the eastern end of the airfield and was airborne at a height of 33 feet for a distance of 550 yards. The new fighter plane landed and returned to the original flight line, where the engineers congratulated the test pilot on his successful initial flight. The chief designer's report on the first flight indicated that the first jump flight had proved the balance of the aircraft, as well as the effectiveness of the ailerons, rudder and elevator controls, as being entirely satisfactory. The leakage of the oleo strut and the incorrect function of the oil cooler were known prior to the commencement of the flight. However, the malfunctions were not serious enough to delay the flight test programme, and repair work was effected subsequent to the stability and controllability tests during the period 2–4 April. At this time the ground crew tuned the engine so that on 4 April, under a light load, flight tests involving balance, stability and controllability were conducted, as well as take-off and landing trials. On the following day under light load conditions, the main and tailwheel undercarriage were retracted and extended while flying, and the various degrees of propeller pitch were tested. Most noticeable was the fact that test pilot Shima flew the aircraft at a higher altitude on a familiarisation flight.

For the first time the prototype plane was flown under normal full-load conditions with an all-up weight of 5,111 lb. Flying in wide circles within the airfield boundary, Shima flew to a maximum altitude of 2,000 feet, whereupon the machine developed minor engine trouble and was forced to land due to continuous slight vibrations. The next day, test flying was taken over by test pilot Aratani, who corroborated Shima's report regarding the effective control response and the continuous vibration. At the same time airspeed indicator calibration tests took place with partial climb and rate-of-descent measurements, as well as tests on the pilot tube apparatus.

The flight test programme for the period 8–12 April was cancelled. During this time aerobatics, dive tests and general performance manoeuvres had been scheduled, including cylinder and oil temperature measurements. However, the opportunity was taken for the ground crews to repair the list of defects that had been detected so far. On 12 April take-off and landing tests took place under full normal loaded conditions. The flight was undertaken with the undercarriage fully retracted. The left leg retracted sluggishly while the right leg retracted correctly. Fortuitously both legs extended correctly. With the undercarriage up at high speed, elevator response appeared heavy, and the vibrations hitherto reported had not been eliminated. The next day the dropping of auxiliary fuel tanks was undertaken successfully under normal full load conditions. Five days later, on 17 April, the chief designer took the decision to change the air screw to a three-bladed propeller. As a result the vibrations were almost entirely eliminated, so that stability and controllability tests were conducted in aerobatic flights. A week later the prototype machine was flying with a gross weight now up to 5,140 lb at a maximum speed of 304 mph, which was almost equal to the original computations. However, the pilot's airspeed indicator was registering some 10 knots lower than was actually attained. But stability and controllability was now deemed satisfactory.

On 1 May 1939 Dr Horikoshi reported the results of the flight test programme to the Naval Bureau of Aeronautics and to the Naval Aircraft Establishment. Those present at the conference included Rear Admiral Wada, chief of the Technical Division of the NBA, Captain Sakamaki, chief of the Flight Testing Division, and Captain Osamu Sugimoto, chief of the Aircraft Division. A proposed schedule of further tests was submitted by Mitsubishi, and the conference decided that the company should complete the test programme. It was further decided that the third prototype plane should be engined with the Nakajima Sakae radial motor replacing the Mitsubishi Zuisei. On the successful completion of the testing programme, it was agreed that Navy pilots would take over any further development schedules. At this stage His Imperial Majesty the Emperor Hirohito was given a full report on the progress of developments to date. Nineteen days later Rear Admiral Wada and staff from the NBA assembled at Kagamigahara airfield to inspect the prototype plane. At the end of May the Mitsubishi Aircraft Company submitted a report on the test flights conducted since 1 April, listing the minor problems, including the now satisfactory state of the stability and aircraft controllability. Recommendations were made concerning modifications to the tail and elevators to improve the control response, with reduction in the rigidity of the elevator control system to improve the feel of the control. A further inspection of the machine took place on 5 June 1939 by Vice-Admiral Hanajima, chief of the Naval Aircraft Establishment, and his staff. A week later the staff of the NBA and NAE assembled at Kagamigahara Airfield for a two-day meeting to confer on the problems arising from the engine troubles that had delayed the development programme twice previously.

In the early part of July 1939, Engineer Commander Yoshitake HIJMN from the Naval Aircraft Establishment made a detailed engine inspection and ran complete operational ground checks before handing the aircraft over to Lieutenant Shigekazu Maki, who flew the plane on a thirty-minute trial flight. The engine test programme was continued by Lieutenant-Commander Chihiro Nakano, flying the prototype fighter for a further forty minutes. In his report of 6 July, Lieutenant Maki recorded the following points:

1. The prototype fighter lands easily.
2. The ease of all-round visibility assists the pilot.
3. Controllability of this machine is greater than that of the German Heinkel He 112 fighter. (Nevertheless, the Luftwaffe had not taken

the He 112 into operational use, preferring the Messerschmitt Bf 109E).

4. As compared with the Mitsubishi A5M fighter the aileron response of the prototype plane was sluggish.
5. Noticeable slow speeds when prototype is rolled.
6. Elevator response was also considered heavy.
7. Sudden control column manipulation causes a sharp response, adversely affecting the pilot's ability to control the aircraft.

Upon studying the report, Lieutenant-Commander Nakano concurred with these views and corroborated Mitsubishi's test pilots' reports. On the following day, 7 July, the machine was flown with the two-bladed propeller, and once again severe vibrations were experienced. As a result, the chief designer, Dr Horikoshi, finally decided upon the three-bladed propeller replacement. A further point now arose in that Lieutenant Maki disapproved of the excessive acceleration experienced while the plane was in a dive.

On 5 August the Nagoya Aircraft Works of Mitsubishi Heavy Industries submitted a formal report entitled 'Report on the Results of the First-Phase Flight Tests of the A6M1 (12-Shi Carrier Fighter) Prototype 1'. The detailed report contained a list of the numbers of flights conducted, the duration of each flight, details of engine test runs, repairs and engineering modifications undertaken. By the end of August the company's responsibilities had ended, and the second official series of test flights were to commence, continuing with engineering changes until about 10 September. Three days later the prototype A6M1 was flown for the last time by Shima, Maki and Nakano before a convened conference of engineering design personnel from the Mitsubishi Aircraft Company and Navy aeronautical engineers. The following day, 14 September, the Imperial Naval Air Service officially accepted Prototype Number One A6M1 carrier fighter. Eventually Lieutenant Maki HIJMN on behalf of the Navy flew the new machine over the Hakone Mountains from Kagamigahara to the naval air station at Yokosuka. The time taken to develop Prototype Number One from design conception to final acceptance by the Naval Air Service was slightly under two years. During the period 1 April to 14 September the test pilots of the Mitsubishi Aircraft Company and the Navy made 119 flights, taking 43 hours 26 minutes while airborne. Engineers had conducted 215 engine tests in 70 hours 49 minutes of development time.

On 18 October 1939 the second prototype passed the company's flight trials and was handed to the Naval Air Service, which accepted the machine on 25 October. Immediately cannon-firing tests took place, in which nine out of twenty rounds scored direct hits on a ground target of 205 square feet. The results were considered excellent! In December 1939 the Mitsubishi A6M2 carrier fighter Model 11 was operational. This model had a wingspan of 39.37 feet, which inconvenienced handling aboard aircraft carriers by causing wing damage to some aircraft, so that it was necessary to hinge some twenty inches of each wingtip to prevent collisions.

At this time the Mitsubishi A6M2 Model 11 appeared in the skies over China for the first time. Fighting in air battles over Chongqing, Chengdu and other Chinese cities, the Zero fighters cut the opposition to shreds. In all, some sixty-four Zeros were available for operational use in two versions. On 20 December Prototype Number Three appeared and was engine tested, as the machine had been fitted with a Nakajima Sakae motor. Flying tests took place on 28 December, in which all performance figures exceeded the original Naval Air Service specification. It has been said by members of the Japanese Imperial Naval Air Service that they could not understand the ignorance of the existence and performance of this plane by the Allies, even as late as 1941. For this machine not only outperformed but out-gunned all the Allied fighter aircraft of the time in the South Pacific area. Previously the performance of the Zero had outshone any of the fighter planes supplied to the Chinese Air Force. Indeed, too many responsible people who should have known better even doubted its existence; their professional ignorance was remarkable!

Meanwhile the development of the naval bomber programme by Mitsubishi ran parallel with the introduction of the Zero fighter scheme. The production of the Mitsubishi G3M bomber had reached a target of twenty-two machines a month by March 1939, and these planes were now equipped with two Kinsei engines rated at 1,200 hp. A civil version of the G3M1 was produced, designated the MC21 freighter, and one such aircraft, registered J-BAC1 and named Nippon-Go, flew around the world between August and October 1939, sponsored by the newspapers the *Tokyo-Nichi-Nichi* and the *Osaka-Mainichi*. The main problems associated with the G3M1 bomber were the Navy's neglect of adequate defensive armament, the lack of self-sealing fuel tanks and armour, and the necessity for fighter aircraft escort. At one stage in the period, high-ranking naval air officers considered the Type 96 G3M1 bomber as possessing superior performance, and a decision was taken to withdraw the effective escort fighter force. Such decisions were a grave delusion, subsequently leading to the development of the A6M1 carrier-based escort fighter. A replacement bomber type, designated G4M1, was completed by September 1939, the design having commenced in late 1937. Having been constructed by the Mitsubishi Dai-San Kokuki Seisakusho, or Mitsubishi

3rd Airframe Works, at Oe-Machi, Nagoya, the new plane was transferred to the naval air station at Kasumigaura for flight testing. This aircraft was flown in October by Katsuzo Shima, the company test pilot who had been hitherto associated with the A6M1 development programme. Upon successfully completing manufacturer's trials, the new machine was flown to the naval air station, Yokosuka, the location of the Dai-Ichi Kaigun Koku Gijitshusho, or First Naval Air Technical Arsenal, for service trials. During October 1939 a total of thirty bombers of the G4M1 type were converted to bomber escort planes with a crew of ten and an increased armament. Upon service testing, the performance of these machines fell short of expectations, which had been predicted by Mitsubishi, and they were reluctantly withdrawn by the Navy. These machines were later converted to transport aircraft and used in the Great Pacific War as parachute troop transports under the designation of G6M1-L2.

During 1939 attempts were made to design and build a four-engine bomber, and Nakajima Aircraft Company produced the G5N Shinzan, or High Mountain. Altogether five prototypes were built, but regrettably they failed flight test programmes. Since they were unsuitable as bombers these machines were converted to transports. At the same time the Douglas DC4 four-engine airliner was purchased from the USA for evaluation purposes as an aid to the study of large four-engine aircraft. At the same time the Navy requested Nakajima and Mitsubishi to prepare designs for a long-range fighter aircraft. Unfortunately the latter company was too heavily committed to be included in this new aircraft design programme, and so the design assignment was accepted by Nakajima. The Kawanishi Aircraft Company had designed a large patrol-type flying-boat designated H6K4. This machine had a crew of nine and carried a bomb load of 3,527 lb or could be used as a torpedo-bomber. With a range of 3,107 miles, a transport version was developed, designated H6K4-L. The aircraft was powered by four Mitsubishi Kinsei 47 radial engines, each rated at 1,070 hp, developing a maximum speed of 211 mph at an altitude of 13,120 feet. A total production of 174 machines was achieved before replacement by an improved type during the middle war years.

The Imperial Navy had designed submarines with the capability of launching reconnaissance aircraft. One such class was the I-9, with sister ships I-10 and I-11. These vessels, developed from the J3 type of submarine, were ordered in 1939 under the 4th Fleet Replenishment Law. Designed with extra telecommunications facilities, they were used as headquarters submarines, and possessed a seaplane hangar with a catapult fitted to the hull casing. Each boat was capable of remaining at sea for ninety days and could travel 16,000 sea miles at 16 knots on the surface. Diving to a depth of 325 ft, they could navigate ninety miles submerged at a speed of 3 knots. Displacing 2,434 tons, they measured $372^3/_4 \times 31^1/_3 \times 17^1/_2$ ft, and were powered by two shaft diesel/electric motors of 12,400 hp/2,400 hp, which gave 23½ knots on the surface or 8 knots fully submerged. Besides housing one catapult seaplane, each boat was armed with one 5.5-inch GP gun, four 25 mm AA cannon and six 21-inch torpedo tubes, and had a capacity to carry eighteen torpedoes. The I-9 was built at the Kure Navy Yard in 1939, with I-10 under construction by Kawasaki Ltd at Kobe in the same year, while I-11 was also built at Kobe in 1941. The I9 class was responsible for sinking nineteen Allied merchant ships during the Great Pacific War, totalling 96,481 tons. During that conflict I-9 was sunk on 11 June 1943, with I-10 lost on 4 July 1944 and I-11 sunk on 11 January of the same year.

The I-15 class of seaplane-carrying submarine was developed from the KD6 cruiser-type submarine as a scouting class. They were constructed with a hull and conning tower highly streamlined, the seaplane hangar being smoothed and rounded, extending forward of the conning tower. The exception was I-17, which had a hangar aft with a 5.5-inch GP gun forward. The range of these boats was 16,000 sea miles surfaced at 16 knots, or ninety-six sea miles submerged at 3 knots, with an operational range of ninety days' duration and a maximum diving depth of 325 ft. Displacing 2,198 tons, their dimensions were 356½ × 30½ × 16¾ ft, powered by two shaft machinery diesel/electric engines producing 12,400 hp/2,000 hp respectively, which gave maximum speeds of 23½ knots surfaced and 8 knots submerged. Manned by a complement of a hundred, they were armed with one 5.5-inch GP gun, two 25 mm AA cannon and six 21-inch torpedo tubes housing a capacity of seventeen torpedoes. In all, some twenty boats of this class were eventually built.

At this time the Imperial Navy had six aircraft carriers fully operational. These were HIJMS *Hosho, Akagi, Kaga, Ryujo, Soryu* and *Hiryu*, with two further carriers under construction – HIJMS *Shokaku* and *Zuikaku*. Since it took the building yards three years to construct an aircraft carrier, a decision was taken by the Imperial Navy to convert auxiliaries and suitable merchant ships to this class of vessel.

The two carriers under construction were planned after the expiration of the naval treaties and were not limited by design constraints. Basically the design was similar to HIJMS *Soryu*, with the bridge placed amidships on the starboard side after the failure of placing the bridge near the bows, as had been the case on HIJMS *Hiryu*. A strengthened flight deck was built, as was a bulbous type of bow similar to the designs of the bow of the battleship HIJMS *Yamato*. These carriers were built for long-range, high-speed operation. Displacing 25,675 tons, they measured $844^3/_4 \times 85^1/_3 \times 29$ ft, and were powered by four geared-shaft turbines producing 160,000 shp, giving a maximum speed of 34 knots. They were armed with sixteen 5-inch AA guns and thirty-six 25 mm AA cannon, as well as later having anti-aircraft rocket launchers installed. Each carrier could launch a total of eighty-four aircraft, and each ship possessed a complement of 1,660 men. *Shokaku* was constructed at the Yokosuka Navy Yard on 1 June 1939, while *Zuikaku* was built by Kawasaki at the Kobe Shipyard on 27 November 1939. Other ships used for carrying aircraft included HIJMS *Nisshin*, a seaplane carrier designed to carry 700 mines and twelve seaplanes, or twenty seaplanes without mines. Two catapults were fitted behind the bridge, but later replaced by two triple 25 mm AA cannon. Built at the Kure Navy Yard on 30 November 1939, displacing 11,317 tons, the ship measured 631½ × 64½ × 23 ft and was powered by six shaft-geared diesel engines producing 47,000 bhp, which gave a speed of 28 knots. Carrying twelve or twenty seaplanes, the ship was armed with six 5.5-inch GP guns and twelve 25 mm AA cannon.

In October 1939 the Mitsubishi G4M1 naval attack-bomber flew for the first time, the design having been commenced late in 1937. A special escort fighter version was developed but failed to reach the naval specification and was discontinued, the planes being converted to training duties. Though the special escort version was unsuccessful, the Mitsubishi Aircraft Company had provided the Imperial Navy with the Japanese Spitfire and had built the long-range bombardment weapon. The stage was nearly set!

During the latter part of 1939, British code-breakers, it is said, had broken the Japanese naval code known as JN25 and other codes. Thus the British government was able to read not only Imperial Navy orders but other communications concerning the Strike South Movement and the European colonies in south-east Asia.

When Winston S. Churchill became British Prime Minister shortly after the outbreak of the Second World War, he was continually briefed with the latest political and military intelligence from abroad. He was reputed to know more than President Roosevelt regarding the spiral of unfolding events, which led to the opening of the Great Pacific War. This situation had arisen due to the rapid technical developments which had taken place during the 1920s and 1930s in the field of radio interception, code breaking and cryptanalysis. How indeed was the British Prime Minister so fortunate?

At the end of the First World War the United States Army had organised a cipher office, which endeavoured to read the Japanese diplomatic codes. Within three years the codes had been broken and messages between Tokyo and Washington were being regularly read, especially at the time of the disarmament conferences.

Meanwhile, during 1919, Arthur Scherbius, in the suburb of Wilmersdorf, Berlin, had built a number of cipher machines, electrically operated, employing a rotor system. By 1923 a partnership had been created to manufacture these cipher machines on a commercial basis in a factory located in the city of Berlin. The media news of the machines was well received at the meeting of the International Postal Union, extensive coverage was given on the radio news and very favourable comments were made by the director of the Institute of Criminology, Vienna. Eventually progress developments followed, and the new code machines were advertised for sale to industry.

In 1926 the German Navy had purchased one of the new code machines, transferred the apparatus to Kiel and in great secrecy had tested the new channel of communications sufficiently well to place large orders with the Berlin company, but with the addition of certain modifications to meet naval specifications. Nevertheless, the Scherbius code machine in commercial form was still available on the open market for industrial purchasers. Two years later the new machine was patented and openly sold to foreign armed forces and industrial concerns. By 1929 secret sales negotiations had been concluded with the Reichswehr for large-scale purchases. However, an important order was fulfilled with William Friedman, the chief of the US Army Signals Intelligence Service. At about this time the Polish General Staff Radio Intelligence Section recorded radio intercepts from German cipher machine transmissions, as did other monitoring services.

Unfortunately, a serious problem arose in 1931 among the politicians and senior operatives of the US Army Signals organisation. The chief of the Cipher Office, Herbert Yardley, lost his government subsidy, with possible abolition of his office, to be absorbed into the Signals Intelligence Service under William Friedman. Protesting, Yardley was offered the post of leading the Signals Intelligence Service by way of compensation. Highly grieved by these developments and other factors, Yardley published extracts in the *Saturday Evening Post* from his unpublished book on the activities of the Cipher Office. In vain Friedman protested vehemently, such that the American Congress passed a special act prohibiting publication of further extracts from another publisher's book entitled *Secrets of Japanese Diplomacy*, or indeed any other publication. Tokyo was instantly stung into feverish activity as codes and ciphers were hastily withdrawn, and work commenced immediately on reconstructing Japanese coding systems.

It was during 1931 that notice was drawn to the head of Section D, Scientific Technical Intelligence and Decoding Service of French Military Intelligence, that a German informant wished to communicate information. After the usual security checks as to background, a meeting was arranged. The German informant was the younger brother of a German general working in the Reichswehr Cipher Bureau. Being a heavy gambler and at the same time hopelessly in debt, the informant felt he had something to sell. Documents were handed to the French officer confirming the Reichswehr's use of a cipher machine – a set of instructions for using the new apparatus, details of construction and the system of machine code keys.

Sometime during December 1931 the French head of Section D visited Warsaw to meet officers of the General Staff Intelligence Branch. The Polish General Staff was modelled on the French system, in which Intelligence was always the second bureau. For some time Polish Intelligence had been endeavouring to break the German machine codes. Since France and Poland had been allies from the days of the Napoleonic Wars, the sharing of information and aid was a natural phenomenon, with strong French influences throughout Poland. The Poles knew the German Navy and Reichswehr had introduced the use of the Enigma coding machines, but partners in a special Polish company named AVA were engaged to purchase a commercial Enigma machine through German business intermediaries. On delivery of the Enigma, a comparison between the machine and radio intercepts was made and proved the machine to be an exact copy. It was recognised that special differences would exist in military Enigma machines as used by the Kriegsmarine and Reichswehr. However, information from the head of Section D, French Military Intelligence, obtained from the German informant included operational instructions and four drawings of the machine, with details of the use of a special switchboard and plugs, and notes about the rotors and rotor inner connections. There was also a special note on how frequently the machine setting had to be reset, the shuffling of the three rotors every three months, and how the ring settings and plugs were changed every day. Finally a sample message was enclosed both in clear and enciphered with relevant keys.

Once again in the following year of 1932, the French head of Section D paid a visit to Warsaw. He brought with him two telegram keys for September and October 1932, which showed how all the Enigmas were operationally set for two months. The Polish intelligence officers now knew how to construct an Enigma machine and how to work out methods of locating the keys. The accumulated information had been shown to British Military Intelligence, which from now onwards was also working on the problems of the Enigma machine. Thus the cryptanalysts of the French, British and Polish intelligence services were working in unison to break the German cipher codes. By September 1932 the Poznan University cryptanalysts were moved to Warsaw to join the General Staff Cipher

Department for greater co-ordination. Thus in November 1932 the rotor connections were known, while the Polish AVA Company began constructing military Enigma machines to the deduced specifications in the shortest possible time. In December, at Christmas time, the first enciphered secret Reichswehr signal was received on the Polish-built Enigma machines. Meanwhile between the years of 1931 and approximately 1934 the Japanese had reorganised the ciphering of diplomatic and military messages. In 1934 a commercial version of the Enigma machine had been purchased and introduced into the Imperial Navy, the Foreign Office, the Army and other ministries, so that by 1937 the reorganisation was complete. However, the Japanese had succeeded through industrial and military espionage in acquiring components of the US SIGABA coding machine M134C built by William Friedman, chief of the US Army Signals Intelligence Service, as well as components of the British Typex machine, the Hagelin's Swedish Model B21 machine used by the Swedish Army in 1926, the French version – the C36 machine of 1936 – and the US code machine M209. To ensure maximum security the Japanese modified the Enigmas and other machines. The so-called Alphabetic Typewriter Number 97 had a system of plugs in a box connected electrically to a separate box of rotors. There were two electric typewriters, one of which typed into the system the message in clear language, while the second typewriter enciphered the message for transmission by radio. A system of radio nets with differing identification keys was used. It was for transmission from this type of machine that the American code-breakers were attempting to encipher the Japanese diplomatic code known to the Americans as 'purple'.

In 1936 the Polish government built a special radio-transmitting station and cipher centre at Pyry, some ten kilometres south of Warsaw, operated by the Polish General Staff 2nd Bureau to maintain permanent radio contact with the French intelligence organisation in Paris and with the Czech intelligence service in Prague.

The successful achievements of the Polish General Staff 2nd Bureau in breaking the German codes and ciphers were beset by the financial and technical costs becoming a drain on resources to the Polish State. At the same time the constantly changing German codes required extra resources of finance and technical equipment to keep abreast of a fast-changing situation. A conference was arranged for 9 and 10 January 1938 to discuss the situation. Neither the British, French nor Polish representatives were able to forward a solution other than acquiring a German military Enigma machine.

Meanwhile, on 24 July 1938, Lieutenant-Colonel Langer of the Polish General Staff invited senior British and French intelligence officers to a two-day conference held in the Radio Transmitting and Cipher Centre at Pyry. The assembled company were given a conducted tour of the station with a brief outline of the history and developments of the code breaking. As a Polish contribution to its allies' efforts in the coming conflagration, the General Staff presented an Enigma military coding machine to each delegation, manufactured by the AVA Company to German military specification, with drawings and plans of the bombes for hunting the code keys used by the enemy transmitters, and groups of specially perforated sheets in twenty-six-sheet bundles for manual tracing of the transmitter keys. (Alan Turing produced the first design for what would become known as the 'Turing Bombe' in 1939. The bombe was a semi-mechanical, semi-electronic machine, 6½ feet tall, that each day could process 158 trillion combinations and crack more than 3,000 messages. It is widely recognised as the forerunner of the modern-day computer. By the end of the Second World War there were 210 bombes at Bletchley Park.)

In August 1938 a group of British Military Intelligence officers arrived back at Victoria Station and handed over the 'luggage' to the deputy head of MI6, and it was possibly sent to the Government Code and Cipher School attached to the Foreign Office in London.

At the time the Government Code and Cipher School was being relocated to Bletchley Park, Buckinghamshire, some forty miles north-west of London, now known as Station X and employing over 10,000 people.

It was at Station X that a code-breaking section was created to break Japanese codes. These and all other important intercepted and deciphered signals were read by the Prime Minister, Mr Churchill.

CHAPTER 13

On to the Brink

In January 1940 the submarine tender HIJMS *Takasaki*, previously launched in 1936, was converted to an aircraft carrier and completed as HIJMS *Zuiho* in December.

On 18 January the third prototype of the Mitsubishi A6M carrier fighter gave demonstrations under the control of Lieutenant Maki HIJMN. Present were Rear Admiral Wada of the Naval Bureau of Aeronautics, Captain Kira, Naval Aircraft Establishment, Rear Admiral Mitsuname, the chief superintendent of the Nagoya Naval District, and Mr Naota Goto, the manager of the Nagoya Aircraft Works of Mitsubishi. The flight trials proved a great success, so that on 23 January the Navy officially accepted the modified aircraft as the Mitsubishi A6M2 carrier fighter aircraft. Next day Lieutenant Maki flew the new plane to the Yokosuka Naval Air Station. Also, in February 1940, the Mitsubishi G4M land-based bomber exceeded performance levels while on flight test with a top speed of 240 knots at a range of 3,000 nautical miles. This machine differed in minor internal details from previous prototypes in that the vertical fin was increased in area and aileron balance tabs were fitted. At this point a serious controversy broke out among senior naval officers in the ranks of the Naval Air Staff. Some senior naval officers wished to order the G4M1 model into immediate full-scale production, while another school of thought, led principally by the Yokosuka Experimental Air Corps, recommended that the needs of the land-based bomber squadrons could best be achieved by continuing production of the G3M2 model, perhaps even refined and further developed. To introduce the manufacture of the new model would require a change in productive output, thus reducing the available number of G3M2 machines for the long-range naval bomber squadrons in China. However, another matter arose in that defence of the bombers on long-range penetration raids would be necessary by using heavily armed twin-engine escort fighters. So the later school of thought prevailed upon the Koko Hombu, or Air Headquarters, to instruct Mitsubishi to continue production of the G3M2 and develop an escort fighter. This machine was to be a twin-engine aircraft using a G4M fuselage with Mitsubishi Kinsei

11 engines, heavily armed and designated G6M1. Unhappily this aeroplane did not achieve the required performance figures, and therefore development was discontinued, the prototype planes being relegated to transport duties.

The Mitsubishi A6M1 Prototype Number Two carrier fighter was undergoing test flights on 11 March 1940 under the control of pilot Okuyama of the Naval Aircraft Establishment. An ex-fighter pilot with seven years' experience, he took off from the airfield at Oppama Naval Air Station to investigate the problems of engine over-revolution during steep diving operations. The plane was flown to operational altitude when Okuyama rolled the little plane over into a steep dive. On the second run the fighter disintegrated in mid-air. Now it had been a regulation of the Naval Aircraft Establishment that the performance of all machines on test flights had to be witnessed. Upon receiving news of the accident, Dr Horikoshi, Mitsubishi's chief designer, departed at once for Oppama to attend the accident investigation enquiries. It was established at the enquiry that the pilot's body had been blown clear of the plunging machine and that his parachute had opened at a height of 1,000 feet. Regrettably the pilot had become separated from the parachute and plunged to his death in the sea. Upon the wreckage being examined, the throttle was found to be fully open, but the pitch lever was only half open, and an estimate of the diving speed had been calculated as 276–288 mph at an engine speed of 2,800 rpm.. The maximum allowable over-speed of the Zuisei engine was 2,700 rpm.! The elevator mass balance arms were broken and lost to the searching investigators, but those components obviously required more attention during the design stage. By 12 March an accident investigation was under way, to which Commander Nakano HIJMN, chief fighter pilot NAE, submitted eyewitness reports. The investigating commission had to study propeller over-speed and the aircraft vibrations during the dives. Now it would appear from the investigation reports that test pilot Okuyama had commenced his first dive from an altitude of 4,900 feet to pull up at 1,600 feet, when all seemed well. Entering a second dive from an altitude of 4,900 feet at an angle of 50 degrees, the pilot failed to pull out at 1,500 feet with the engine noise ever increasing in intensity, and a loud explosion followed as the plane disintegrated into a multitude of pieces scattering over the ground. As a result of this accident the Navy instituted research work to determine:

1. The main wing spar life under repeated stress, and under varying combinations of load.
2. The life of the main spar with sharp corners and riveted holes, also under repeated stress load.

3. Life of the main spar when manufactured of alloy other than ESD and under load.
4. Life of other critical structures under repeated load stress which contributed to high performance of later plane types.

In April 1940 the Navy issued basic requirements for a new land-based fighter plane – the Mitsubishi J2M1. It was to be a single-seat fighter with a maximum speed of 375 mph at 19,700 feet, with a rate of climb to operational altitude of 5 minutes 30 seconds and a landing speed of 80 mph or less after an endurance of 0.7 hours at full power. The take-off in overloaded conditions was not to exceed 948 feet overall. Armed with two 20 mm cannon and two 7.7 mm machine-guns, the plane was to be equipped with a Type 96 transreceiver radio and with instruments and oxygen equipment similar to that fitted to the A6M1 fighter plane. Surprisingly the pilot's seat was armoured with eight mm metal! The fuselage was to be spindle shaped, with the largest cross-section at 40% of fuselage length. The engine end of the cowling was slenderised to relieve compressibility. The cabin would be flared into the fuselage to form a continuous surface with a low, curving windscreen. The wing was to have a laminar-flow aerofoil and conventional aerofoil for the wingtips, as proposed by engineer Tsutomu Fujino of the Mitsubishi Wind Tunnel Section. The low-pressure points would be towards the centre of the chord, and not one-third from the leading edge. This shifted the maximum thickness to the centre of the chord. The idea was first pioneered by a NACA scientist. By calculation the radius of a vertical turn and dive recovery height would exceed by 50% to 60% the space required by the Zero plane. To reduce the space required, flaps were designed to increase lift with minimum drag and increasing wing area to improve manoeuvrability. As a result, fowler flaps were used. Total mass production was envisaged for the manufacture of this new fighter, with split assembly and sub-assembly production. Forged instead of machined steel parts with simplified machining and sheet-metal fabrication would be used. A spacious fuselage gave ease of equipment installation. However, the pilot would have a restricted field of vision. In the development of this design the Mitsubishi Aircraft Company was the only contender for the 14-Shi interceptor design, with the same design team assigned as had been assembled for the Zero project. An exception was Teruo Tojo, who transferred to the company's Army Air Corps section. Additional engineering staff included Shiro Kushibe, Sadao Kobayashi and Massao Kawabe, all working on stress problems, while the flap and tail unit was engineered by Kujoshi Horikawa. A consideration of the engine options indicated that the Mitsubishi Kasei gave the Interceptor a better rate of climb, but the maximum speed was lower by 11.5 mph. Alternatively the Aichi Atsuta motor could be installed, which was a Japanese version of the German Daimler-Benz DB601A built under licence. The cowling was streamlined by extending the propeller shaft, and by using an induction fan air could be circulated around the engine.

Meanwhile, despite the disintegration of the Mitsubishi A6M2, the Navy would be happy to accept the new prototype subsequent to the correction of the cause. Indeed, Navy pilots flying over the Chinese theatre of operations had requested immediate shipment of the new fighter planes, and as a result the Navy felt obliged to accelerate the test programme during May 1940. While the test pilots continued the experimental work, a group of A6M2 fighters had arrived in China, where the Navy commenced organising the aircraft into combat groups with the necessary supporting ground organisation.

In June 1940 the United States Navy announced the 3rd Vinson Plan, which authorised the construction of three additional aircraft carriers and the manufacture of 1,500 naval aircraft. The Japanese reaction was swift, as the Imperial Navy initiated its 5th Expansion Programme. Unfortunately the Japanese government had no means of preventing what was to be an escalating arms race, since the military were now in full control. The 5th Expansion Programme demanded the construction of 160 squadrons of aircraft, totalling 5,499 machines, as well as three aircraft carriers, two seaplane carriers and 159 other vessels. In the period 10–24 June 1940, Lieutenant Shimokawa HIJMN, Fighter Division Leader of the Yokosuka Air Corps, conducted a series of twenty-four tests, firing 2,396 rounds of explosive 20 mm cannon shells under various flight attitudes – accelerating steep turns of 3G to 5G, steep descending turns of 3.5G to 5.5G, as well as loops, level and straight flights, including climbing and diving. Minor defects in the systems were eliminated, while Mitsubishi modified the cartridge shell and the link ejector chutes, thereby increasing their reliability. The next day, 25 June, high-altitude test flights took place with Prototype Number Four of the Mitsubishi Type A6M2. Ascending to an altitude of 32,808 feet in 28 minutes 7 seconds, the fighter reached 19,685 feet in 9 minutes 18 seconds, then climbed to 26,247 feet in 14 minutes 30 seconds, and finally, at 32,792 feet, the ascent had to be concluded due to inadequate fuel pressure. On subsequent ascents the problem was solved by the use of 92-octane fuel, which stopped fuel vaporisation blocking the fuel lines.

In the summer of 1940 the Mitsubishi G6M1 escort fighter project was cancelled since the aircraft did not possess the necessary performance to achieve Navy acceptance. But the G4M1 did receive recognition, being designated Type 1 attack-bomber. However, reprieve for the G6M1 came when the various prototypes were converted to training planes, designated G6M1-K. Other machines of this type were converted to G6M1-L2 transports, being used to convey Japanese marine parachute troops at a later stage during the Great Pacific War by the 1,006th Naval Air Service Kokutai.

By mid-July 1940 the majority of the practicability and acceptance tests had been completed, but the remaining tests were simple and would only require a further few months of work. Already the Naval Air Service commanders in the Chinese theatre of operations were pressing for immediate delivery of the new fighters. As a result the Navy decided to break with tradition and transfer a group of the new Zero fighters to China. On 21 July two squadrons of Mitsubishi A6M2 fighters, comprising some fifteen machines, flew from Japan across the sea to mainland China under the command of Lieutenant Tamotsu Yokoyama and Lieutenant Saburo Shindo. Two bombers of the Mitsubishi Type G3M1 acted as mother ships over the ocean for navigation purposes. Included with the bomber crews were Engineer Lieutenants Shoichi Takayama of the Aircraft Division, Lieutenant Osamu Nagano of the Engineering Division and Lieutenant Sotoji Unishi of the Ordnance Division, all officers of the Naval Air Research and Development Centre, formerly the Naval Aircraft Establishment. Flying to the Imperial Naval Air Station at Hankou, they set up training courses for mechanics and armourers in the maintenance of the new fighter planes. The over-flight to China had been successfully accomplished without any major incident. The pilots reported a tendency for the cylinder temperatures to rise, but ground adjustment by maintenance staff quickly cured the ailment. Auxiliary fuel tank drops were successfully made with speeds in excess of 207 mph. During the last days of July 1940, the Navy officially accepted the A6M2, just at the time when the naval pilots had commenced familiarisation flights. In commemoration of the 2600th year of the Japanese calendar, the Mitsubishi carrier-based fighter A6M2 was designated Navy plane Type O carrier fighter Model II. On 19 August 1940 the first operational flight took place when fifteen A6M2 fighters escorted thirty-two G3M2 bombers flying out of Hankou to attack the Chinese temporary capital at Chongqing. No Chinese fighter opposition was encountered, and two subsequent aerial attacks were made by the Hankou-based naval aircraft. Imperial Navy statistics show that between December 1938 and August 1940 a total of eighty large-scale air battles developed with groups of Chinese fighters between twenty and forty strong in attack formations. Fierce air battles had developed. The first air combat that involved the new Zero fighters took place on 13 September,

155

when thirteen A6M2 planes dived upon twenty-seven Chinese Russian-built I-15s and I-16s. In the resulting mêlée all the Russian-built fighters were shot down in flames without a single loss of an Imperial Navy fighter. During the course of the action Flight Sergeant Koichiro Yamashita claimed the shooting-down of five Chinese fighters. In the course of this aerial mission the Imperial aircraft flew 575 miles to Chongqing from Hankou, and engaged the Chinese interceptors in the short period of thirty minutes. As a result of the persistent attacks in the air war, the Chinese government fled the city of Chongqing sometime during late September 1940.

In September 1940, Admiral Yamamoto, the Commander-in-Chief of the Combined Fleet (Rengo Kantai), officially opposed war with America. He realised the significance of the American industrial potential as a machine capable of mass producing all the weapons of war totally unopposed. This potential was far greater than that possessed by Japan. Further, he had negotiated increases in Japanese naval strength at the International Disarmament Conferences of 1930 and 1934. His understanding of the strategic situation involving Japan and American naval strength had led him to organise unorthodox naval weapon systems now acknowledged as the most efficient in the armed forces. Admiral Yamamoto's flyers from the Misty Lagoon crewed the Navy's long-range bomber squadrons and were rated as the finest bombers of the time. Success was achieved by the simple process of practising until every bomb hit the target fair and square. In the previous May he had observed to the Chief of the Naval Staff that a crushing torpedo attack by surprise could be a great success to defeat the Americans. In the following months he had worked out a comprehensive plan of attack. But the US Navy picket ships were patrolling off the Asian coast, prowling around and monitoring Japanese naval radio traffic. As a result, in October 1940 Admiral Yamamoto introduced the new Admiral's Code, which remained uncracked for some considerable time to come.

By October 1940 the flight testing of the G4M1 bomber was completed at Nagoya, and the Navy accepted the Type 1 land-based bomber Model II. Unhappily operational equipment had reduced the speed from 240 knots to 231, but this was 16 knots faster than the 12-Shi specification required. At the same time the G6M1 was finally cancelled after thirty aircraft had been produced. The machine was designated the Navy Type 1 wingtip convoy fighter, armed with a 20 mm cannon in nose, tail, ventral gondola and side blister, with bomb bay faired over and internal fuel capacity reduced to 3,640 litres, the aircraft proved to be far too heavy. The calculated result was that once the escorted bombers had dropped their bombs and consequently became lighter, the convoy fighters would not have the top speed to continue an escort mission. It was a disappointing failure, but the first production models of the G4M1 bomber were becoming available to operational squadrons.

Commander Minoru Genda HIJMN, previously Assistant Naval Attaché in London, returned home to Tokyo sometime in October. Subsequently he was appointed operations officer of the First Carrier Division, but while on leave he took the opportunity to visit a cinema, where he saw newsreel films that included pictures of four US carriers steaming in formation. Now it was the practice in the Imperial Navy for aircraft carriers to operate on the high seas in groups of two-ship formations. He concluded that several carriers could operate together to form squadrons for mass aerial attacks. The idea emerged of using mass formations of dive-bombers and torpedo-bombers protected by large fighter squadrons. Discovery by an enemy would be possible and inevitably highly probable, but defence would be maintained by concentrated anti-aircraft defence and the large concentration of defending fighter planes. When presented, the idea was accepted by the commanding officer of the Imperial Navy's First Carrier Division. However, it had to be acknowledged that to obtain absolute surprise the use of radio guidance techniques would be impossible since silence would have to be maintained. In any case, from dispersed carriers a longer form-up of attacking aircraft would be necessary, which would use greater quantities of fuel with a consequent reduction of aircraft flight range.

In the course of the month the Heijo class of aircraft carrier was launched. Originally laid down as commercial liners, conversion was completed as the design allowed since the hulls were still on the stocks. HIJMS *Heijo* was formerly *Izumo Maru*, and HIJMS *Junyo* was previously laid down as *Kashiwara Maru*.

On 24 October 1940 a naval delegation led by Admiral Yamamoto paid a visit to the city of Osaka – the mercantile centre of Japan. Now the samurai despised capitalistic materialism, and it was the tradition that a samurai would never handle money and never indulge in financial speculation. This duty was always handled by a financial manager specially appointed to deal with the financial necessities of his master. However, if the naval expansion plans were to be completed, then the mercantile class would have to provide the extra financial funding necessary. Refusal would lead to a loss of construction of the required ships, a retreat to the inland sea and exposure of the coast of the Home Islands to harassment by US aircraft carriers, with aerial attacks on the city of Tokyo. The industrialists agreed to fund the expansion programme after some careful deliberation, and the building schemes were completed eighteen months ahead of schedule. Then in November all fleet messages were transmitted in Admiral's Code, which was not broken by the Americans and so ensured the success of the later Pearl Harbor attack. Later in November, Admiral Yamamoto communicated the plan of the forthcoming Pearl Harbor aerial attack to Prince Takamatsu, Commander HIJMN and a Naval General Staff officer, who later presented the scheme to the Emperor. Meanwhile Rear Admiral Onishi Takajiro advised that the plans be checked out by Commander Genda HIJMN as a precaution, and this was satisfactorily done.

Then on 12 December 1940 a high priority signal was received by the chief of the Third Section of the Imperial Naval General Staff, Tokyo, from the Japanese Naval Attaché in Berlin. Despite having been recorded by American intelligence radio listening posts, the contents could not be decoded owing to transmission in Admiral Yamamoto's new Admiral's Code, and they remained so until August 1945.

The message concerned a report on the minutes of a British Cabinet meeting held on the previous 15 August, which had been obtained by the German Naval Intelligence Service. The main points of the communication were as follows:

1. Singapore, upon attack, would be defended by the Army and Air Force units. The fleet would not be sent to the Far East.
2. Britain would not go to war if French Indo-China and Siam were occupied by Japanese forces.
3. Hong Kong would be abandoned, but resistance would be maintained as long as possible.

The circumstances surrounding this high-priority signal of 12 December to the chief of the Third Section of the Imperial Naval General Staff was a fortuitous piece of good luck.

During the course of July 1940 the British Chiefs of the General Staff met in London to consider the war situation in Europe and the Far East. The Royal Navy was fighting the German and Italian Navies and was denied the assistance of the French fleet in the battles of the North Atlantic and the Mediterranean. Therefore, naval forces could not be spared for any aggressive threat originating in the Far East. The pre-war plans of 1937 had envisaged that in the eventuality of hostilities towards Malaya and Singapore the British main fleet would be dispatched from home waters. Now, in the summer of 1940, at the height of the Luftwaffe's assault on British armaments centres and airfields, no such fleet of ships existed. The Royal Air Force was heavily committed, with Fighter Command looking to the aerial defence of south-east England and industrial areas in other parts of the country. At the same time

Bomber Command was being built up and organised for the forthcoming bomber offensive against central Europe, with the new Avro Lancaster and Handley Page Halifax four-engine night-bombers. Indeed, to cope with the air defence of Malaysia and Singapore it was estimated that the British overseas air forces would have to be increased fourfold, but surplus modern aircraft and equipment were not available. In conclusion, the British Chiefs of Staff in a secret twenty-three-page report to the Prime Minister and to the War Cabinet advised the pacification of the Japanese government with economic concessions, and avoidance of war at any cost. Even if Japan should invade Thailand and occupy French Indo-China, war must not be declared. Britain had to play for time at all costs.

On 5 August 1940 the Prime Minister and members of the War Cabinet ordered the Chiefs of Staff to dispatch a copy of the secret report to the Far East Command. The safest method of conveying the contents of such a secret document was via the secure telegraph service from London to Singapore. Such a service forwarded only a summary of the original document, and the full text of the secret report would be dispatched via steamer at a later date.

By 24 September 1940, the Blue Funnel Line steamer SS *Automedon* sailed from Victoria Dock, Birkenhead, bound for Singapore with cargo and passengers. The cargo included confidential mail and a copy of the secret report in a weighted canvas bag left in the chart room behind the ship's bridge. In the eventuality of an emergency the canvas bag could be thrown overboard by the captain or watch officer quite conveniently. On 11 November SS *Automedon* was in the Indian Ocean *en route* for Malaya and Singapore when, at 0800, the watch officers sighted what was considered to be a Dutch freighter. This vessel was in fact the Kriegsmarine commerce raider *Atlantis* under the command of Kapitän Bernhard Rogge. She was armed with six 5.9-inch guns and four torpedo tubes, and in the course of her cruise had sunk twelve Allied merchantmen. Quickly running up the German naval ensign, *Atlantis* opened fire and scored direct hits on the bridge and radio room. Meanwhile *Automedon* had veered to starboard so as to bring her only 4.7-inch gun into action, despite its antiquated mounting, which had been manufactured at the time of the Russo-Japanese war. The first salvo from *Atlantis* had smashed the bridge and killed Captain Evans and the officers of the watch. The second salvo had blown up the radio room so that the 'Raider attacking' warning could not be transmitted to friendly shipping in the immediate vicinity. After further salvoes of gunfire *Automedon* was stopped dead in the water. A German naval boarding party was sent across to the stricken ship, with Kapitän Rogge's orders for the transfer of passengers and cargo to *Atlantis* within the following few hours. Six hours later Rogge ordered *Automedon* to be scuttled and the boarding party to return, as all the passengers and cargo were aboard the German ship.

The boarding party had discovered in the cargo holds 550 cases of whisky as well as 2.5 million packaged Chesterfield cigarettes. Leutnant Ulrich Mohr and his party had also found the ship's strong room, and blown the door open to discover fifteen bags of secret documents. But the greatest prize was the discovery of the weighted canvas bag among the tangled wreckage of the bridge and chart room, obviously containing important papers. The importance of the contents of the strong room and the canvas bag were not realised until carefully scrutinised by the German naval officers and Kapitän Rogge's attention was drawn to the information acquired. The documents that had been discovered included Royal Navy cipher codes, fleet orders, gunnery instructions and MI6 Secret Service reports, but most important was the twenty-three-page report entitled 'Situation in the Far East in the Event of Japanese Intervention against US – Chiefs of Staff Committee, War Cabinet, London'.

In the evening of 5 December Kapitän Rogge deposited the intelligence documents with Rear Admiral Wennecker, the German Naval Attaché in Tokyo, who promptly signalled his naval superiors in the Kriegsmarine, Berlin, seeking permission to show the Japanese naval authorities the contents of the various papers. On 10 December a Tokyo radio message to the Japanese Embassy in Washington announced the appointment of Ambassador Nomura. The coded message also stated the wish to organise a plan for propaganda and intelligence gathering within the United States. The Embassy officials were to seek the co-operation of Japanese banks and businessmen, who subsequently agreed to channel all intelligence information through Japanese diplomatic channels. Permission was received from Berlin on 11 December for the British documents to be revealed to the Japanese. On the following day Rear Admiral Wennecker had a meeting with Admiral Kondo, Vice-Chief of the Imperial Naval General Staff, at Naval Headquarters, Tokyo. To begin with, Admiral Kondo tended to view the papers as mere forgeries in the game of international deception, until Admiral Wennecker described in detail the circumstances of the encounter between *Atlantis* and SS *Automedon*, with the subsequent discoveries by the boarding party. Upon the realisation of the importance of the captured documents, in particular the report by the British Chiefs of Staff, Admiral Kondo expressed the opinion that he had not appreciated just how serious the weakening of the British Empire had become. The significant document was placed in the hands of the commander of the Imperial Combined Fleet – Admiral Isoroku Yamamoto. He was most impressed with the report, for previously he had concluded that an American, British and Dutch combined plan of defence had existed. He had further opposed the idea of Japan waging war on both the Americans and British simultaneously because he believed that Japan did not possess the industrial capacity to wage a war on two opponents at the same time. Upon completion of the study of the report of the British Chiefs of Staff, planning of the Pearl Harbor attack commenced in detail, for Admiral Yamamoto had revised his estimates of the short-term success for Japanese aggression. Meanwhile a secret message was transmitted on 30 December to the Admiralty in London from Naval Headquarters, Singapore, reposted to the Commander-in-Chief, China, West Indies and Cape Town from a captain on the staff at Singapore. In this message it was stated that a Norwegian seaman taken prisoner as a result of the capture of *Automedon* had witnessed the transfer of sixty small mailbags to the raider *Atlantis*. Also, there was a statement by an officer on board the raider that some $6,000,000 in notes had been seized and the entire contents of the strong-room aboard *Automedon* had been taken aboard *Atlantis*. Unfortunately a local shipping agent in Singapore had confirmed that a large quantity of unissued Malayan currency was among the cargo. As a result of this secret signal, the Admiralty on 3 January 1941 ordered all British merchant ships to put into port and collect new signal-code books. It would seem that the British Admiralty did not realise that other naval ciphers had been compromised, nor did they appear to realise that the report by the Chiefs of Staff was now in the hands of the Japanese naval planners. For Admiral Yamamoto was able to communicate on 7 January to the Navy Minister the opinion that upon the destruction of the American naval forces in the central Pacific in a decisive battle, the British and Allied forces south of the Philippine Islands would be defeated.

A signal dated 30 January 1941 from the Japanese Foreign Office, Tokyo, to the Embassy in Washington instructed the co-ordination of all intelligence-gathering activities in the USA by the Japanese intelligence organisation with those of the German and Italian espionage services, as previously discussed and agreed. A further coded message from the Foreign Office, Tokyo, to the Japanese Embassy in Washington, from the Foreign Minister Matsuoka in person, advised that the setting-up of the intelligence service was to be recognised as due to the serious situation between Japan and America, for the specific purpose of being prepared for the worst! Obviously the planning for the Pearl Harbor operation and the great advance southwards was gaining momentum.

Nevertheless, the captured secret Chiefs of Staff report had blown away any pretence at a possible British defence in south-east Asia in the event of hostilities with Japan. For the report headed 'War Cabinet – Chiefs of Staff Committee' stated categorically:

> At the same time as the collapse of France, the development of a direct threat to the United Kingdom and the necessity for retaining in European waters a fleet of sufficient strength to match both the German and Italian fleets have made it temporarily impossible for us to dispatch a fleet to the Far East should the decision arise. In consequence neither of the two underlying assumptions of the 1937 review is any longer tenable and we have therefore thought it necessary to review again our Far East defence problems as a whole. We have done this at length in thoroughness to …

This frank and decisive report, having fallen into the hands of the Japanese Naval General Staff in Tokyo, indicated that someone in London had made a monumental blunder!

Sometime between April and August 1941, a double-cross agent operated by British Military Intelligence was informed by his German Abwehr superior control officer that a high-ranking officer of a Japanese Naval Mission to Berlin had wished to obtain information concerning the previous British raid of November 1940 on the Italian battleships in Taranto harbour. Unfortunately the Japanese could only obtain these highly secret details of the air raid by Fairey Swordfish torpedo-bombers by an introduction from senior officers of the Abwehr to the Italian naval authorities. A visit to Taranto was duly arranged, and a close study of the aerial attack was undertaken by the Japanese party. Thus Admiral Yamamoto's conceived and planned attack on the American warships and installations at Pearl Harbor was brought a stage nearer. This information was confirmed when a decoded message was later intercepted by the Ultra machine. In the meantime the Japanese Foreign Minister Matsuoka had signed a non-aggression pact with the Soviet Union on 13 April 1941, which had the effect of guarding and neutralising the Japanese northern flank, making a southward movement a distinct possibility. At the same time intelligence gathering by Japanese businessmen and newspaper correspondents in America was intensifying, for on 9 May, and again on 11 May, radio signals were intercepted originating in Los Angeles and Seattle, and transmitted to Tokyo by Japanese intelligence agents. On decoding, the subject matter of these messages concerned the recruitment of sympathetic agents, surveys of US aeroplane manufacturing plants, production statistics and details of connections made with second-generation Japanese American citizens recruited into the US armed forces. Further radio intercepts of Japanese intelligence communications took place on 14 July, when a transmission from Canton, China, to Tokyo scheduled the next conquest:

After Indo-China, the Netherlands Indies. In seizing Singapore the Imperial Navy would play the principal part, with one Army division seizing Singapore and two Army divisions to seize the Netherlands Indies. Air Forces to be located in Canton, Spratly Island, Palau, Singora, Timor, Indo-China. The submarine fleet would be based in the Mandates, Hainan, Indo-China. We will crush British and American military powers.

Further interest in the Hawaiian Islands, and Pearl Harbor in particular, came in August 1941, when the German Abwehr ordered the captured double-cross agent in British hands to go to America. The objectives of this mission were to organise a new spy ring independently of the existing German intelligence organisation in the USA, to obtain details of war supplies being shipped to England, and information concerning the sailing of ships and convoys across the Atlantic, and to acquire details of the Liberty shipbuilding programme, as well as the repair schedules of Allied warships in US yards. Most important was a separate questionnaire concerning detailed information required about installations in Hawaii, and Pearl Harbor especially. Such a separate questionnaire would not have originated from the Kriegsmarine in Berlin, whose strategic interest in that area was nil. The answer lay in the activities of the Japanese Naval Intelligence Service, which required the double-cross agent to visit the Hawaiian Islands.

On the following 25 September, a radio message was decoded, originating in Tokyo to the Japanese Consul in Honolulu, requesting reports on details of installations in and around the harbour by means of dividing it into five areas for research purposes. Again communications were opened on 29 September from Tokyo to the Consul in Hawaii, with a designated code system for reporting the layout of the military objectives at Pearl Harbor. Precise and detailed planning of the Imperial Navy's attack on the United States warships in the Hawaiian Islands was now proceeding in haste!

Meanwhile talks had been initiated during July 1940 by the Japanese Foreign Minister, Mr Matsuoka, in both Berlin and Tokyo, involving the Foreign Ministries of Germany and Italy. These talks concerned questions of German and Italian policy towards the status of the European colonies in south-east Asia, and above all else the attitude of both Germany and Italy to the possibility of Japanese intervention at some time in the future in that sphere of influence.

By 27 September 1940 an agreement as to future policy had been achieved by the three governments, embodied in a tripartite pact. The representatives of the three governments agreed a policy of defeat and dismemberment of the British Empire, but without involving the government of the United States of America in war, which was at all costs to be excluded.

Despite the signing of the Tripartite Pact, the leaders of the Japanese Navy held certain reservations, and were not prepared to move militarily before full preparations had been completed, particularly if active operations against the United States were ever contemplated.

It was during the course of 1940 that the American code breakers had constructed a decoding machine named Magic, which was capable of breaking the Japanese diplomatic code. Thus, it is alleged, every signal originating from the Foreign Ministry in Tokyo to Japanese embassies around the world could be intercepted and read.

That such a secret and sensitive British Cabinet paper fell into the hands of German Naval Intelligence and was passed to the opposite Japanese number was disastrous for the British. It could not but encourage the military planners to lay their plans of action precisely, confident in their estimation of the crumbling British position.

Meanwhile, in December 1940, the aircraft carrier HIJMS *Zuiho* was completed. She was a flush-deck carrier with no island structure, converted from the former fleet oiler HIJMS *Takasaki*. This light carrier displaced 11,262 tons, and with an overall length of 662 feet she attained a maximum speed of 26.2 knots, which was achieved using steam turbine machinery, replacing the former diesel engines, and she could launch a total of thirty-one aircraft.

Eventually, by the end of December, a total of 153 new Zero fighters, in twenty-two sorties, had shot down fifty-nine Chinese planes and had destroyed a further 101 in the course of ground attack operations, without sustaining any losses to themselves. At the same time the Mitsubishi Aircraft Company had delivered twelve new G4M1 bombers to the operational squadrons, and had constructed the first mock-up of the J2M1 fighter plane, which had been inspected by Naval Air Service officers.

The strategic situation in the Far East was now changing, and only the 12th and 14th Naval Air Corps were stationed in China. The carrier force hitherto cruising off the coast was drastically reduced and withdrawn in preparations for the forthcoming offensive operations within the year. At this time the Navy's air units comprised twenty-five air corps, of which fifteen were operational and ten training, which had been organised since 1938.

Operational units possessed

	Squadrons	Aircraft
Carrier-based fighters	5	80
Dive-bombers	1	16
Torpedo-bombers	4.5	72
Twin-engine bombers	7.5	107
Flying-boats, medium & large	6	28
Flying-boats, small	1.5	18
Sea reconnaissance	5	60
	30.5	**381**
Ship-based units		
Carrier-based fighters		225
Dive-bombers		192
Torpedo-bombers		342
Sea reconnaissance		329
		1,088
Training units		
Carrier-based fighters	2	32
Torpedo-bombers	3.5	56
Reconnaissance	2	32
Sea reconnaissance	2	24
Flying-boats, small	0.5	6
Primary trainers	4	120
Secondary trainers	5	120
Advanced trainers	2	32
Research aircraft	1.5	24
	22.5	**446**
Total	**53**	**1,915**

The Sino-Japanese Incident was beginning to develop step by step into the Great Pacific War, and during this period production of the 20 mm cannon was becoming seriously short. As a result the Dai-Nikon Heiki Company organised six further factories for the production of this weapon. But even these measures proved inadequate, so that the Navy was compelled to commence the production of these aerial cannon at the Tokakawa Arsenal and the Tagajo Arsenal, and by 1945 35,000 of these weapons had been manufactured. At the start of the Great Pacific War the German Rhine-Metal Borsig machine-gun was being manufactured in Japan for the Naval

Air Service. These weapons comprised the 7.9 mm Type 1 machine-gun and the Type 2 13 mm machine-gun for use on a movable mounting, together with a modification of the US Browning weapon Type 3 of 13.2 mm calibre for both fixed and free mountings.

Details of Naval Air Service 20 mm cannon armament

Type	99 Model 1	99 Model 1	99 Model 2	99 Model 1	99 Model 2	99 Model 2
Fixture	Fixed	Movable	Fixed	Fixed	Fixed	Fixed
Model	MK3	MK3	MK3	MK4	MK4	MK5
Calibre mm	20	20	20	20	20	20
Shell weight, oz	4.34	4.34	4.34	4.34	4.34	4.34
Muzzle velocity	1,970	1,970	2,460	1,970	2,460	2,490
Weight, lb	51	66	73.7	59.4	82.6	84.7
Length, in.	52.5	65.1	74.4	52.5	74.4	73.1
Length of bar	20.6	20.6	49.2	20.6	49.2	49.2
Shell fired	M	M 45R	M 100R	BT	BT	BT
Rate of fire	520	520	490	550	500	750

M = Magazine   BT = Belt   R = Rounds

With the possibility of a future war coming ever closer, it was considered necessary to hastily construct twelve airstrips at the major plants of the various aircraft companies. These included:

> Nakajima Aircraft Company at Koizumi
> Kawasaki Aircraft Company at Akashi
> Kawanishi Aircraft Company at Naruo
> Mitsubishi Aircraft Company at Mizushima

The Navy continued its policy of importing foreign aircraft for evaluation purposes. During the course of 1940 these included the German Junkers Ju 88, Heinkel He 100 and Heinkel He 119. Though the Imperial Navy was not interested, quite mistakenly, in transport aviation, it did acquire the American Douglas DC3 design. The machine was put into production by both the Nakajima and the Showa Aircraft Companies. A total of 487 machines were built, designated Type O transport L2D, but the number produced was inadequate for the transportation requirements of the Imperial Navy. Other types relegated to transport duties included conversions of the Type 96 to transport L3Y Type 96, Type 97 to transport H6KL Type 97FB, Type 2 fighter bomber to transport H8KL Type 2 and Type 1 bomber to Type 1 bomber trainer.

Difficulties were now arising within the aircraft manufacturing industry when the Mitsubishi Aircraft Company was approached by the Navy with a specification for a 16-Shi carrier fighter to replace the Mitsubishi Zero fighter. Unfortunately the Mitsubishi Company was encountering problems with the 14-Shi land-based interceptor owing to the lack of a suitable engine and other difficulties. So it was decided to postpone the new plane by twelve months. Meanwhile the Navy asked the Kawanishi Aircraft Company to produce a fighter float-plane. During the Sino-Japanese war a need arose to utilise beach airbases during the periods of gaining air superiority in local theatres of operation until land airbases had been constructed or captured from an enemy. A specialised anti-submarine aircraft was developed by the Kyushu Aircraft Company, designated Q1W1 Tokai, for patrol operations against submarines. This machine had a span of 52 feet 6 inches and was powered by two 610 hp Hitachi Tempu 31 radial engines, giving a maximum speed of 200 mph at an altitude of 4,396 feet, over a range of 914 miles. With a crew of three, the aircraft was equipped with radar apparatus and anti-submarine detection gear, having an offensive armament of one 20 mm cannon, one 7.7 mm machine-gun, together with a bomb load of 1,100 lb.

A special reconnaissance float-plane was designed by Yokosuka, designated E14Y1, which was a small submarine-based reconnaissance catapult float-plane, first appearing in 1940. Manufactured by the Kyushu Aircraft Company and armed with one 7.7 mm machine-gun, it had a loaded weight of 3,197 lb. In the course of the Great Pacific War this machine was reported over the coasts of America and Australia. At the Yokosuka Navy Yard, Doctor Masuo Yamana commenced the design of twin-engine aircraft with dive-bomber capabilities and a crew of three, designated P1Y1. At the same time the Yokosuka R2Y1 Keuin (Beautiful Cloud) was acquired from Germany. This machine was the Heinkel He 119 that at sometime during 1945, in the course of the Great Pacific War, was used to destroy US bombers on raids over American Pacific airbases. The Heinkel He 119 V7 and V8 were sold to Japan in 1940. These aircraft were high-speed, high-altitude reconnaissance aircraft powered by twin-engine V Daimler DB606 motors. A central fuselage, in which the cabin, nose and fuselage were combined, housed a pilot, co-pilot and bombardier. A central propeller shaft between the pilots connected the engines with a 14 ft diameter propeller. The tail and wings resembled the Heinkel He 111 configuration, first tested in Germany by Erprobungsstelle Rechlin, some fifty-five miles south-east of Rostock, having been designed by Siegfried and Walter Gunter, project engineers for Ernst Heinkel Flugzeugwerke AG, of Rostock, Marienhe, north Germany. The aircraft achieved a maximum speed of 373 mph as compared with the British Vickers Armstrong Supermarine Spitfire MK1, which attained 355 mph. However, a range of

1,678 miles could be reached at a speed of 248 mph, at an altitude of 14,750 feet. Nevertheless, on 30 March 1939 a new world airspeed record had been claimed for the Heinkel He 119 of 463.92 mph when flown by Flugzeugführer Hans Dieterle, a 22-year-old test pilot. The Heinkel He 119A was a three-seater reconnaissance aircraft, while the He 119 V7 and V8 were high-speed bombing aircraft carrying 2,205 lb of bombs and equipped with a rear-firing armament.

The year 1940 was most notable for the signing of a Mutual Defence Pact between Germany, Italy and Japan to give aid and assistance so that Japan could conquer the Dutch East Indies as well as the Malayan Peninsula for the oil and rubber supplies, which she so desperately required for the prosecution of the war in China.

Sometime during the course of 1940, Winston Churchill decided to send 300 British Military Intelligence agents to America. Under the command of Sir William Stephenson, a millionaire and First World War air ace who had shot down twenty-seven enemy aircraft, and was the inventor of the picture wiring system, the American network set up headquarters in the Rockefeller Center in New York, using a 'commercial front' called the British Security Co-ordination Company. From here British Military Intelligence had a liaison with the Federal Bureau of Investigation and contacts in the White House in Washington. The purpose of the network was to discredit American isolationists and the anti-war faction of the US establishment, as well as exposing the activity of the Nazi and Japanese Military Intelligence services operating in the USA. To achieve these ends the British agents used covert propaganda warfare and 'dirty tricks' techniques. These included intercepting mail, telephone bugging, safe cracking, code cracking, kidnapping important people and operating a rumours department. Exposure of the enemy was achieved by feeding unlimited evidence to the courts and supplying embarrassing information to the State Department in Washington concerning the activities of German and Japanese agents. One such operation involved forging a letter supposedly from the Bolivian Military Attaché in Berlin, Major Elias Belmonte, to the Bolivian Chief of Staff, hinting at a possible coup to overthrow that government and to replace its members with pro-German sympathisers who would rescind the commercial contracts with America and Britain. This was a serious matter, for America depended upon Bolivia as a main source of supply for wolfram, from which tungsten, so important in the manufacture of armaments, was produced. It is alleged that President Roosevelt believed the evidence, which had been forged. Naturally a private undercover war flared up between the British Intelligence Service and its German and Japanese counterparts. The Imperial Navy had set up an American headquarters in Mexico City, where a radio-monitoring station was operated in conjunction with a similar establishment at Baja California, Mexico, where the radio signals from American shipping leaving the west coast were monitored and radio sources were located for transmission to Imperial Headquarters, Tokyo, for intelligence evaluation purposes.

Most important for the Imperial Navy planners of the Naval General Staff was a shift in the strategic situation, which took place during March 1940. The withdrawal of the Royal Navy's 4th Submarine Flotilla from the South China Sea removed a serious threat to the movements of the Imperial Navy when the movement south took place at a later date. Foolishly the British Admiralty had left wide open the defence of the Malayan Peninsula, as well as the city and port of Singapore. There was now nothing to stop the Japanese air and naval forces invading south-east Asia, and Japan was not slow to perceive her opportunity.

Meanwhile, in Europe the so-called 'Phoney War' came to an abrupt end when on 10 May 1940 the German advance on the Western Front in northern France commenced. The strategic armoured thrust by the German armies through the Ardennes and through Belgium by use of Blitzkrieg tactics caused a defensive collapse among the French and British forces. The German tanks and armoured vehicle formations were welded into a single offensive weapon with the dive-bombers, and were supplied by a highly organised air transport service. The German advance was assisted by detailed aerial reconnaissance flights and the attacks of long-range bombers destroying facilities, communications and supplies deep in the Allied rear.

As the co-ordinated German forces swept westwards, then southwards and splitting south-eastwards, the resistance of the French Army and the British Expeditionary Force disintegrated and the retreat became a general rout. With both the German and Allied units racing for the Channel ports, the bulk of the British personnel were evacuated from the port of Dunkerque, at the cost of the total loss of all British equipment. At the same time many of the French military personnel were successfully evacuated to England to continue the conflict to liberate France from German occupation at a later date. Back in London, General de Gaulle set up a Free French government in exile, to rally French citizens to continue the cause of liberation. But on 22 June the Battle of France was over, and Marshal Pétain had accepted the German terms and conditions of surrender, and signed the armistice agreement. This new French government was promptly moved to Vichy. The French Vichy government now controlled half the metropolitan territory of France and controlled all the overseas colonial possessions.

During July 1940 the Japanese government opened talks in Berlin with its German counterpart regarding the policy of the then German government as to the European colonial possessions in the South Pacific. At the time the German government had no real interest in these far-off possessions, and did not have adequate policies as to the future of those territories. Ostensibly a free Japanese hand could be assumed. On 27 July an Imperial Liaison Conference took place in Tokyo. Interested parties of the Strike South Movement were present, together with the members of the government and the propagators of the Co-Prosperity Sphere policy in south-east Asia. The questions concerning the expansion in south-east Asia were discussed, and an agreement to talks with European powers as to the possibility of increasing raw material imports into Japan. If all diplomatic efforts should fail then recourse to military action might be a necessary option.

In the following month, August 1940, Japanese diplomats opened discussions with representatives of the French Vichy government regarding the establishment and use of bases by the Japanese Army and Navy in territories of French Indo-China.

By September 1940 an agreement had been reached with the French Vichy government whereby Japan was permitted the establishment and use of military bases and naval stations, while respecting French sovereignty over the territory in the Province of Tonkin. Furthermore, Japan was allowed to move troops through Indo-China in the event of a war involving a third party. Unfortunately, talks on a similar basis with the Dutch proved abortive. It was considered that military action would be necessary to obtain the petroleum products from the then Dutch East Indies, now Indonesia.

These negotiations were concluded by 27 September, at the same time as a tripartite agreement was signed between Japan, Germany and Italy for mutual co-operation in the event of war.

The activities of the Japanese diplomats and the Vichy French authorities had not been lost on the Canberra government of Australia. Some nine hundred miles east-north-east of the Australian coast lay the French colony of New Caledonia, with its capital of Nouméa. The Minister of External Affairs – Mr John McEwen, in the administration of Mr Robert Menzies, the Australian Prime Minister – judged quite correctly that should Vichy France allow Japanese forces to occupy New Caledonia, then the shipping lanes from Australia to America would be irrevocably blocked by hostile forces.

With the agreement of the Federal Prime Minister and without the knowledge of the Cabinet, Mr McEwen charted a Norwegian freighter, which secretly conveyed Free French forces under the command of one of General de Gaulle's nominees to peacefully occupy Nouméa and neutralise the pro-Axis Vichy French colonial administration.

At the time the Norwegian freighter arrived, HMAS *Adelaide*, under the command of Captain (later Rear Admiral) H.A. Showers RAN, weighed anchor in Nouméa Harbour. Quickly the island was occupied by Free French forces, denying the Japanese the facilities and bases needed for any future conquest of Australia.

By November 1940 the military occupation of southern Indo-China had been successfully completed, despite the belief that a negotiated settlement with the United States could be achieved, since America supplied large quantities of scrap metal for Japanese industry and over 60% of all petroleum requirements. When President Roosevelt and the American administration froze Japanese assets in the United States, worth some $33,000,000, in retaliation, the future suddenly became uncertain and hostile. A serious economic problem had now loomed above the horizon and would need to be resolved urgently by the Tokyo government. Thus the gauntlet had been thrown down!

CHAPTER 14

Charge!

In 1941, the lessons from the earlier air battles over China suggested the necessity of incorporating self-sealing fuel tanks, armour and other devices into the design of naval aircraft. The matters were considered by other nations as indispensable necessities, but the Japanese military failed to insist upon such provisions despite offers being made by the aviation industry, nor were research projects initiated into such ideas. This was a matter of total negligence, which was to lose the air war for Japan and end in an aerial defeat.

In January 1941, the annual performance contest between the Army and Navy was due to take place. The Navy entered two leading pilots – Lieutenant-Commander Yokosuka from the Yokosuka Air Corps, and Lieutenant Shimokawa, a fighter squadron leader – to determine the best fighter plane and pilot. The machines being used included the A6M2 Type O, the Army Ki-43 Type 1, the Ki-44 Type 2 and Ki-27 Type 97 fighter. The summary report from the Yokosuka Air Corps stated:

1. Speed – A6M2 was as fast as the Ki-44 and faster than the Ki-43.
2. Climb – From 42 feet to 13,124 feet A6M2 faster than Ki-43, Ki-43 lower all-up weight but higher hp.
3. Zoom performance – A6M2 faster than Ki-43.
4. Turning circle – A6M2 superior in climbing turns.
5. Dogfighting – A6M2 superior in simulated operational manoeuvres than the Army plane.
6. General – wandering in level flight due to excessive trim and over-responsive rudder at very high speed.

To obtain the overall performance capabilities, the chief designer of Mitsubishi conferred with the two naval pilots and ascertained the following points:

1. The A6M2 was superior to the Ki-43 with better controllability and capability.
2. The A6M2 was more manoeuvrable than Ki-44.
3. The Ki-27 can turn inside the A6M2, which possesses superior dogfighting qualities.

4. Wandering in level flight does not hinder firing accuracy but enhances mildness of stall and ease of pull-out.
5. The inability to roll at low speeds results from a large wingspan required for short turning radius and rapid climb.
6. Navy reported a lack of sensitivity in correcting roll and yaw during landing glides – uncorroborated by test pilots.
7. Undesired spinning delayed by making slight inward slip during
a turn.
8. The A6M2 speed inferior to Heinkel He 100, with max speed of 464 mph using a Daimler DB601 steam-cooled engine of 1,070 hp with wing surface radiators.
9. The maximum speed of A6M2 could be increased by 17.25 mph at expense of manoeuvrability.
10. The He 100 was over-sensitive on the elevator.
11. The A6M2 controllability and stability was satisfactory.

Undoubtedly the Mitsubishi A6M2 fighter plane had shown exceptional flying characteristics, and was accepted as a first-class machine. However, during this month the Navy visited Nagoya to inspect the changes ordered in the inspection of the first mock-up of the new J2M1 interceptor plane. But a new aircraft carrier joined the 4th Carrier Division. This was HIJMS *Shoho*, formerly HIJMS *Tsurugisaki*, a high-speed oiler, launched originally in 1935 at the Yokosuka Navy Yard, completed as a submarine tender in 1938, and due to some indecision as to the role the ship was to play with the combined fleet, converted to an aircraft carrier role in January 1941. Diesel power was replaced by steam turbines, with a donkey engine exhaust funnel at the stern. Unfortunately the carrier could not be armoured, as this would impede the manoeuvrability of the ship. Displacing 11,262 tons, she measured 712 × 59 × 21¾ ft, and was powered by two geared-shaft turbines of 52,000 shp, which gave a top speed of 28 knots. Armed with eight 5-inch anti-aircraft guns and fifteen 25 mm cannon, she was equipped to accommodate thirty aircraft. Eventually she was lost at the Battle of the Coral Sea on 8 May 1942.

Commander Kazunari Miyo was an aviation operations officer on the Naval General Staff who on 27 January 1941 attempted to convince his General Staff superiors of the importance of forcing the United States into a decisive naval battle. He proposed the use of six- or eight-engine bombers raiding the US Pacific Headquarters at Pearl Harbor. This he calculated would ensure the USA fleeing the Pacific for the mainland or fighting a naval engagement near the Marshall Islands on Japanese terms. News concerning these discussions came to the ears of Dr Schreiber, the Peruvian envoy in Tokyo, who informed First Secretary Edward Crocker of the US embassy, Tokyo, and resulted in Ambassador

Crew cabling the information to Washington. The information was routed to the US Naval Intelligence Service, which did not think any move appeared imminent! At this time the Pearl Harbor attack plans were still under secret independent review on the orders of Rear Admiral Takijiro Onishi, Chief of Staff of the 11th Air Fleet, for the Emperor. On 1 February Admiral Yamamoto wrote an unofficial letter to Admiral Onishi, outlining the Pearl Harbor plans and asking for a feasibility study. Having previously turned over the plans to Commander Genda HIJMN, the originator of the long-range fighter operations in China, and whose influence went far beyond his modest rank, Admiral Onishi discussed the plan with Captain Kanto Kujoshi, who closeted himself in his cabin for several days and emerged with Plan Kuroshima. Two days later Admiral Yamamoto revealed the plan to Captain Kanji Ogawa, chief of Naval Intelligence, who had three assigned agents already working in Hawaii – a German by the name of Otto Kuhn, a Buddhist priest and two other persons, all of whom had supplied unimportant information. It was decided to send to Hawaii Ensign Takeo Yoshikawa, who had attended the Naval Academy at Etajima, had become swimming champion on the Etajima–Miyajima Shrine ten-mile race and had achieved 4th rank at Kendo. At some time he had studied Zen Buddhism to attain spiritual discipline, had attended one term as a code officer on a cruiser and had attended training courses at torpedo, gunnery and aviation schools. He had temporarily retired from the Navy due to stomach ailments, but returned as a reserve officer in the Intelligence Branch, where he had served in both the British and American sections concerned with shipping movements and studying various types of equipment. In the previous spring of 1940 he had been asked by the chief of Section 5, Captain Takeyoshi, to consider serving in the Hawaiian Islands as a secret agent. He became a consular official and was sent to study international law and English at the Nippon University and at the Foreign Ministry, where he researched US politics, economics and the work of Section 5.

Just over one week later, Commander Genda arrived at a conclusion concerning the Pearl Harbor attack plans. The operation would be difficult to mount and the attack would be risky but had a reasonable chance of success. For two weeks Genda had shut himself up in his cabin aboard the aircraft carrier HIJMS *Kaga*, while it lay at anchor in Ariake Bay, Kyushu. Under Commander Genda's guidance the dive-bombers and torpedo-planes had been trained over Kagoshima Harbour in southern Japan, since it most resembled the Pearl Harbor layout. Meanwhile the Mitsubishi A6M2 Model 21 appeared, commencing with serial number 127. New balance tabs were installed on ailerons linked to the retracting-gear mechanism, which reduced the stick force required for high-speed lateral control. A total of 740 aircraft of this type were produced by Mitsubishi.

In March 1941 Mitsubishi produced the G7M1 Taizan (Great Mountain) bomber project in response to the 16-Shi experimental naval attack-bomber specification. The Nakajima Company produced the G8N1 Renzan (Mountain Range) four-engine bomber. As neither of these proposals was acceptable to the Navy, the Mitsubishi Aircraft Company was instructed to develop the G4M2 design as a stop-gap measure. The engineers on this development were Honjo and Hikeda, whose objective was to improve the range, speed, armament and the aircraft's general stability. A new bomb aimer's window was designed, and two Mitsubishi MK4P Kasei 21 engines with water methanol injection were installed and rated at 1,800 hp with four-bladed propellers. An auxiliary fuselage fuel tank was added to the design, giving an extra 6,490 litres of aviation spirit, while a new wing with laminar flow characteristics was added to the design. The defensive armament was revised, with a new power-operated gun turret aft of the cabin being installed with a Type 99 20 mm cannon. On either side of the nose was mounted a Type 92 machine-gun of 7.7 mm calibre with new glazing to the nose windows. The horizontal tail surfaces were increased in span and area as well as the tail tips now being rounded. At this time twenty-five aircraft of the G4M1 type had been delivered to the 1st Kanoya Kokutai at Kagoshima on Kyushu. This unit was assigned to the 21st Koku Sentai.

But the air war over China continued unabated. While escorting bombers attacking Chongqing in central China on 20 March, twelve Type O A6M2 fighters engaged thirty Chinese interceptors. In the course of the twenty-minute action, twenty-seven Chinese planes were shot down, crashing in flames to the Japanese guns. Thus control of Chinese air space was virtually assured to the Imperial Naval Air Service by the introduction of the A6M2 carrier fighters. It was now possible to reintroduce the Type 97 torpedo/horizontal bombers and Type 99 dive-bombers in offensive operations over land targets. At times as many as 130 aircraft were being used by the Imperial Naval Air Service on operations over Chunking.

While the air war was in escalating progress over China, Ensign Yoshikawa HIJMN boarded the Japanese liner *Nitta-maru* in Yokohama Harbour, bound for Honolulu, where he disembarked on 27 March, to be met by the Japanese Consul-General, Nagao Kita. At a salary of $150 per month and $600 expenses for six months, Yoshikawa set about his intelligence-gathering activities. He visited all the islands of the Hawaiian group as a tourist, making two automobile trips and taking air excursions as a visiting holidaymaker. Twice a week he would drive around the island of Oahu in six-hour trips, visiting Pearl Harbor every day. Meeting the Japanese community, he quickly realised that they had little information for his purposes. By a stroke of good luck he was able to visit Hickham Field Airbase, which he was able to tour. Once a week he would report the information gathered to the Consul-General in Honolulu, who would cable the information to Imperial Headquarters in Tokyo.

In April 1941 other Naval Intelligence agents had penetrated the Pearl Harbor installations and were established in Hawaii. Their purpose was to watch the shipping movements, and a monitoring service was organised to this end. The information collated included US naval radio call-signs, messages, supply requisitions, pier locations and anchorages of US warships. During the course of all these activities a monitoring station was installed at Baja California, in Mexico, to track US shipping movements off the west coast of America. A monitoring station was also established in Mexico City to monitor messages and movements of the US Atlantic Fleet. It was in this month that the commanding officer of the 1st Air Fleet accepted the mass carrier tactics as proposed by Commander Genda. Under this scheme all six fast Japanese carriers were organised into a single task force, cruising in a rectangular box formation with ships 7,700 yards apart. This type of formation was to be used against enemy ground and port installations in the Pearl Harbor attack, subsequently in the Indian Ocean attacks, and at the Battle of Midway against land bases.

Against an enemy fleet an inverted V formation was proposed, with three by two groups of ships at a distances of 150–250 nautical miles along the arms of the V, as well as 300–500 nautical miles between the arms of the V encircling the enemy fleet. But the inverted V battle formation was never to be used.

By 7 April Admiral Nagano replaced Prince Fushimi as the Navy Chief of Staff, just as General Sugiyama replaced Prince Kan'in as the Army Chief of Staff, a principle in which the Imperial family withdrew from the direction of government organisation, so that no aggressive intent or action could be attributed to Imperial direction.

On 10 and 11 April Operation Z was planned, when Rear Admiral Kusaka, former captain of HIJMS *Hosho* and *Akagi*, and having been commanding officer of the 24th Air Squadron on the island of Palau, was requested to report to Admiral Shigeru Fukudome, chief of the Naval Operations Bureau of the Naval General Staff. Admiral Kusaka was 48 years old, sturdy, energetic and with a candid face, whose father had been a business executive. He had graduated from the Naval Academy in 1913, having studied naval aviation. In his time he had crossed the Pacific Ocean in the German airship Graf Zeppelin. Admiral Kusaka was shown an operations plan written by Admiral Onishi and was ordered by Admiral

Shigeru Fukudome to make the plan work if required during a war situation.

By 11 April Rear Admiral Kusaka reported to Vice-Admiral Chuichi Nagumo on board the flagship HIJMS *Akagi*. The Vice-Admiral, though short and slight, knew little of aviation matters but was a torpedo expert. He informed Admiral Kusaka that he was in charge of planning Operation Z, though neither officer had been a flyer. The senior staff officer, Commander Tamotsu Oishi, and the senior air operations officer, Commander Minoru Genda, were instructed by the Rear Admiral to draw up a workable scheme for an aerial attack on Pearl Harbor, and Admiral Kusaka himself proceeded with the scheme after a visit by Admiral Yamamoto. From now on the overall plan was under the direction of Commander Tamotsu Oishi, and the technical planning of the aerial attack was conceived by Commander Genda. Admiral Yamamoto had considered such an attack ever since 1940, when he had watched newsreels of US carriers on manoeuvres.

Four days later, on 15 April, the Japanese became most upset when it was learned that President Roosevelt had authorised reserve officers and enlisted men to voluntarily resign from the US Army Air Corps and Naval and Marine Air Services to join Colonel Chennault, the commanding officer of the American Volunteer Flying Group openly training in Burma, to fight the Japanese on behalf of China. All arrangements were made through the Central Aircraft Manufacturing Company of China, which had been authorised to contract a hundred US pilots as well as several hundred ground crews. The Japanese considered this a hostile and provocative act, considering that neither Japan nor America was at war.

The next day Sub-Lieutenant Yasushi Nikaido, a fighter squadron leader from the carrier HIJMS *Kaga* of the 1st Carrier Division, was performing aerobatics in an A6M2 Model 21, serial number 140, over the naval air station at Kisarazu on the eastern shore of Tokyo Bay. Upon looping with sharp turns at high speed, Sub-Lieutenant Nikaido surprisingly observed the outer wing skin had wrinkled, and upon placing the Zero in a 50-degree dive with throttling-back from 12,000 feet, the indicated air speed (IAS) rose to 333 mph. With increasing speed the pilot lost control but managed to recover from the power dive, when a tremendous shock could be felt throughout the fighter plane as the speed approached 345 mph at 2,300 rpm. The pilot almost blacked out, but once again managed to regain control of the plane. Sub-Lieutenant Nikaido noticed the wing ailerons were now missing and the upper right wing surface had been sheared off. The pitot head had broken off and the airspeed indicator had jammed at 184 mph. Nevertheless he was able to make a good emergency landing on the airfield.

Imperial naval regulations required that any aircraft accident attributed to faulty design or construction must be reported to the headquarters of the aircraft's unit and to:

1. The Naval Bureau of Aeronautics
2. Naval Research and Development Centre
3. Yokosuka Experimental Air Corps
4. All naval aviation units equipped and operating this type

The captain of HIJMS *Kaga* submitted a full report on the circumstances of the accident, and Lieutenant Manbei Shimokawa, fighter squadron leader and corps instructor with the Yokosuka Naval Experimental Air Corps, was assigned to continue further investigations.

In the hangars stood two A6M2 Zero fighters, serial numbers 50 and 135, which engineers critically examined, particularly the wing surfaces, to try and find a cause of the accident. They were unable to find any reasonable cause, so Lieutenant Shimokawa decided to make flight tests with these machines on 16 April 1941. Now Lieutenant Shimokawa was an excellent test pilot and was most careful, for he was a product of the pre-war Japanese aviation policy in that a few highly skilled pilots were trained rather than many mediocre flyers. So on 17 April a second test flight was made in aircraft serial number 135.

Lieutenant Shimokawa took off and steadily climbed to 13,000 feet, where he rolled the Zero fighter over and commenced to dive at an angle of 55–60 degrees and appeared to make a good recovery at 5,000 feet. To watchers on the ground a white object appeared to rip off the left wing, while a black object was hurled from the fuselage. The plane made two turns to the left to regain the dive and, accelerating, plunged into the sea at high speed, with a high-pitched motorised roar. The pilot was killed instantly, as the wreckage sank to ten fathoms, some 1,000 feet off Natus Shima Island.

Two very good pilots had been lost in the progress of the development programme – assistant test pilot Okuyama in the disintegration of the second prototype of the A6M1, and Lieutenant Shimokawa in an A6M2 Model 21 serial number 135. This was a serious loss to the Naval Air Service, particularly as they were highly skilled pilots.

Curiously enough, on the morning of 17 April, the Lieutenant had test-flown aircraft serial number 50, simulating Nikaido's original flight. Entering a 50-degree dive at 368 mph, he successfully pulled out to level off at 4,000 feet, with no appreciable wrinkling of the wing surfaces.

Both planes were in the same production batch as Nikaido's machine, and the two flights had been observed by the ground crew, several pilots and engineers. Shimokawa had been warned by Lieutenant-Commander Yoshitomi, fighter group leader, to pull out of a dive if the wrinkling of the wing surfaces appeared, but it had been too late.

Fortunately the aircraft had been salvaged virtually intact, with only the ailerons and tailplanes missing. Engineers of the Naval Aircraft Research and Development Centre investigated the accident and concluded that:

1. Ailerons over-balanced at high speed due to hinge failure of the balance tabs leading to the aileron.
2. Ailerons had been ripped off the wings due to violent flutter. Vibration tests were made on the A6M2, and wind tunnel flutter tests were conducted on model wings, which enabled Mitsubishi engineers to estimate critical speed of aileron rotating wing flutter as 575 mph.

For the NARDC a restriction to a 5G pull-out was imposed on all A6M2 aileron-balancing-tab-equipped machines with a maximum speed of 278.5 mph IAS when power diving. Results of the investigations suggested that aileron rotating wing torsion flutter could occur at speeds above 345 mph. Concluding that severe flutter caused the accident, a conflict of opinion resulted between the use of an inaccurate theoretical formula suggesting 575 mph and practical tests with calculations of 345 mph.

Corrective measures were introduced, which included increasing the torsional rigidity by increasing the skin thickness of the wing outer skin, increasing the aileron mass balance and finally setting the critical IAS of the aircraft at 414 mph.

Between January and April 1941 the Naval Air Service introduced the Air Fleet organisation, or Koku-Kantai. Thus the 11th Air Fleet consisted of eight land-based air corps, while the Carrier Fleet consisted of eight aircraft carriers. The Joint Air Corps was abolished for operational units, but retained for training formations, and an intermediate level of command was introduced as the Air Wing, or Koku-Sentai. Thus the Naval Air Service from now on was organised into squadrons, air wings, air corps and air fleets.

In May 1941 three liners of the NYK Line were converted to aircraft carriers –*Kasuga Maru*, *Nitta Maru* and *Yawata Maru*. The latter two liners were rebuilt at the Kure Navy Yard, while the former ship had been towed to the Sasebo Navy Yard for conversion. When rebuilding had been completed the three ships were renamed, respectively, HIJMS *Taiyo*, *Chuyo* and *Unyo*. Each displaced some 17,830 tons and measured 73¾ × 26¼ ft, and was driven by two geared-shaft turbines developing 25,200 shp at a maximum speed of 21 knots. Armed with eight 4.7-inch anti-aircraft guns and eight 25 mm cannon, they could house twenty-seven aircraft and were manned by 850 officers and men, except HIJMS *Taiyo*, which was crewed by 747 men. Originally built by Mitsubishi Heavy Industries at Nagasaki in 1939/40, each ship of this class was to be lost to United States submarines in 1943 and 1944.

During the course of a conference held on 25 April 1941 by the British military Chiefs of Staff in London, a request was made by the Vice-Chief of the Naval Staff to his opposite number in the Air Force urging the immediate dispatch of Hawker Hurricane fighter planes to Malaya and Singapore. The Vice-Chief of the Air Staff in his reply indicated that the American Brewster Buffalo fighters were quite adequate for the aerial defence of Singapore and the archipelago, since Japanese aircraft were not considered to be of the latest type and lacked high performance.

Unhappily for this decision, the first Mitsubishi A6M1 Zero fighter aircraft was shot down in the air war over China on 20 May 1941, the month following the London conference. The remains of this machine were carefully examined, and the technical details of the armament, tankage and other items of equipment were conveyed to Singapore.

It was not until 26 July that the technical analysis of the Zero was communicated to Headquarters Air Command Far East, Singapore, and to the Air Ministry, London. The Air Staff should have amended the assessment of the air war situation of the Far East in the light of these latest technical revelations, which the Air Staff failed to appreciate.

Meanwhile the British Air Attaché in Chongqing had either obtained access to the Japanese plane or obtained further information that had enabled him to calculate the combat performance of the new machine, which subsequently proved reasonably accurate.

The Combined Intelligence Bureau, Singapore, on 29 September 1941, transmitted all this information to interested formations. Originally established as a branch of the Royal Naval Intelligence Service for gathering naval intelligence on shipping, it had later been broadened into the gathering of secret information for the Army, and subsequently for the Royal Air Force, by agreement with the government in London. For these purposes the Combined Intelligence Bureau was ill equipped, for it possessed neither the specialist operators nor the secret networks for so complex and comprehensive a service of intelligence acquisition. Consequently, despite valiant efforts by dedicated officers, the resulting information was inadequate, incorrectly analysed and communicated too late to be of real value for such specialised purposes.

Furthermore, Air Command Far East unfortunately did not possess an air intelligence branch with which all commands within the Royal Air Force normally function. Consequently, the result of this failure denied the

British fighter squadron commanders in Malaysia access to vital detailed information concerning Japanese aerial dispositions, tactics, numerical strengths and technical information affecting the outcome of the air war over Malaya. Any information which was received during the course of the air battle remained among the mass of unsifted intelligence documents and reports deposited unanalysed in the vaults of the Singapore headquarters until the surrender revealed the extent of this cataclysmic historical disaster for the British armed forces in February 1942.

While the first Zero fighter was lost in the air war over China on 20 May 1941, during the same month the maiden flight took place of the Nakajima J1N1 Gekko in response to a naval specification for a long-range escort fighter design, work having commenced sometime during 1938. The initial flight test was to prove the design unsuitable, requiring further development work. The Nakajima J1N1-S was powered by two Nakajima Sakae 21 radial engines rated at 1,130 hp each, which gave a maximum speed of 315 mph at 19,030 feet, with a maximum ceiling of 30,580 feet and a range at cruising speed of 1,584 miles. The aircraft had a wingspan of 55 feet 8^1/$_2$ inches, a length of 39 feet 11^1/$_2$ inches and a height of 13 feet 1^1/$_3$ inches, and attained an all-up weight of 15,983 lb. With a crew of two, the machine was armed with four Type 99 20 mm cannon. Development continued until July 1942, when the type was placed in full production, and by 1943 the J1N1-C version was introduced, with full reconnaissance equipment, as well as a Type J1N1-F with a rotating gun turret. Later the J1N1-S model was produced, being the first Japanese twin-engine fighter to be equipped with aircraft interception-type radar. A total of 479 aircraft of the J1N1 Gekko were manufactured, of which some were equipped as bombers, with a 2,432 lb offensive load.

Commencing in June 1941, the Mitsubishi Aircraft Company was to build 343 A6M2 Model 32 carrier fighters. However, at about this time a sudden problem arose, which no one had ever anticipated would take place. At the Nagoya Works of the Mitsubishi Aircraft Company, engineer Sone broke down and collapsed, while engaged on the Mitsubishi J2M1 development project. His doctor eventually ordered him to take a prolonged rest because of physical and mental exhaustion, although only a quarter of the detailed production blueprints had been prepared. Even the chief designer, Dr Horikoshi, was ordered to rest on medical grounds. This situation presented a serious problem for the company because it represented a shortage of technically trained personnel in the development of the company's aviation projects who could assist the work of the senior engineers. It was therefore resolved that Mijiro Takahashi should be appointed chief designer, with Hirotsugu Hirayama, the director of the Prototype Construction Shop, acting as chief assistant. A shortage of highly skilled engineers and the over-burdened work load on the existing engineering staff would delay many future military engineering plans.

During the summer of 1941, the aircraft carriers HIJMS *Shokaku* and *Zuikaku* formed the 5th Carrier Division, each ship capable of carrying eighty-four aircraft. They were damaged at the Battle of the Coral Sea in 1942, and withdrawn from active operations for dockyard repairs, the anti-aircraft armament being increased to seventy 25 mm cannon. However, neither of these ships appeared at the Battle of Midway, due to a shortage of trained pilots, but HIJMS *Shokaku* sank at the Battle of the Philippine Sea on 19 June 1944.

By mid-1941, His Imperial Majesty the Emperor Hirohito had the Imperial Naval General Staff study war prospects with the United States and Britain. The Emperor was not satisfied with their report and insisted on a further study, which eventually was to confirm the first report. Japan could wage war successfully on the United States and Britain for a period of eighteen months, but not beyond June 1943. After this date, Japan would have to seek negotiations with Washington and London.

Meanwhile the planning for Operation Z continued, and by late June 1941, Admiral Kusaka had convinced Admiral Onishi that the plan for the Pearl Harbor attack had many serious flaws. A visit to Admiral Yamamoto was agreed, but he insisted upon drawing up a viable tactical plan of operation. Basically the problem facing Admiral Kusaka was how to bring the attacking fleet within operational distance of the American base at Pearl Harbor without being detected. The staff navigation expert Lieutenant-Commander Toshisaburo Sasabe studied the nationalities and types of ships sailing across the Pacific during the previous ten years, to determine the various shipping lanes. Toshisaburo was able to report that no shipping line travelled in latitude 40 degrees north in the months of November and December because of the rough nature of the ocean at that time of year. Therefore, an attacking fleet coming from the north would stand least chance of detection by American forces.

Normally the US fleet conducted manoeuvres to the south-west of the Hawaiian Islands, and US fleet leaders had automatically assumed that any Japanese naval aggression would originate from their bases in the Marshall Islands. Such factors were crucial, particularly regarding the position at which the ships of the attacking fleet would require to refuel on the high seas.

Imperial Naval Intelligence agents had discovered that the US Naval Catalina PBY flying-boats patrolled an area 500 miles north of the Pearl Harbor base and 500 miles south of the US base at Dutch Harbor in the

Aleutian Islands far to the north. This would leave an undefended gap in between completely unpatrolled by American forces. The Japanese strike force, therefore, would head due east from the Kurile Islands within the unpatrolled area to a point 800 miles north of the Pearl Harbor anchorage. From there, in the dark, the fleet would steam southwards so that at first light the armada of attacking naval planes would take off to their assigned targets.

Commander Mitsuo Fuchida, the squadron leader from HIJMS *Akagi*, became aerial strike leader. As the flight commander at 39 years of age, a veteran of the China war, he had logged over 300 hours of flying time. Unfortunately, some of the carrier captains had objected to his appointment, but Admiral Kusaka in an executive decision enforced the appointment.

The aerial attack of Operation Z was conceived by Commander Genda, the RAF College-trained staff planner who envisaged the primary target as 'Battleship Row', the two lines of battleships moored off Ford Island, within Pearl Harbor. The outside row of battleships could be conveniently attacked by the torpedo-bombers, but the inside line would have to be struck by the dive-bombers and horizontal bombers. Admiral Kusaka believed this was not possible since the Naval Air Service did not possess an accurate bombsight apparatus. However, constant aerial practice with the Type 99 bombsight, a copy of a piece of German aviation equipment, proved the solution. Nevertheless, Naval Intelligence agents had been unable to penetrate the cloak of US security, and thus plans and a model of the US Norden bombsight had not been obtained, and its existence was not known at this time.

Unhappily, at this juncture it was recognised that the Naval Air Service did not possess a bomb capable of penetrating the thick armour plating of a battleship. The problem was resolved by Commander Genda, Commander Fuchida and naval aviation engineers by using 15-inch battleship shells fitted with tail fuses, a strengthened casing and streamlined tailfins.

Flying training was scheduled to take place over Kyushu Province in the south of Japan, famed for its active volcanoes and an area conducive to men of war-like spirit. The fighter pilots would be based at Saeki Naval Air Station, while the dive-bombers were located 150 miles further down the coast at Tomioka Naval Air Station, from where night attacks were practised using a towed raft making a heavy wake. The horizontal bombers and torpedo-bombers conducted practice-runs near the south of Kagoshima Bay, flying over the mountains that rise to 5,000 feet behind Kagoshima city, then dropping down low over the city to skim over the harbour at 25 feet towards a pier and zooming right to avoid Sakurajima, an active volcano. Complaints poured in from outraged citizens at the inconvenience of the noise and interruption of the daily civilian routine. Each plane, with a crew of three – pilot, observer and radioman/gunner – flew over mountains, city department stores, the railway stations, telegraph poles, factory chimneys, dropping to 25 feet above the breakwater, zooming right, skimming the bay waters and away. Meanwhile torpedo experts at Yokosuka Naval Arsenal were adopting torpedoes for shallow-running purposes by fitting wooden fins and rudders.

On 30 June 1941 a severe difference of opinion reared its ugly head when the Japanese diplomat Mr Matsuoka expressed a wish to see the government postpone the Strike South campaign, predicting ultimate disaster. He proposed a six-month postponement of the project for the occupation of Indo-China, but Lieutenant-General Tsukada, a member of the Emperor's inner cabal, pointed out that the Indo-China occupation plans had already been sanctioned by His Imperial Majesty. The Home Minister Hiranuma, president of the National Foundation Society, wished to resurrect the Strike North idea. However, the Chief of the Naval Staff, Admiral Nagano, stated that all preparations had been made for a Strike South attack, and an alteration to the projected plans would require fifty days to replan a new scheme.

A new government came into power on 18 July 1941 under a new Prime Minister, Prince Konoye. The investiture ceremony took place in the presence of the Emperor at the Imperial villa at Hayama. Admiral Toyoda Soemu was given the Foreign Minister's portfolio, while Tanabe Harumichi was appointed Home Minister. Three days later a liaison conference was held under the direction of the Chief of the Naval Staff, Admiral Nagano, who made the following points:

1. Japan would have a chance of winning an immediate war with the United States.
2. Nine months later the situation would become progressively worse.
3. The United States was dragging out negotiations so as to build up defences.
4. Time was not on Japan's side.
5. War should be prosecuted fully from declaration.
6. War for the Navy inevitably arose due to the question of oil acquisition in the south.

By 29 July the Admiral was urging the Emperor to declare war on the United Kingdom whenever necessary, so as to seize the Malayan Peninsula. Meanwhile Japanese troops had occupied the southern provinces of French Indo-China under an agreement between France and Japan. The United States retaliated by enforcing diplomatically an economic embargo policy. These military and economic pressures on Japan were made in agreement with the support of Britain, the Netherlands and China. The Imperial Navy and the Japanese government were now surer than ever that war was inevitable, and they agreed that Japan should not become the aggressor.

Between 25 and 30 July, the 21st Koku Sentai was transferred from its airbase in Japan to Hankou Air Station in China. Offensive operations were flown against Chengdu and Chongqing in central China. On 1 August the 1st Kokutai was detached from the Imperial Naval Air Forces in China to return to Kanoya in Kagoshima Province. Here intensive crew training was undertaken so as to prepare for the forthcoming war against the Western powers.

During the second week of August 1941, His Imperial Majesty the Emperor Hirohito reviewed the naval war planning. Previously, on 2 July, he had taken the final decision not to strike north. In the spring and summer of 1941, Admiral Yamamoto had ordered practice exercises to take place off the Kyushu ports of Kagoshima, Kanoya, Saiki and the Shikoku port of Sukomo. The dive-bombers and torpedo-bombers had followed maps and briefings with high success. By early August 1941 Admiral Yamamoto and his staff had arrived in Tokyo to explain the Admiral's convictions to the Emperor's personal naval representative. The Admiral initially protested, urging the avoidance of an all-out war with the United States of America – ideas only known to the Emperor, Prince Takamatsu and Rear Admiral the Marquis Komatsu Terukisu. Nevertheless, if war was to come, Admiral Yamamoto urged the destruction of the US base at Pearl Harbor, together with the sinking of the ships of the US Pacific Fleet. A demonstration of the Admiral's plan took place at the Naval Staff College to the fleet captains and commanders. Prince Takamatsu and Admiral Teruhisa were convinced of the feasibility of the operation.

Meanwhile, during August 1941 the national oil situation came under review, since Japan was dependent upon the USA for its oil and some raw materials, especially for aviation fuel. An inventory of stocks held in the Home Islands had to be made, as the USA was intervening to prevent oil supplies from reaching Japan from as far afield as Indonesia and South America. Oil stocks in 1941 were 300 million US gallons of aviation fuel and 1,150 million gallons of crude petroleum. These figures were to drop by 1943 to 210 million US gallons of aviation fuel and 600 million US gallons of crude petroleum, due to consumption during the course of two years of active military operations.

On 9 August 1941 President Roosevelt met Prime Minister Winston Churchill aboard USS *Augusta* off the coast of Newfoundland. It became apparent that Mr Churchill had access to more information concerning Japanese expansionist intentions than President Roosevelt, despite the American invention of the Magic machine, which had enabled them to read Japanese consular coded messages. Nevertheless, the US Navy had established a chain of radio direction-finding stations based on Pearl Harbor at station Hypo, under the command of Commander Rochefort USN, enabling Imperial naval units manoeuvring off the coast of Japan to be located and identified. In parallel with the naval wireless network the US Army maintained monitoring stations in the islands. At station MS5 Honolulu, the monitoring staff intercepted consular messages from the Foreign Office, Tokyo, to the Japanese Consul, and sent the texts directly to US Army Intelligence in Washington. This information was not given to the senior military in Hawaii. The messages were dispatched over the normal commercial cable network to the Japanese Consul, requesting information relating to the disposition of warships in the US fleet anchorages at Pearl Harbor.

Despite the blockading of the Chinese coastline by the Imperial Navy, the Chinese armed forces were being supplied with weapons and materials from the USA and Great Britain routed through Burma. Normally supplies would be landed from shipping at the port of Rangoon and thence by rail to Mandalay and by lorry along the Burma Road, over the Chinese border, to Yunan. Further supplies were coming in from the USSR through the province of Sichuan. Imperial Navy bombers attacked Chinese roads, railways and bridges in an endeavour to stop supplies getting through. During the course of eight months of 1941, over fifty aerial attacks on the various routes were undertaken by Type 99 Aichi dive-bombers co-operating with Type 96 Mitsubishi G3M1 twin-engine bombers. In the period November 1938 to August 1941, over 410 long-range aerial operations were conducted by aircraft of the Imperial Navy's 1st and 2nd Joint Air Corps in association with the Koshun Air Corps. Targets included the Burma Road supply route as well as Chinese airbases and installations located at Chongqing, which was raided some one hundred times. Other targets attacked were located at Chengdu, Hengyang, Guilin, Kweiyang, Kunming, Lanzhou and Yichang. Initially aerial attacks were conducted by formations of eighteen to thirty-six bombers, but the numbers increased as the air war increased in intensity.

In September 1941 the Japanese Honolulu consular spy made his report to Tokyo regarding the US fleet dispositions and installations as requested. His report was intercepted by monitoring station MS5 and immediately communicated to Washington. With the Japanese armed forces station in

Indo-China and President Roosevelt's embargo on strategic oil and material supplies, Washington was anxious to know the opinion of the Japanese senior officials and the effects being felt by the relevant agencies. Consequently the monitoring unit stationed at 717 Market Street, San Francisco, part of the 12th Naval District organisation under the command of Captain McCulloch USN, and reputed to have a direct line of communication with Washington, forwarded an analysis to the presidential staff. In the meantime, on 24 September, a further signal, from Tokyo to the Japanese Consul General, was intercepted by MS5 Honolulu. This message requested information on the installations in and around Pearl Harbor in greater detail, by dividing the harbour area into five sections for analysis. Section A would cover the area east of Ford Island, where the battleships and aircraft carriers would be anchored, with Section B west of the island and Section C where the smaller ship anchorage lay. Sections D and E would cover areas where other small craft would be anchored.

At the time of the request to the Honolulu spy network requesting further information on the US warship anchorages, the planning and preparations for Operation Z continued during autumn 1941. Captain Fumio Aioko HIJMN, a torpedo expert stationed at Yokosuka Naval Base, had solved the problem of running torpedoes shallow within the confines of Pearl Harbor. The solution lay in the use of wooden fins as aerial stabilisers and fitted to the torpedoes carried by the Nakajima torpedo-bombers. Tests were conducted in Kagoshima Bay, where an 80% success rate was achieved, as the torpedoes ran shallow enough for the Pearl Harbor attack. Immediately the Yokosuka Naval Arsenal mass-produced the stabilising fins in sufficient quantities for the operation.

The Nipponese naval strength in the Pacific during autumn 1941 consisted of ten battleships, ten aircraft carriers, eighteen heavy cruisers, seventeen light cruisers, a hundred destroyers and sixty-four submarines.

The aircraft carrier force comprised the First Air Fleet, under the command of Vice-Admiral Nagumo HIJMN, with the 1st Carrier Division under his personal command, including the carriers HIJMS *Akagi* and *Kaga*, plus three destroyers. The 2nd Carrier Division was under the command of Rear Admiral T. Kuwabara controlling the carriers HIJMS *Hiryu* and *Soryu*, plus three escorting destroyers. The 4th Carrier Division was commanded by Rear Admiral K. Kakuta, with HIJMS *Ryujo* and two escorting destroyers, while the 5th Carrier Division was commanded by Rear Admiral C. Hara with HIJMS *Shokaku* and *Zuikaku* escorted by two destroyers. The aerial strength of the First Air Fleet comprised fighter planes, dive-bombers and torpedo-bombers.

Aerial strength of the First Air Fleet

Carrier	Fighters	Dive-bombers	Torpedo-planes
Akagi	21 Zeros	18 D3A	27 B5N
Kaga	21 A6M	27 D3A	27 B5N
Hiryu	18 A6M	18 D3A	18 B5N
Soryu	18 A6M	18 D3A	18 B5N
Shokaku	18 A6M	27 D3A	27 B5N
Zuikaku	18 A6M	27 D3A	27 B5N
Hosho	11 A5M	8 D3A	8 B5N
Zuiho	16 A5M	12 D3A	12 B5N
Ryujo	16 A5M	18 D3A	18 B5N
Total	**157**	**173**	**182**
	= 512 aircraft		

In the course of 1941, ten merchant ships were converted to seaplane tenders, each with an armament of two 5.9-inch DP guns and two 13 mm anti-aircraft cannon. Each ship in this class carried eight seaplanes and was equipped with two catapults on the aft deck.

In autumn 1941, Sir John Masterman was the chief of Section XX of the British Military Intelligence Department Five, with the task of infiltrating the ranks of the British-based German intelligence organisations. These included the German armed forces' Abwehr and the Schutzstaffel Sicherheitsdienst. German agents were detected, caught, interrogated and 'turned around' (persuaded to operate on behalf of British Military Intelligence). By such operations the British were able to learn from the German intelligence centres that an attack on the US base at Pearl Harbor by the Japanese would be the number-one target if America was invaded in the war. This information was conveyed to the leaders of the US administration, but its military significance was not grasped.

At this period, the chief designer of the Mitsubishi Aircraft Company became seriously ill during the course of designing the new Mitsubishi A7M Reppu land-based interceptor fighter plane. Unfortunately he did not recover until the early months of 1942. No doubt the cause of the illness lay in the tremendous pressure under which the design staff were operating at this time. These factors could be interpreted as indicating a shortage of trained engineers within the aviation industry.

On 2 September 1941, Admiral Yamamoto and his staff attended a conference under the chairmanship of Rear Admiral the Marquis Komatsu at the Naval Staff College situated on the edge of the Shiragane Imperial Woods, some three and a half miles due south of the main Imperial Palace. This was a rehearsal of the naval operations for Army Unit 82, the Strike South land-based planners. Demonstrations were given of the naval plans for supporting the troop landings in Malaya, the Philippine Islands, the islands of Wake, Guam, Borneo and Java. Unhappily, Admiral Yamamoto's Pearl Harbor attack was only considered a sideshow, but subsequent to six days of continual criticism more ships and aircraft were allocated for the attack on the American naval base. Present were thirty-nine admirals, ships' captains, commanders and other officers who would lead on the field of conflict.

During a further meeting on 4 September, the Prime Minister, Prince Konoye, was advised by the military leaders that the oil reserves in Japan were running out. A deadline of 16 October was set as the final date by which the US oil embargo had to be lifted. This decision gave Prince Konoye just six weeks in which to reach a diplomatic settlement with America.

With this task in mind, Prince Konoye proposed to hold a personal meeting with President Roosevelt to try and settle the impasse. Unfortunately the President of the United States was advised by his Secretary of State, Mr Cordell Hull, that prior concessions would be required from the Japanese government, including the withdrawal of Imperial armed forces from the territories of nationalist China, before discussions between the two governments could take place. Such a demand by the US government made negotiations impossible to achieve, for the Japanese would be agreeing under threat to a condition that they could never concede. So on 16 October 1941, Prince Konoye tendered his resignation to the Emperor, and General Tojo, hitherto Minister of War, was appointed Prime Minister.

Earlier, at an Imperial conference on 6 September, at which Admiral Nagano was present, ministers of the government were briefed one by one. Japan was entering negotiations with America and Britain but prudently preparing for war at the same time, and the chief of the Cabinet Planning Board, Suzuki, urged immediate acquisition of the islands of the Dutch East Indies as an alternative source of supply of petroleum products, but no one pointed out the utter hopelessness of Japan's negotiating position. Meanwhile the Navy withdrew many of its naval aircraft from China to prepare for a war with the more sophisticated Western air forces. It had been calculated originally that the numbers of aircraft required for a campaign in the Pacific theatre of operations would be:

For the Malaya campaign	150 aircraft
For the Philippines campaign	300 aircraft
For Operation Hawaii	400 aircraft
For the Marshalls campaign	50 aircraft
Seaplanes required by the combined fleet	75 aircraft
Combat aircraft available in Japan in res	275 aircraft
Total allocation for war operations	**1,250 aircraft**

Sometime during October 1941, the liner *Taiyo-maru* sailed from Japan on a cruise to the Hawaiian Islands, destined for Oahu. On board the ship were two agents of the Imperial Naval Intelligence Service, Commander Toshihide Maejima, a submarine expert, and Lieutenant Commander Suzuki, an expert on aviation affairs. Unbeknown to the passengers, the liner's course navigated due eastwards from the Home Islands to a point some 800 miles due north of Hawaii, from where the ship turned south for her intended destination of Oahu. The liner had tracked the exact course to be taken by the forthcoming aircraft carrier attack fleet in preparation for the Pearl Harbor mission. The tasks set for the two naval officers included surveying the positions of the various targets, recommendations as to the types of bombs to be used, the whereabouts of suitable emergency landing-strips, whether Lahaina Harbor on Maui was used as a US fleet anchorage, and a precise study of the sea state and weather conditions during the course of the voyage to Oahu. On 1 November *Taiyo-maru* passed Diamond Head. Lieutenant-Commander Suzuki carefully studied the harbour entrance through powerful binoculars from a position on the liner's bridge, concluding that the harbour mouth was very small. Once having landed, he accosted the post officials, as well as Navy officers, and learned the depths of the water, position of the minefields and the whereabouts of the steel net at the Harbor entrance, and noted a British warship on a liaison visit, which was equipped with radar apparatus. A list of ninety-seven questions was compiled for investigation, but the two officers were advised to stay on board the liner and within its comparative security away from the watchful eyes of the US counter-intelligence agents, lest any mishap should take place. The task of conducting the intelligence survey was handed to Ensign Yoshikawa and his secret service network, now well established among the islands. Meanwhile Lieutenant-Commander Suzuki confined himself to taking photographs from the bridge of *Taiyo-maru*, including the Pearl Harbor installations and the facilities to be found at the Hickham Field Airbase. Eventually Japanese consular officials visited Suzuki on board the liner with all the information required hidden on sheets of notepaper within the pages of daily newspapers.

Meanwhile the Imperial Navy announced its 5th Expansion Programme for Naval Aviation, upon the eventual completion of which, the Navy would be provided with:

132 operation squadrons	2,696 aircraft
Ship-based & transport aircraft	2,041 aircraft
156 training squadrons	3,671 aircraft
Grand total	**8,408 aircraft**
Trained flight crews	31,500 men

Thus the Navy regarded aviation of prime importance in the forthcoming Great Pacific War.

On 1 November 1941 a liaison conference took place in the Imperial headquarters building north of the Fukiage Gardens, not far from the Imperial Palace. Present were Admiral Nagano, Chief of the Naval General Staff, and General Tsukada, Army Vice-Chief of Staff, who both agreed on the importance of continuing the negotiations with the Barbarians, but wished for 13 November to be the final deadline for the commencement of hostilities. Two days later, 3 November, Admiral Nagano went to the Imperial Palace to answer questions on the forthcoming naval strategy as asked by the Emperor in company with General Sugiyama, Army Chief of Staff. During the course of this meeting His Imperial Majesty announced the target day for the commencement of hostilities as 8 December 1941.

On 5 November operators of the US radio interception intelligence decoding machine Magic picked up an Imperial diplomatic message in the Purple code advising the Japanese Ambassador in Washington, Admiral Nomura, that a revised date of 25 November had been agreed for the restoration of the supplying of oil from the USA. Returning to the Imperial Naval Staff Headquarters building on the edge of the Hibiya Park, just outside the Imperial Palace gates, Admiral Nagano met Rear Admiral Kuroshima Kumahito, who had brought Imperial Order No. 1 from Admiral Yamamoto's flagship lying in Hiroshima Harbor on the Inland Sea. In forwarding the order, Admiral Yamamoto had threatened to resign if the projected plan was not accepted. The Chief of the Naval General Staff, upon reflection, gave his consent, and Admiral Kumahito flew south to report to the commander of the Combined Fleet. The date for the meeting of the Supreme War Council, under the direction of His Imperial Majesty The Emperor, was 4 November. Those present included the Army and Navy Ministers with their respective chiefs of staff, the commanding officers of the Kure and Yokosuka Naval Bases, two former Navy Ministers and General Count Terauchi, the Strike South Army commander. Also present were General Doihara and Admiral Shinozuka

Yoshio of the Imperial Secret Service, and Imperial Princes Kan'in, Fushimi, Asaka and Higashikuni. Admiral Nagano informed the members of the Supreme War Council that war would prevent the stranglehold of America, Britain and the Netherlands. Should war start in early December with the present military strengths, the first phase could be won, resulting in the defeat of the Barbarian forces and a reduction of their strongholds in the south-west Pacific area. Since the outcome of war depended upon the execution of phase one, the result had to be achieved by surprise and swift action. Such a war for Japan could be won in one or two years' time, but the country would face uncertainties if the war was prolonged to a three- or four-year period. The United States was too big to be invaded and occupied, and since Germany might invade and conquer Britain, the United States would be more receptive to a negotiated peace.

The following day, 5 November, Admiral Yamamoto issued Combined Fleet Top Secret Operation Order No. 1 of Operation Z. This was a document of 151 pages outlining the first phase of hostilities, covering the operations at Pearl Harbor, with the simultaneous assaults on Malaya, the Philippine Islands, Guam, Wake, Hong Kong and the South Sea Islands. All unit commanding officers and squadron leaders assembled on board the new super battleship HIJMS *Yamamoto* in Hiroshima Bay for a personal briefing by Admiral Yamamoto himself. This was the day of departure, Admiral Suzuki summarised his findings as the final courier arrived at three o'clock in the afternoon with diplomatic pouches from Ensign Yoshikawa, containing the latest naval intelligence and maps obtained by the naval spy network operating in the Hawaiian Islands. On 7 November the Emperor arrived at Hayama, the site of the Imperial seaside villa, where little protocol was observed and where the Imperial yacht was moored, equipped for marine biological studies and connected with a ship-to-shore radio telephone. The Imperial Chamberlain Marquis Kido had remembered the Emperor speaking of an Imperial vacation at Hayama. But at such a bleak season of the year? Then the Chamberlain did not have a further audience with the Emperor until 9 November, when the Emperor immediately departed once again, not returning to Hayama until 13 November. It was said that the Imperial presence was on board the aircraft carrier HIJMS *Akagi* at sea, conducting war games with senior officers at which Admiral Yamamoto announced to trusted high-ranking officers that the date of the Pearl Harbor attack would be 8 December. Co-ordinating war studies continued, but the Emperor took the opportunity to say farewell to so many Army and Navy commanders, both on board the aircraft carrier and at the Army base at Numazu some forty miles down the coast from Hayama, or a hundred miles south by sea. At this time Admiral Yamamoto issued the second secret fleet order regarding the commencement of hostilities, tentatively fixed for Sunday 8 December 1941. The full moon at this period would facilitate the launching of the aircraft from the carriers, while a Sunday-morning attack would ensure that the US Pacific Fleet, having come into harbour on the Friday night, would not depart for sea again until Monday morning. Admiral Nagumo issued the 1st Operational Order to the carrier fleet on 10 November. Should the diplomatic negotiations with the USA be successful, then the Pearl Harbor operation would be cancelled. Meanwhile the carrier strike force was to rendezvous at latitude 42 degrees north, longitude 170 degrees east, staying in readiness until further orders were issued, while the carrier fleet captains were instructed by Admiral Nagumo to complete all preparations by 20 November. At the same time the 1st Kokutai of the Naval Air Service, equipped with Mitsubishi G4M1 twin-engine attack-bombers, was redeployed at Tainan in Formosa in preparation to strike at US installations in the Philippine Islands.

An insight into the real feelings of Admiral Yamamoto may be gauged from a private letter written by him to a senior naval officer who was a very close friend, in which Yamamoto confessed that he held diametrically opposite views as to the feasibility of the war with the United States, but added that he had no choice but to lead at the Emperor's command.

By 14 November Admiral Yamamoto, commander of the Combined Fleet, had arrived at Saiki Bay in north-east Kyushu, where the rest of the task force was now assembling. At Yokosuka Naval Base, Admiral Nagumo in HIJMS *Akagi* was taking on a consignment of special torpedoes for shallow-water running and equipped with the special wooden fins as devised the previous September and subsequently tried out successfully in Kagoshima Bay. This was one of many technical developments in the realm of torpedo technology among others. The Imperial Navy, having assembled the ships of the carrier strike force under the command of Vice-Admiral Nagumo in Saiki Bay, the task force would sail northward to a rendezvous off the Kurile Islands. At the time the Royal Navy monitoring stations at Hong Kong and Singapore had intercepted the Japanese naval signals, so that the British code breakers were able to read the contents of the communications. As the Imperial warships were sailing northwards, the Dutch Army intelligence radio network in Java was able to track the ships and intercept the radio traffic. The coded signals revealed that if war was necessary a voice message would be transmitted as part of a weather forecast within a news bulletin, and repeated at the conclusion of the newscast. These messages would indicate with which country Japan was breaking diplomatic relations. Thus 'north wind cloudy' – break with Russia, 'north wind rain' – break with the USA, and 'north wind clear' – break with Great Britain. These messages were intercepted by the Royal Navy monitors at Singapore, sometime about 20 November. Two days later, further intercepted signals revealed that the carrier strike force had been given orders to sail. At the same time Japanese Foreign Office coded signals had instructed the Ambassador in Berlin to formally approach the German government with a view to seeking support in the event of Japan going to war. This information was conveyed to the US government by Britain.

On 26 November President Roosevelt presented an impossible ultimatum to the Japanese, who could not in honour accept such terms. As a result of the Japanese rejection, hostile warnings were dispatched to the US military and naval commanders in Hawaii. Search as they might, the radio direction-finding chain under the command of USN Honolulu could not detect the Japanese fleet in its home waters. By chance radio signals were intercepted from the cruisers, battleships and oil tankers some 1,000 miles out in the Pacific. Unfortunately for Vice-Admiral Nagumo, these transmissions were in breach of Admiral Yamamoto's orders for radio silence to be maintained while the strike force was at sea. Nevertheless, Washington was now listening to every Japanese diplomatic communication and any other radio transmission that could be decoded emanating from Japan. The Imperial Navy had produced the Type 93 Long Lance torpedo specially designed for use against heavy naval surface units and warships. This weapon had a calibre of 600 mm, or 23.62 inches, with a 3,418 lb warhead, and was propelled by an engine burning a mixture of petrol and bottled gaseous oxygen producing 550 hp.

At a military meeting chaired by President Roosevelt on 26 November, it was agreed that US aircraft carriers would scull from Pearl Harbor to provide fighter defensive cover for a force of forty-eight Boeing B-17 Flying Fortress bombers *en route* to the Philippine Islands from the west coast of America. Based on Midway Island and Wake Island, US naval fighters would protect the bombers across the Pacific Ocean.

Meanwhile President Roosevelt had sent his son Colonel James Roosevelt to see Sir William Stephenson, head of British Military Intelligence, New York, with a presidential message. This was a communication specially to be relayed to Winston Churchill advising that diplomatic negotiations had failed and that US military leaders considered hostilities would commence within fourteen days.

During this period up to 26 November, a total of five aircraft carriers had left their base one by one, night by night cruising to Hitokappu Bay in the far north of Japan at Etorofu Island in the Kuriles group. Here in the cold of autumn, amid a wild, desolate, empty coastal landscape, only seals and walruses played on the beaches. At some time in the past a quay had been constructed with a lonely warehouse, which now contained drums of oil fuel for the ships. The drums were used to fill the carrier tanks and the remainder were stacked upon the decks. Radio silence was to be maintained from now onwards, while Admiral Yamamoto would transmit information to the carrier strike force in the new Admiral's Code, which was relayed in voice code so as to confound the US Navy's cryptographers. The old radio officers were based in Japan and commenced a radio warfare operation in which bogus messages were sent from one to another to confuse American Naval Intelligence. The new radio officers were aboard the carrier strike fleet receiving transmission of the Admiral's Code, which relayed the latest intelligence information.

While the carrier fleet was steaming across the ocean, the Japanese Foreign Ministry advised its Ambassador in Berlin on 27 November that diplomatic negotiations would be discontinued within two or three days.

However, a surprise meeting took place on 29 November between Prince Takamatsu and the Lord Privy Seal Marquis Kido, together with some armed forces officers, in which they agreed to appeal to the Emperor on the grounds that they had no confidence in the Navy's ability to win a sea war. A hurriedly arranged conference took place at which the Navy Minister and Chief of the Naval General Staff presented themselves to His Imperial Majesty. After an intensive two hours it was agreed that war would automatically follow and that the diplomatic negotiations were to be a mere sham. The United States had on a previous occasion demanded the complete withdrawal of all Japanese forces from the Chinese mainland and denounced the Mutual Defence Pact, which Japan had entered into with Italy and Germany.

On 30 November 1941 the Japanese Ambassador in Berlin advised Hitler that hostilities were imminent.

In the meantime war preparations continued, with the Nakajima Aircraft Company commencing production of the Mitsubishi A6M2 Model 21 carrier fighter. But the Mitsubishi Aircraft Company had been steadily increasing the production of the Mitsubishi G4M attack-bomber since the beginning of 1941.

The production figures were:

January to March	13
April to June	41
July to September	53
October to December	75
Total	182
B/F	12
Grand total 1941	**194 machines**

These bombers were to operate from:
Hankou, China
Jolo, Philippines
Kanoya, Japan
Kisarazu, Japan
Tainan, Taichu – Formosa (Taiwan)
Takao, Japan
Thudaumot, French Indo-China

G4M Units and Operational Bases, 1941

Unit	Bases	Attached to
1st Kokutai	Kanoya, Hankou, Kanoya,Tainan, Jolo,Kagi	21st Koku Sentai, April 1941
Kanoya Kokutai	Taichu, Thudaumot	21st Koku Sentai, October 1941
Kisarazu Kokutai	Kisarazu	Yokosuka Naval District, October 1941
Takao Kokutai	Takao	23rd Koku Sentai, October 1941

The Mitsubishi G4M2 Navy Type OB01 Model 22 of 1941 was a twin-engine naval land-based bomber and reconnaissance monoplane, sometimes used as the carrier of the piloted Baka bomb. The wing was a cantilever mid-wing centre section, with two outer sections and detachable wing types. The wing was a two-spar structure, with smooth stressed skin flush riveted, electrically operated flaps in which the ailerons were statically balanced Frise type with controllable trim tabs. The fuselage had an elliptical-section, semi-monocoque construction of aluminium alloy. The tail was a cantilever monoplane type of aluminium alloy frame, with metal-covered fixed surfaces with the elevator and rudder covering in fabric. The landing-gear was operated by electric motors with the main wheels retracting forward into the engine nacelle and the tailwheel retracting rearwards into the fuselage. The power plants were two Mitsubishi Kasei 21 motors, 14-cylinder radial air-cooled two-speed supercharged engines with water injection rated at 1,350 hp at 10,000 feet, 1,260 hp at 20,000 feet and supplying 1,500 hp for take-off. Fuel was retained in ten fuel tanks within the leading edge of the wings inboard of the engine nacelles. Propulsion was provided by two four-bladed Mitsubishi Hamilton air screws.

The machine was flown by a crew of seven, who would comprise a navigator, bomb aimer, two pilots, an upper turret gunner, one or two waist gunners and a tail-end gunner. The Model 22 was armed with one 7.7 mm machine-gun in the nose, a 20 mm cannon in the tail turret. The Model 24 version had an increased armament comprising one 7.7 mm machine-gun in the nose, a 20 mm cannon in the upper turret, two 20 mm cannon in the waist positions and one 20 mm cannon in the tail turret. These heavy-calibre weapons were operated by movement of the trigger butts, which activated electric motors, thereby moving the weapon in the vertical and horizontal planes as desired. The bomb bay was capable of housing either one 1,760 lb torpedo or a bomb load of 4,840 lb as required. No bomb-bay doors were fitted when the machine was bombed up, but a reflector plate was located at the end of the bay to deflect turbulence. A door fairing was attached when no offensive weapons were being carried. With a span of 81 feet 10½ inches and a length of 65 feet 7 inches, the machine had a loaded weight of 27,500 lb and a cruising speed of 250 mph, at an altitude of 19,680 feet.

All this rearmament and movement of military material had been sanctioned by the Ministry of Finance, and since the Army and Navy controlled political power, the Ministry was hardly in a position to refuse the necessary fiscal funds, but in more normal times the required military budget would never have been accepted.

At this time Admiral Nagano, Chief of the Naval General Staff, at the Emperor's direction dispatched a naval mission to Switzerland before the outbreak of hostilities, equipped with coding machines and cipher clerks. The purpose of the naval unit in Switzerland was to keep open the lines of communication with Tokyo in the unlikely event of opening negotiations with the US government and intelligence service. As it turned out, the American government was not interested in a negotiated peace and had become confident in eventual victory. Nevertheless this special naval unit remained on duty in Switzerland for the duration of hostilities.

In the course of 1941 aviation developments continued at a greater pace, with the prospect of war breaking out at anytime. The Aichi Aircraft Company produced the A16A1 Zuian, a three-seater machine originally built for reconnaissance duties but later developed into a dive-bomber to replace the E13A1 machine. It was powered by a single Mitsubishi Kinsei 54 radial engine producing 1,300 hp, giving a maximum speed of 279 mph at an altitude of 18,305 feet, for a range of 600 miles. Armed with two 20 mm cannon, two 7.7 mm machine-guns and one 13.2 mm heavy-calibre machine-gun, the plane carried an offensive weight of two 550 lb bombs. A total of 259 aircraft of this type were manufactured. The Kawanishi Company designed a replacement for the Model H6K2 flying-boat. This was a four-engine machine powered by Mitsubishi Kinsei 22 radial engines that gave the new 124 feet 8 inches wingspan H8K2 model a top speed of 283 mph, at an altitude of 15,485 feet, with a loaded weight of 54,013 lb. Carrying an offensive load of 4,410 lb of bombs, the Model H8K1 was armed with five 20 mm cannon and four 7.7 mm machine-guns, and could fly a total distance of 4,000 miles. Two years later the improved Model H8K2 appeared, including a special transport version known as the Model H8K2-L in a batch production of only thirty-six machines, but the series was manufactured to a total of 167 flying-boats. Meanwhile Mitsubishi had continued the development programme of the A6M carrier-based interceptor fighter otherwise known as the Zero. This new version was the A6M3, of which 343 examples were manufactured, but more important was the building of the Mitsubishi G7M Taizan bomber, the replacement for the G4M2 series of twin-engine attack-bombers, of which 376 had now been produced. Eventually it became apparent that the Mitsubishi Company was unable to design the aircraft to meet the Naval Air Service's specification due to a lack of suitable engines with a high enough power output. In the design stage the Navy and the company became involved in conflicts over the performance requirements, and eventually the project had to be cancelled.

At this time the second phase of the Sino-Japanese War was coming to an end, to later enter a new phase. The Army was now fighting a defensive war to consolidate previous territorial gains, and no further expansion operations would be undertaken until the Great Pacific War commenced. Thus, from 1938 to 1941, air operations were undertaken to destroy Chinese cities in an endeavour to convince the Chinese that further resistance was useless.

In December 1941, the Imperial Naval Intelligence Service had acquired sufficient information for the Naval General Staff to assume that on the commencement of hostilities the US Navy war plans envisaged the American Pacific Fleet steaming across the ocean to attack the Marshall Island bases to distract the Japanese from occupying the oil-rich Dutch East Indies and thereby land American reinforcements in the Philippine Islands. Unhappily the strategic move that opened the Great Pacific War – the crippling attack on the US base at Pearl Harbor – effectively stopped the movement of all capital ships to execute the American plan.

In December 1941 the G4M1 bomber groups were pitifully small in numbers as compared with the bombing groups of the more sophisticated Western air forces, and certainly as compared with the RAF Bomber Command squadrons or the bombing groups of the US Army Air Corps, though for the Imperial Naval Air Service the concept of long-range strategic bombing operations was fundamentally correct. As a result the

Kanoya Kokutai were re-equipped with the new Mitsubishi G4M1 bombers, while the Kisarazu Kokutai retained the older G3M2s, as did the Takao Kokutai, which also received a small number of the new machines. A total of 120 of the new G4M1 bombers were operated by the Kanoya Kokutai out of 170 then in service at the Taichu Naval Air Station, Formosa. This operational group maintained a detachment of the new machines at the Thudaumot Naval Air Station in French Indo-China, and a further detachment was based on the Tainan Naval Air Station in Formosa. However, to date only two A6M2 Zero fighters had been lost over China as a result of intensive ground fire.

It had been decided at a conference on 1 December 1941, on the orders of President Roosevelt, that Admiral Stark USMN would deploy three small US warships in the South China Sea, flying the Stars and Stripes. If these warships were attacked then America would have an excuse to enter the war, despite the Washington isolationist lobby, which would be politically isolated by popular American public opinion.

In Tokyo on 1 December, high-level discussions took place between the Japanese government, the Army and the Navy, which were convinced that diplomatic negotiations had failed to achieve the Japanese government's aims and that war with America, Britain and Holland would be unavoidable. The Navy's task was still to send a carrier task force to Hawaii, to send land-based air units into the Central Pacific and mobilise thirteen land-based air corps, ten aircraft carriers and nine seaplane tenders.

These high-level discussions took place at an Imperial conference presided over by Emperor Hirohito, which was but a mere formality. Admiral Nagano reaffirmed the Navy's confidence, Field Marshal Togo reviewed the results of eight months of negotiations with the Americans, and the Prime Minister, General Tojo, reported on the morale of the nation.

That night, all Japanese Navy ships changed callsigns as planned by Admiral Yamamoto.

On the following day, 2 December, radio transmissions were intercepted emanating from the cruisers, battleships and oil tankers of the carrier strike force ploughing through the breakers some 1,000 miles out in the Pacific. These radio signals were also received by General de Porten, Dutch Army Radio Intelligence Service in Bandung, Java. The texts were handed to the American liaison officer and received by General Marshal, Chief of the Joint Chiefs of Staff, Washington. Subsequent to delivering three intercepts in which Pearl Harbor was the possible target, Marshal was not interested! Nevertheless, Washington was now listening with greater intensity to all Japanese diplomatic radio transmissions.

On the same day the American liner SS *Lerning* sailed from Golden

Gate, San Francisco, on a weekly service to Honolulu, carrying more service personnel than civilians. When some three days from Hawaii, the radio officer picked up unidentified radio signals from a position north-west of Honolulu, and later signals were identified as originating from JCS Yokohama. On the second night the ship's radio officer continued tracking the transmissions and recording the signals with a view to handing over a report to US Naval Intelligence, Hawaii, immediately on the docking of the liner at her usual berth on 4 December. The report was directly forwarded to Washington.

Meanwhile the Royal Navy monitors and code breakers in Singapore had been kept busy when the coded radio transmissions from Admiral Nagano had transmitted Great Navy Order 12 to all Japanese ships and shore-based commanders, advising them that the attack was to be on 8 December 1941. This information, and other details contained within the order, were quickly communicated to the Admiralty, London, for restricted circulation within the British government.

By 2 December, direction-finding stations had been activated between Alaska and San Francisco to intercept the wireless transmissions originating in the Central Pacific, and the co-ordinates with any decoded messages were reported to Captain McCulloch's monitoring unit at 717 Market Street, San Francisco. On 2 December the co-ordinates indicated that the originators of the radio traffic were just east of the International Date Line, but still keeping well north across the Pacific on an easterly course. No. 717 Market Street was plotting the co-ordinates on a Great Circle Chart, which information was urgently required by Washington. At the same time Tokyo was advising Japanese embassy staff in Washington and London to destroy all ciphers and decoding machines.

On Thursday 4 December, the day the radio officer of SS *Lerning* filed his report of the unidentified radio transmissions originating in the Central Pacific, the Japanese carrier strike force was some three days from Pearl Harbor. Fortunately, as already noted, the American aircraft carriers had left Pearl Harbor to fly fighter plane squadrons to Wake Island and Guam to cover the flight of a formation of Boeing B-17 Flying Fortress bombers navigating across the Pacific Ocean to Clark Field, north of Manila in the Philippine Islands. At the same time, in Hawaii arrangements had been made by the Japanese for one of their consular officials to occupy a lonely beach-house and display lights to off-shore Japanese submarines, signalling the latest disposition of the US ships at anchor for onward transmission to Imperial Headquarters in Tokyo.

All monitoring units, whether the Royal Navy monitors in Hong Kong or Singapore, the Royal Australian Navy monitoring unit located at Park Orchards near Melbourne or the Washington radio intercept organisation,

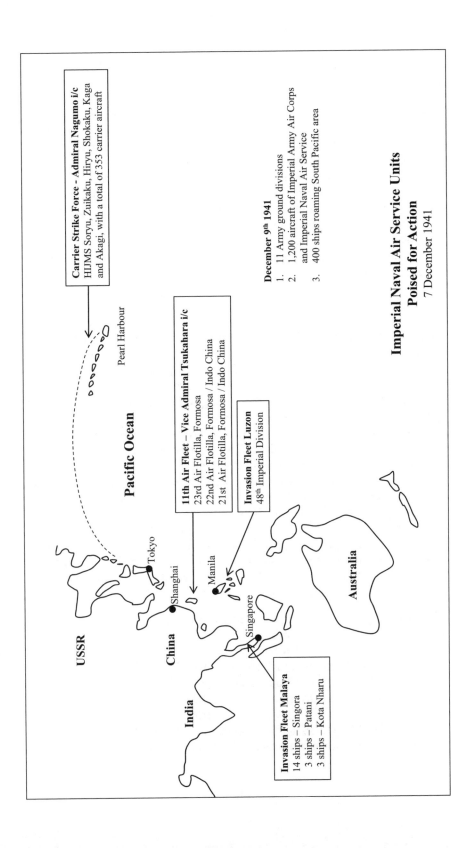

Carrier Strike Force - Admiral Nagumo i/c
HIJMS Soryu, Zuikaku, Hiryu, Shokaku, Kaga and Akagi, with a total of 353 carrier aircraft

Pearl Harbour

Pacific Ocean

11th Air Fleet – Vice Admiral Tsukahara i/c
23rd Air Flotilla, Formosa
22nd Air Flotilla, Formosa / Indo China
21st Air Flotilla, Formosa / Indo China

Invasion Fleet Luzon
48th Imperial Division

Tokyo

Shanghai

Manila

Singapore

USSR

China

India

Australia

Invasion Fleet Malaya
14 ships – Singora
3 ships – Patani
3 ships – Kota Nharu

December 9th 1941

1. 11 Army ground divisions
2. 1,200 aircraft of Imperial Army Air Corps and Imperial Naval Air Service
3. 400 ships roaming South Pacific area

**Imperial Naval Air Service Units
Poised for Action**
7 December 1941

were all alerted to listen out for the Japanese news broadcast that would contain the coded message indicating which country Japan would break with and attack. 'North wind rain' – relations with the USA in jeopardy! Not, as had been presupposed, relations with Britain in jeopardy, much to the surprise of the monitors.

Four days later, on 6 December, a Consolidated Catalina PBY flying-boat of RAF Coastal Command discovered and sighted a large Japanese convoy – the Southern task force off the coast of Indo-China. The aircraft proceeded with the patient and cunning task of shadowing this group of ships as it steamed towards the Malay Archipelago.

On the same day a US Naval Intelligence operative, a Mrs Dorothy Edgers in Washington, had deciphered a Japanese radio transmission giving details of light signals to appear from a building in Honolulu that would give information concerning the number of US warships in Pearl Harbor. Regrettably Commander Kramer USN, the officer in charge of decoding the Japanese Purple-coded message, did not consider the information of sufficient significance to warrant a more detailed study.

On Saturday afternoon, 6 December, the Japanese carrier strike fleet was plotted from radio intercepts by 717 Market Street as being 500 miles north of Pearl Harbor. During Saturday night the Japanese consular spy in Hawaii sent his last messages by means of lights and bonfires to waiting off-shore Japanese submarines for final transmission to Tokyo, concerning the ultimate US Navy dispositions. These messages were intercepted and shown to President Roosevelt, who exclaimed, 'This means war!'

By Sunday morning, 7 December, 717 Market Street and the chain of direction-finding stations had plotted the Japanese force as 200 miles north of Hawaii. In that latitude and longitude the air crews were awoken at approximately 0330 hours, to commence preparations for the initial air attack on Pearl Harbor, which would cripple the US Pacific Fleet for a period of six months and cause the United States to be brought into the Second World War. The disposition of the warships and aircraft of the Imperial Navy were as follows:

1. The First Battle Fleet. Base – Hiroshima Bay
   C-in-C Admiral Yamamoto
   Seaplane tender – HIJMS *Chiyoda*
2. The Third Blockade and Transport Fleet. Base – Formosa
   C-in-C Vice-Admiral Takahashi
   Seaplane tender Division 12   HIJMS *Kamikawa Maru*
                                        HIJMS *Sanyo Maru*
3. The Third Blockade and Transport Fleet. Base – Truk
   C-in-C Vice-Admiral Inouye, 4th Mandate Fleet

Air Flotilla 24   HIJMS *Goshu Maru*
                                            HIJMS *Kamoi*

3rd Base Force   HIJMS *Sanuki Maru*

Under direct command seaplane tender HIJMS *Kiyokawa Maru*

4. The Fifth Northern Fleet. Base – Maizuru or Ohminato

    C-in-C Vice-Admiral Hosogaya

    Seaplane tender   HIJMS *Kamikawa Maru*

5. Aircraft Carrier Strike Force. Base – Kure

    C-in-C Vice-Admiral Nagumo

    Squadron 1 Striking force   HIJMS *Akagi, Kaga*

    2 Striking force   HIJMS *Hiryu, Soryu*

    3 Training force   HIJMS *Hosho, Ryujo*

    4 Southern force   HIJMS *Taiyo, Zuiho*

    5 Striking force   HIJMS *Shokaku, Zuikaku*

6. Combined Air Fleet. Base NAS Kanoya, Japan

7. The Eleventh Air Fleet. Base – Formosa

    C-in-C Vice-Admiral Tsukahara

    Aircraft transport   HIJMS *Kamogawa Maru*
                                             HIJMS *Kayo Maru*
                                             HIJMS *Lyon Maru*

    21st Air Flotilla base – Indo-China A/C transports
                                             HIJMS *Katsuragi Maru*

    22nd Air Flotilla base - Indo China A/C transports
                                             HIJMS *Fujikawa Maru*

    23rd Air Flotilla base - Formosa A/C transports
                                             HIJMS *Komaki Maru*

8. The Ninth Base Force

    Southern Expeditionary Fleet. Base – Indo China

    Seaplane tender   HIJMS *Sagara Maru*

9. Southern Force – Philippines

    C-in-C Vice-Admiral Nobutake Kondo. Comprising:

    2 Battleships

    1 Light aircraft carrier – HIJMS *Ryujo*

    12 Heavy cruisers

    9 Light cruisers

    53 Destroyers

    16 Submarines

    Seaplane tenders and servicing ships

The target areas in the Philippine Islands were within range of the naval air stations in Formosa and French Indo-China. The 22nd Air Flotilla comprised 132 twin-engine attack-bombers, twenty-four fighter planes and six reconnaissance aircraft ready and poised for the initial assault.

At this time over 400 Zero fighters were in operational squadrons, and of these 108 took part in the Pearl Harbor attack.

Now Operation Z was gaining momentum hour by hour, as the Kido Butai, or aircraft carrier fleet, raced through the dark night to the aircraft launching-point some 200 miles north of Pearl Harbor. Hitherto the voyage from Japan had taken place during radio silence, in severe weather conditions, poor visibility, rain and excessively dark nights. Not a single ship had been sighted during the entire voyage, and the fleet had been surrounded by an air of isolation and doom. The morale of ship crews and aircraft personnel had become somewhat depressed, as the fleet ploughed its way through the white-capped foaming masses of watery mountains and valleys. At the early hour of 0330 on Sunday 7 December 1941, Hawaiian Time, the pilots and air crews were roused from their bunks, dressed, wrote last letters home, left fingernail clippings and locks of hair for the families and donned clean loincloths, as well as 'thousand-stitched' belts. Each stitch in a belt represented a prayer for good luck and a good flight, and had been made by the mothers, wives or sisters of the men taking part.

Earlier, the Southern Assault Force had approached the darkened Malayan coastline, with a main force of fourteen ships heading for Singora, three ships approaching Pattani and three more transports arriving off Kota Bharu, dropping anchor at about midnight under the star-studded heavens. A vivid electrical storm lit up the green-palmed golden beaches, upon which the white-capped breakers leisurely collapsed into masses of glistening foam. At about 0115 the synchronised guns of the escorting warships had roared into life, as brilliant blood-red flashes stabbed into the night, raising tower-like masses of explosions among the coastal fortifications. Landing-craft spilling over with heavily armed troops raced for the beaches. The Great Pacific War had commenced. The hour was 0545 Hawaiian Time, though the Pearl Harbor attack was not scheduled until 0800, as agreed previously by Commander Genda and Commander Miyo of the Naval General Staff during the planning stage.

The Kido Butai, or carrier fleet, slipped past the launching-point as the faint light of day glimmered and shot across the sky from horizon to horizon, 200 miles north of the American naval base at Pearl Harbor. The crews, having been finally briefed, were seated in their planes, engines roaring, with the carriers rolling some 12–15 degrees. The command 'Carriers turn into the wind' was given as the big ships simultaneously and laboriously turned, pitching in the heavy swell. On the decks of the six carriers were the first aerial assault wave comprising forty-three fighters, forty-nine high-level bombers, fifty-one dive-bombers and forty torpedo-bombers. From the yard-arm of the flagship the red triangular pennant with the white circle, which indicated 'Get ready for take-off', was run up to half mast and now fluttered furiously in the high breeze sweeping down the decks of the carriers. Within a short while the pennant was quickly hauled to the mast top, which signalled 'All planes take off!' With a thunderous roar the first wave of 183 aircraft rose from the carrier decks into the fresh sea breeze of the Sunday morning under the air command of Commander Fuchida HIJMN, and headed for Pearl Harbor.

Prior to the attacking planes taking off, Pearl Harbor had been reconnoitred at 0735 by a catapult reconnaissance plane from the heavy cruiser HIJMS *Chikuma*, which had radioed confirmation that the enemy fleet was concentrated in the roads at Pearl Harbor. Likewise a catapult reconnaissance plane from the heavy cruiser HIJMS *Tone* had flown over Lahaina Roads to report the absence of any American warships. Both these cruisers had operated at a distance of 150 miles north of Hawaii as an advanced guard for the carrier strike force. Both the reconnaissance machines had arrived over the target half an hour prior to the attack, and transmitted radio reports on wind speed and direction and cloud cover over Hawaii. The reconnaissance report observed the absence of the US carriers from Pearl Harbor, which fortunately for the Americans had sailed previously to locations of safety.

At 0749, Hawaiian Time, Commander Fuchida HIJMN with the first wave arrived over the target area, gave the command to attack and signalled Admiral Nagumo in the flagship with the coded word 'To, To, To', which were the first syllables of the Japanese Totsugeki, meaning 'Charge!' This word was used to indicate 'The first wave attacking.' Thus, on 7 December 1941, America was brought into the Great Pacific War.

On 8 December 1941, the United States declared war on Japan. The British Prime Minister, when informed in London that the Japanese had attacked the American fleet at Pearl Harbor, is reputed to have said 'that he slept the sleep of the thankful – and saved!'

CHAPTER 15

The Glorious Seventh

Oahu Island! At 0735 Commander Fuchida in the lead plane sighted land through broken cloud flying at 9,000 feet, with the fighter escort above at 14,000 feet, and observed the glistening white surf and the green carpet of tropical vegetation. The air armada was approaching its target from the north, so the lead planes swept along the western side of the island so that the torpedo-bombers could commence their runs from the southerly ocean side. Then at 0749 Commander Fuchida fired a black signal rocket to instruct the dive-bombers to climb to 12,000 feet as the torpedo-bombers spiralled down to wave-top height, ready for the command to attack. Meanwhile the horizontal bombers descended to 3,500 feet as Commander Fuchida fired a second rocket commanding the fighter escort to fly with the bombers to strafe the American positions. Eventually, at 0749 on the R/T, Fuchida gave the command 'Attack, Attack, Attack!' All 182 planes of the first wave converged on their pre-selected targets and commenced their bombing runs. In a crescendo of exploding bombs and the roar of high-speed motors, destruction rained on Army Air Corps runways at:

Wheeler and Hickham Fields
Marine Air Corps runways at Ewa
Navy Air Corps runways at Ford Island
Catalina basin at Kaneohe

At these airfields the American planes had lined up as an anti-sabotage precaution. In the harbour and roads at Pearl were 90 vessels:

8 Old battleships
1 Old battleship used for target practice
2 Heavy cruisers
6 Light cruisers
30 Destroyers
5 Submarines
9 Minelayers
10 Minesweepers
2 Repair ships

3 Destroyer tenders
2 Cargo ships
1 Submarine
1 Hospital ship
6 Seaplane tenders
2 Oilers
2 Gunboats

At first, little or no American anti-aircraft fire or defending fighters responded to the initial Nipponese attack.

The second wave of Japanese carrier planes attacked Pearl Harbor at 0845, including seventy-eight bombers, fifty-four torpedo-bombers and thirty-five fighter planes. Commander Fuchida followed the last of the attackers back to the aircraft carriers at 1000, wishing to lead a third wave of planes to assault the American naval base. However, Admiral Nagumo refused the request on the point that further aerial attacks might reveal the position of the carriers at sea and precipitate an American counter-attack. Having suppressed any further argument as to the advantages of a third attack on Pearl Harbor, the Admiral ordered a course to be set for the home base in Japan, and the ships of the fleet turned back into the rising fog and disappeared as grey shapes as silently as they had come. Behind them the Imperial flyers left five battleships, three destroyers, one minelayer and a possible aircraft carrier sunk, 200 US aircraft destroyed and 2,008 sailors, 109 Marines, 218 soldiers and 68 civilians killed. The shock in the USA was traumatic!

In this initial attack the Imperial Naval Air Service launched 353 aircraft, comprising Aichi D3A1 dive-bombers, Nakajima B5N2 torpedo-bombers and Mitsubishi A6M2 fighter planes. These aircraft for a loss of twenty-nine planes and fifty-five men were launched from the aircraft carriers HIJMS *Soryu*, *Zuikaku*, *Hiryu*, *Shokaku*, *Kaga* and *Akagi*.

On the morning of 7 December 1941 the naval air stations in Formosa of the 11th Air Fleet were fog bound and all air traffic was grounded. The naval air commanders feared an attack by American Boeing B-17 Flying Fortress bombers, flying from Clark Field and other US airfields in the Philippines some 600 miles away.

Unfortunately, although General MacArthur had been informed by the US War Department at 1500 the previous day that an attack by the Japanese was highly possible, he had made a tactical mistake in refusing permission for the Boeing B-17 Flying Fortress bomber group to make a dawn raid on Imperial airbases in Formosa, which attack the Japanese had feared.

In the meantime only Japanese Army Air Corps bombers took off to bomb targets north of Manila. However, about mid-morning the fog commenced to lift and the commander of the 11th Air Fleet ordered the start of offensive operations. Altogether 109 naval aircraft took off to bomb and strafe targets near Luzon, the main group of planes flew to the US airbase at Clark Field, while the lesser group headed for Manila. At 1225 a formation of twenty-seven Mitsubishi G4M1 attack-bombers roared over Tarlac, twenty miles north of Clark Field, with a second formation, of G4M1 bombers, flying close behind escorted by thirty-five Mitsubishi A6M2 fighters high above, all in V formation.

During the course of the morning the American fighter pilots had been airborne on patrol, realising that an air attack would be imminent at any time, but due to the bad weather little air activity had been noticed. So returning to their airfields the US fighter pilots landed to refuel and to stand down to take lunch in the officers' club when at 1235 the aerial bombardment was unleashed. Bombs crashed, exploding with a roar among the airbase facilities, fuel dumps and ammunition magazines burst into searing flame, parked aircraft disintegrated into jagged wreckage amid the tangled mass of fire hoses and the forlorn efforts of the medical rescue services. The Japanese A6M2 fighters came in low, cannon and machine-guns blazing, strafing the US B-17 bombers and P40B fighters. Then a further forty-four Zero fighters, having strafed a nearby US airbase, suddenly appeared to strafe Clark Field. In the general confusion three US P40B fighters took off only to be jumped by the Zeros and shot out of the sky. The Japanese Naval Air Service formations, their mission accomplished, turned for their home bases in Formosa. Behind, all the P40B fighters and the entire medium-bomber force and observation planes were left furiously blazing and forever broken. All but three of the Boeing B-17 Flying Fortress bombers were destroyed, and the fuel tanks, having been hit by tracer bullets, blew up in large balls of red-hot fire. MacArthur's US Far Eastern Air Force had been obliterated. The American pilots who had survived the aerial onslaught were very surprised at the speed, manoeuvrability and rate of climb of the Mitsubishi A6M1 fighter. During the autumn of 1941, General Chennault had dispatched full details of this Japanese naval aircraft to the US War Department in Washington, but the information had been filed and never issued to the US combat air units. In this document General Chennault had outlined tactics for P40B pilots to combat the highly manoeuvrable A6M2s, but foolishly these ideas had been ignored, with disastrous results.

The next day the air war continued unabated when thirty-four Mitsubishi Type G3M2 assault bombers of the Chitose Naval Air Corps based on Wotto Island in the Marshall Islands attacked the American base on Wake Island. Flying over the blue expanse of the Pacific, they hit the airbase and harbour installations in a devastating aerial bombardment in preparation for a landing by troops of the Japanese Army.

Meanwhile Japanese troops continued to land from transports on the north-east coast of Malaya, reinforcing those units which had already gone ashore.

Happy hunting

Twilight gently slipped over the island and city of Singapore. Despite the war going on in northern Malaya, many tried to carry on in an endeavour to forget the seriousness of the situation. The business quarter lit up in the last hours of commercial intercourse, the theatres and restaurants came to life once again, all twinkling and bustling with pleasure-bent humanity. But two large capital ships and four escorts quietly slipped their cables and departed from the harbour and roads of Singapore. It was 1735, and the departure had not gone unnoticed. Agents of the Imperial Naval Intelligence Service in the harbour area dutifully reported the event to the intelligence centre in Tokyo by long-range radio. Imperial Headquarters, Tokyo, had the the Emperor's authority to order a full-scale effort to hunt the ships and destroy them. On departing Singapore, Admiral Phillips RN had requested RAF fighter escorts, but the RAF airfields in north Malaya were already in the hands of the invading Japanese Army, so that HMS *Prince of Wales* and *Repulse* would be at sea without a protective air umbrella.

On departing Singapore harbour, Admiral Phillips had received an official radio message from Changi Radio Station, which advised him that on Wednesday fighter aircraft cover for the warships would not be available. A more prudent naval commander might have altered his operational plan in the light of such unsatisfactory information.

Airfields in south Indo-China were alerted by Imperial Headquarters, Tokyo, and over a hundred pilots trained at Kasumigaura, near Tokyo, were stood to for offensive operations. These men were the finest bombing crews of the Imperial Naval Air Service, they were the flower of the service. Besides aircraft of the Naval Air Service, Imperial Headquarters had deployed submarines in the lower half of the Gulf of Siam to detect the British warships and shadow them.

Upon leaving the harbour at Singapore, the British squadron, under the command of Admiral Phillips, known as Force Z, set course eastwards for the coast of Borneo, and by dawn was anticipating cruising off the Anambas Islands. It was known that Japanese Naval Air Service reconnaissance planes were out from Saigon looking for the ships. However, Admiral Phillips calculated that he would be beyond the effective range of the Japanese bombers while off the coast of Borneo, and the ships turned north, expecting the enemy search area to be off the Malayan coast. Fortuitously the battle squadron was engulfed by an overcast sky, with frequent rain squalls shrouding the ships in grey mist and heavy seas, making radar navigation a virtual necessity as the warships sailed on into the following day.

At 1410 on 9 December a Japanese submarine on patrol made a sighting of the battle fleet, which was reported as a hundred miles due south of the tip of Indo-China. Through a navigation error, this information was inaccurate, and the battleship HMS *Prince of Wales* and her accompanying ships were in fact 140 miles south-east of this point.

The attack-bombers of the Imperial Naval Air Service's 21st Air Flotilla were bombing-up in preparation for an aerial attack on the city and docks of Singapore. Upon receipt of the reconnaissance information, the commanding officer of the Thudaumot Naval Air Station de-bombed his aircraft and rearmed with torpedoes. Eventually four twin-engine bombers took off in an endeavour to locate the two British capital ships. At 1700 the four reconnaissance bombers spotted HMS *Prince of Wales* on the horizon and made a situation report to air flotilla headquarters. After some forty-five minutes, fifteen bombers armed with torpedoes, accompanied by eighteen bombers carrying heavy-calibre armour-piercing bombs, took off to attack the British ships. The aerial attack was inconclusive, for by 1758 the gun crews aboard the British ships had been stood down. Now the destroyer HMS *Tenedos* was detached from the fleet to act as a decoy, and during the night was attacked by a Japanese submarine as well as by enemy aircraft.

At 2015 the Nipponese submarine shadowing Force Z lost contact with the British battleships, and a small force of Mitsubishi G3M2 torpedo-bombers continued the hunt. Once again contact was lost with the illusive British ships, and the G3M2 planes returned to Thudaumot Naval Air Station at 2359. In the meantime Force Z had been signalled that Japanese Army formations had landed at Kuantan, some 200 miles north of Singapore. At the same time Admiral Phillips on *Prince of Wales* ordered a high-speed feint, steaming for one hour towards Thailand, followed by a course doubling back south-eastwards. But the Japanese reconnaissance bombers had looked along the projected course of the British battle squadron in vain. Some fifty miles to the east the destroyer *Tenedos* broke radio silence with a transmission to decoy the Japanese naval forces further eastwards. During the night, at 0100, the battleships changed course and made for the Malayan coast. Admiral Phillips was unable to break radio silence, otherwise his position would have been revealed, and therefore no fighter aircraft cover could be requested from the Royal Air Force. In the pre-dawn darkness a shadowing Japanese submarine fired five torpedoes, but missed the ships, which disappeared into the early-morning light. Once again the Imperial naval forces had lost contact.

The hunters were now resting and refuelling at a base far to the west of the British ships. Admiral Phillips decided to make a run for Singapore, to arrive by noon the following day. Meanwhile a course was set for Kota Bharu at maximum speed, to arrive by dawn, but the Japanese were fully prepared. Turning south, the British battle fleet made for Kuantan half way up the Malay Peninsula, and on arriving at 0800 on 10 December, the destroyer HMS *Express* surveyed the harbour with care, as intelligence of the Japanese dispositions was unreliable, but the Japanese transports were apparently nowhere to be found.

Meanwhile the Imperial 21st Air Flotilla continued to search for the two battleships as eleven reconnaissance bombers took off from Thudaumot Naval Air Station to search beyond the 400 nm safe range. During the morning the decoy destroyer HMS *Tenedos* was attacked by nine G3M2 attack-bombers, but no serious damage was sustained. Meanwhile the British battle fleet moved out to sea to gain space for manoeuvrability, realising the outcome of the sighting. Unfortunately, at 1015 on 10 December, a force of Mitsubishi G3M2 bombers returning from a bombing raid on Singapore by chance sighted HMS *Prince of Wales* and HMS *Repulse*, and reported the position to the air flotilla headquarters at Thudaumot, in French Indo-China.

Immediately a mixed force of Mitsubishi G3M2 and G4M1 attack-bombers took off armed with bombs and torpedoes. The battle commenced at approximately 1100 with a bombing raid by a nine-group formation. All the bombs were near-misses but sufficiently near to indicate that the ships were not under attack by Italian bombers, as had been the case when HMS *Prince of Wales* was on convoy guard duty in the Mediterranean, while bound for Alexandria *en route* for the Far East. This was something more sinister; these were the graduates of the Misty Lagoon naval air station, the flower of the naval air arm demonstrating their efficiency and prowess. During the air battle the Kanoya Air Corps dispatched twenty-six Mitsubishi G4M1 attack-bombers, while the Bihoro and Genzan Air Corps flew fifty-nine Mitsubishi G3M2 bombers from the naval air station at Thudaumot. Bombing runs were made at an altitude of 10,000 feet, dropping 250 and 500 kg high-explosive bombs, with the torpedo-equipped aircraft dropping 18-inch Type 91 torpedoes. These Japanese naval aircraft formations had won the Yamamoto combined fleet prize for high-level bombing and precision bombing during 1940.

Initially HMS *Repulse* was attacked from an altitude of 10,000 or 12,000 feet, and sustained one direct hit on the battleship's catapult and innumerable near-misses, while HMS *Prince of Wales* was attacked by five torpedo-bombers all at once from five different directions, when three torpedoes missed, one struck the rudder and the second blew in the communication room, and the ship was caused to reduce speed from 30 knots down to 15.

At 1200 Lieutenant Iki Haruki HIJMN, flight commander, with three fully loaded Mitsubishi G3M2 attack-bombers, circled around awaiting the arrival of a further eighteen naval bombers, which had returned to Thudaumot for reloading. The lieutenant surveyed the tactical problem; twenty-four out of sixty-one Japanese planes had spent their torpedoes, yet neither ship had sunk, for they were sailing close in, with effective heavy defensive fire power.

Shortly afterwards, at about 1220, Lieutenant Iki led fourteen torpedo-bombers in a split formation simultaneously from two different directions, which resulted in innumerable direct hits on HMS *Repulse*. The battlecruiser was now mortally wounded and completely crippled, and at 1225 Captain Tennant, commander of *Repulse*, reluctantly gave the order to abandon ship. Those not trapped within the hull below scrambled over the side into the dark grey waters of the South China Sea amid the shot and smoke of the anti-aircraft armament and the oppressive roar of the Mitsubishi bombers overhead. At 1233 the battlecruiser gently rolled over as Captain Tennant abandoned the bridge and was dragged under. Rapidly rising to the surface, he discovered the sea was deserted but for the mass of swimming survivors and floating wreckage. The great ship had gone, bombed and torpedoed. Over 513 ratings were drowned, though 796 survivors were picked up.

Those aboard HMS *Prince of Wales* witnessed the sinking with horror, but realised it was only a matter of time before they too would receive the *coup de grâce*. Further Mitsubishi G3M2 bombers were regrouping to continue the battle. Once again the same tactics were used: flying over the wave tops very low in line abreast at the command of the flight leader they dropped all torpedoes simultaneously. The Japanese machines were so low that the British anti-aircraft gun crews could easily distinguish the crewmen in their positions as the planes flew over the ships. At 1315 the end came; HMS *Prince of Wales* was sinking, and the order to abandon ship was given. Once more those sailors trapped within the hull patiently awaited their watery doom, and the great ship heeled over as those lucky enough scrambled over the side. Some, in panic overlooking the abandon-ship drill, went to the wrong side as the hull turned turtle, and were cut down by the giant propellers as they still revolved. The shattered battleship slid below the waves, taking 327 souls with her. By tradition it is said within the Royal Navy that no admiral who has lost a sea battle ever returns to Portsmouth Harbour. Admiral Phillips RN went down with his flagship.

As the Mitsubishi G3M2 attack-bombers regrouped to fly back to Thudaumot Naval Air Station, it is alleged that it was noticed from one of the escorting destroyers that one particular attack-bomber appeared to detach itself from the air flotilla. It seemed as though this machine had decelerated, flying at a lower altitude towards the escorting destroyer, and very carefully and slowly flew overhead, almost swimming gracefully above the warship. Suddenly those on deck were amazed to see a message streamer fall from the attack-bomber in a multi-coloured plume, and descend onto the destroyer's deck. The message streamer came down in a multi-coloured heap of tail plumes, while the attack-bomber almost leisurely rose in altitude to form up with the now disappearing air flotilla. Soon all was silence, the planes mere dots on the horizon. The order was given to recover the message capsule, which upon being opened was found to contain a message written on an Imperial Navy message-pad sheet that read, 'We have completed our task, you may carry on!' This was the high-water mark of Japanese *noblesse oblige* – it was the pilots of the Misty Lagoon, the expert flyers specially trained under the direction of Admiral Yamamoto. By 1330 the Japanese naval bombers had signalled the news of the sinking of the two British warships, which was immediately received by the Emperor at Imperial Headquarters, Tokyo.

Treachery?
The Imperial Naval Intelligence Service operated a number of highly trained foreign networks in its endeavour to accumulate intelligence news and assessment. One network operating in the United States of America was led by a Señor Angel Alcazar de Velasco, a 69-year-old Spanish spy chief, who returned from the espionage world in Madrid during 1973, it was disclosed by US government documents released during September 1978. This Spanish diplomat operated a network of twenty-one agents of various nationalities without hindrance until detected by the Canadian counter-intelligence organisation, when based at the Spanish Consulate in Vancouver.

Apparently the movements of HMS *Prince of Wales* and HMS *Repulse*, as well as other Royal Navy units, had been very carefully monitored on their voyage from England, through the Mediterranean Sea, into the Indian Ocean and finally to Singapore.

Some of this information was obtained by Velasco via his intelligence network from semi-official leaks originating in naval circles in Washington DC. It is suggested that the movements of British naval units in the Far East were semi-officially leaked in a positive US government policy to reduce British sea supremacy and hasten the final eclipse of the British

213

Empire. This information, it is alleged, was handed over by an American naval officer to his girlfriend, an agent who worked for the network, and was deliberately planted to obtain information of Allied ship movements. This information had been passed to the intelligence centre in Tokyo for evaluation, resulting in the subsequent sinking of the two British capital ships with a loss of over 800 lives. Other information acquired by this Spanish network of the Imperial Naval Intelligence Service included information on convoys leaving the west-coast ports of America, resulting in over 60% of US convoys being attacked and a very high percentage of shipping tonnage being sunk by Japanese submarines. Further information collected concerned the development of the American atomic bomb, but Tokyo was sceptical about this piece of data, which could not be corroborated at the time. But the fact remains that HMS *Prince of Wales* and HMS *Repulse* were sunk allegedly as much by American treachery as by Japanese bombs and torpedoes.

Air war over the Philippine Islands
At dawn on the morning of 8 December 1941, thirteen Type 97 bombers and nine Type 96 carrier fighters attacked targets in Davao, Mindanao, from the carrier HIJMS *Ryujo*, destroying two Consolidated PBY flying-boats on the water for the loss to American anti-aircraft fire of one carrier bomber. During the afternoon eighty-nine Mitsubishi A6M2 carrier fighters and a mixed force of 108 Mitsubishi G3M2s and G4M1s based in Formosa with the Navy's 1st Takao Air Corps and 3rd Tainan Air Corps attacked US airfields located at Iba and Clark Field on the island of Luzon.

At 1000 on 8 December, Mitsubishi A6M2 fighters and Mitsubishi G4M1 bombers took off from Tainan on the island of Formosa, an island which was captured by the Japanese together with the peninsula of Korea in 1896 under the Emperor Meiji.

Defensive patterns had been flown by Boeing B-17 pilots during the morning. They had been ordered to land and stand down for refuelling, while the pilots and crews were sent to lunch at midday owing to a lack of air activity. At approximately 1230 the first wave of the A6M2 carrier fighters arrived, circling at 22,000 feet over Clark Field, which was located some fifty miles north of Manila. Unfortunately at 1240 a communication breakdown took place and the radar surveillance unit failed to inform the commanding officer at Clark Field of the approaching naval bombers. Below, parked on the airfield, were some sixty aircraft, including twenty-one out of thirty-six Boeing B-17 bombers lined up ready for take-off. The G4M2 naval attack-bombers arrived at 1245, approaching Clark Field from the north, and immediately commenced their bombing runs over the pre-selected airfield installations and aircraft. A devastating rain of explosives descended upon Clark Field, which became shrouded in bomb bursts as refuelled aircraft burst into flames with a roar and buildings crumbled, trapping personnel. Fire and rescue units became overwhelmed with the ferocity of the bombardment as the bombers steadily droned overhead. The same events took place at Iba airfield and Nicholas Field, where a total of eighteen Boeing B-17s, fifty-three P40B fighters and some thirty other US machines were destroyed on the ground, together with wrecked airfield facilities. This action resulted in fifty-four US fighters surviving in the whole of Luzon, while fourteen Boeing B-17 bombers that had been on detachment to Del Monte Field in Mindanao in the south Philippines were unable to escape, but subsequently would fly in support of the defence of the Netherlands East Indies. The Japanese lost seven Zero fighter planes, and forty-six days later Borneo fell to the Imperial forces. On the same day the first heavy air raid was made on the city of Singapore by Mitsubishi G3M2 attack-bombers of the Bihoro Air Corps, from Indo-China.

By the morning of 9 December, the Japanese Army and Navy were freely roaming the entire South Pacific area. A total of eleven ground divisions of the Army, supported by 1,200 aircraft of the Imperial Army Air Corps and Imperial Naval Air Service with 400 ships, had been activated.

On the next day, at 0100, 400 Imperial marines landed from nine transports onto the beaches of the US base on Guam in the Mariana Islands. This base had been used by the US Pan American Clipper Service as a stopover point on the Hawaiian Manila air route. The battle lasted 3½ hours, with the island defended by 430 US Marines and Navy personnel, supported by 180 Guamanian Insular Guards, who were eventually forced to surrender. At the same time, 4,000 troops from the 2nd Taiwan Regiment of the 48th Division waded ashore near Aparri on the north coast of Luzon in the immediate vicinity of a US airstrip. After a brief engagement, over 200 Filipinos commanded by a US lieutenant were forced to retreat into the hills. A second landing of Imperial troops took place on the beach near Vigan on the north-west tip of Luzon by the 48th Division, whose target was a nearby airstrip. But mastery of the air had not been fully achieved by the Naval Air Service, as five Boeing B-17 bombers, having escaped the holocaust of Clarke, Iba and Nicholas Field, managed to bomb and leave three Japanese transports burning furiously, having beaten off attacks by Zero fighters.

In the meantime Imperial Supreme Headquarters, Tokyo, announced the annihilation of the US Pacific Fleet, which allowed the Japanese Navy freedom of the seas for more than six months. However, the Imperial Navy overlooked the fact that the United States had three aircraft carriers, twenty-one cruisers, fifty-eight destroyers and twenty-one submarines still operational. All the battleships sunk at Pearl Harbor were old and due for eventual replacement. The Imperial Navy achieved tactical success but failed to seize the initiative and exploit the strategic advantage. On 23 December, Admiral Nagumo's carrier strike force arrived in Hiroshima Bay, minus the six vessels that had been earmarked for the successful invasion of the US base on Wake Island. These six ships, including the aircraft carriers HIJMS *Hiryu* and *Soryu*, eventually returned to Hiroshima Bay.

During December, and in readiness for the outbreak of the Great Pacific War, a ship-borne reconnaissance aircraft went into production. This was the Aichi E13A1 float-plane, which had been used to reconnoitre Pearl Harbor and Lahaina Roads before the initial air attack and would subsequently be operated for reconnaissance operations at the Battle of the Coral Sea, as well as during the Midway and Solomon Islands operations. It was powered by a single 1,080 hp Mitsubishi Kinsei 43 radial motor giving the machine a top speed of 234 mph at an altitude of 7,150 feet, with a wingspan of 47 feet 6¾ inches. The parent company subsequently produced 132 machines, with the associated Kyushu Company building a further 1,100 aircraft of this type. The submarine tender HIJMS *Taigei*, mother ship to three submarines, returned to Japan during the month of December. This vessel was withdrawn from operations on 18 December and was converted to an aircraft carrier of 13,360 tons. The conversion was not completed until November 1942, when the ship emerged with overall dimensions of 707½ × 64¼ × 21¾ ft. It was powered by two geared-shaft turbines, developing 52,000 shp, with a maximum speed of 26½ knots. Armament comprised eight 5-inch DP guns and thirty-eight 25 mm anti-aircraft cannon, with a complement of thirty-one aircraft. During the course of this ship's active life structural defects began to appear, the cause of which was attributed to the electrically welded hull.

In January 1942, HIJMS *Shoho* (Propitious Phoenix) joined the fleet. This aircraft carrier was the former 12,000-ton *Tsurugisaki*, a fleet oiler. Conversion commenced in January 1941, and when completed the ship became a light carrier of 11,262 tons. The new ship had an overall length of 662 feet, with no island structure, and was powered by steam turbines replacing diesel engines, which gave the ship a maximum speed of 26.2 knots. The hangar accommodated a total of thirty-one aircraft.

On 11 January 1942, a force of twenty-seven Mitsubishi G3M2 bombers, which had been converted to troop-carrying transport planes, flew from Davao to Manado in the Celebes, dropping 324 parachute troops of the 1st Special Imperial Marine Corps – the Tokubetsu Rikusentai – into Manado airfield and capturing the nearby city within two days for the loss of thirty-seven marines.

The carrier strike force was on the move once again, with HIJMS *Hiryu* and *Soryu* leaving the Inland Sea on 12 January, while the remaining four carriers of the force were revictualling at Truk in the Caroline Islands from 14 to 17 January. The two carriers, which had left the Inland Sea on the 12th, arrived on the 17th at Palau to take on supplies, and promptly steamed southwards on 21 January.

Air strikes in the South Sea Islands

Carrier-borne dive-bombers Aichi D3A1s, horizontal and torpedo-bombers Nakajima B5N2s and carrier fighters Mitsubishi A6M2s attacked the British base at Rabaul on 20, 21 and 22 January, meeting little or no aerial opposition, flying from the aircraft carriers HIJMS *Akagi* and *Kaga*.

At the same time Naval Air Service carrier-borne bombers flying from HIJMS *Shokaku* and *Zuikaku* hit Allied targets at Lae, Salamaua, on the north-east coast of New Guinea, and once again Allied air defence was non-existent.

On 23 January Imperial Naval Marine Corps units waded ashore from landing-craft on the beaches at Kavieng in New Ireland, covered by Mitsubishi A6M2 carrier fighters, and also at Rabaul in New Britain.

Meanwhile HIJMS *Shokaku* and *Zuikaku*, having completed air-covering the troop landings, returned to the Imperial base at Truk. However, both HIJMS *Hiryu* and HIJMS *Soryu* cruised westwards and attacked targets at the Dutch Port of Ambon. The Naval Air Service set up an air station at Rabaul, which became the headquarters for the Yokohama Air Corps and the Chitose Air Corps for operations over the southern seas. The foundation had been laid for the erection of the Rabaul Air Command.

On 1 February, anti-aircraft fire and a force of Mitsubishi A5M2 fighters shot down four US Douglas Dauntless dive-bombers attacking the naval air station at Roi, north of Kwajalein. The Americans massed thirty-seven Dauntless dive-bombers and nine Douglas Devastators from the carriers USS *Enterprise* and *Lexington*. Though the naval air station was damaged, mist over the airfield prevented further American air attacks. In the nearby lagoon at Kwajalein, a 6,500-ton Japanese freighter was sunk by nine Devastator torpedo-bombers and ten Dauntless dive-bombers, as well as a submarine chaser, and a cruiser was damaged. At Maloelap Naval Air Station the Americans discovered some thirty or forty Mitsubishi G4M1 attack-bombers, but were prevented from successfully attacking by a standing patrol of Mitsubishi A5M2 fighter planes. During the ensuing mêlée the US Grumman Wildcat fighters were plagued with jammed machine-guns and lacked the use of incendiary bullets. So the air station was bombarded by US cruisers and destroyers, though Dauntless dive-bombers destroyed a limited number of parked aircraft.

USS *Enterprise* was attacked by five Mitsubishi G4M2 bombers, which failed to score one hit, and in sheer desperation a G4M2 made a suicide dive at the carrier, crashing ingloriously into the sea nearby. Then at three o'clock in the afternoon, two more G4M1s attacked *Enterprise*, but without scoring any results.

Admiral Nagumo received a signal that a US carrier raid was in progress on the Marshall Islands, and the aircraft carriers HIJMS *Shokaku* and *Zuikaku* were ordered to return to Japan to counter any raids on the Home Islands.

By 4 February, the carriers HIJMS *Hiryu* and *Soryu* stopped at Palau for a time, and continued into the area of the Java Sea. Here Aichi D3A1 dive-bombers and Nakajima B5N2 bombers attacked Allied surface forces in the Java Sea, damaging a heavy and a light US cruiser as well as a Dutch cruiser. The A6M2 carrier fighters shot down four US Consolidated PBY flying-boats. After these aerial engagements the carrier planes flew on to the naval air station at Kendari in the southern Celebes to land, refuel and return to the decks of the carriers whence they came. The two aircraft carriers returned to the island of Palau, to be joined later by the carriers HIJMS *Akagi* and *Kaga*, to make a four-carrier strike force under the command of Admiral Nagumo, which cruised at speed through the islands of the Dutch East Indies.

Air battle over Port Darwin
The four-carrier strike force advanced to within 220 miles of Port Darwin in north-western Australia. Admiral Nagumo launched an aerial attack on the port on 19 February 1942, with 188 carrier-based planes, consisting of Aichi D3A1 dive-bombers, Nakajima B5N2 torpedo-bombers and Mitsubishi A6M2 carrier fighters. At the same time fifty-four twin-engine land-based Mitsubishi G4M1 attack-bombers flying from the naval air station at Kendari bombed the port in conjunction with the carrier-based planes. The raid commenced at 0930 and built up to a miniature Pearl Harbor attack in ferocity and intensity. Facilities at Darwin Airport were attacked and ten US Army Air Corps P40 fighters were shot down, against the Imperial Naval Air Service's loss of one bomber. In the harbour an Australian troop ship was sunk, and a freighter loaded with ammunition exploded in a spectacular fireball and disappeared. A seaplane tender was sunk with three bombs and one near-miss, while a US destroyer capsized in a hail of five exploding bombs, was holed and went to the bottom. Other ships in Port Darwin harbour that were blown up and sunk included two corvettes, two freighters, one freighter damaged and beached, one hospital

US President Franklin D. Roosevelt signing the declaration of war against Japan the day after the attack on Pearl Harbor, 8 December 1941.

US President Franklin D. Roosevelt in conference with General Douglas MacArthur, Admiral Chester Nimitz and Admiral William D. Leahy, Hawaiian Islands, 1944.

Admiral Chuichi Nagumo, in Vice Admiral uniform. Appointed Commander in Chief of the First Air Fleet, the main carrier battle group, 10 April 1941.

Admiral Isoroku Yamamoto, Naval Marshal General and the Commander-in-Chief of the Combined Fleet, died following an ambush by US Lockheed P-38 *Lightning* fighters, 18 April 1943.

The Emperor Showa Hirohito, the 124th Emperor of Japan, reigned from 25 December 1926 until his death on 7 January 1989, the longest reign of any Japanese Emperor.

L-R Rear Admiral David W. Bagley, Commander, Hawaiian Sea Frontier Fleet, Admiral Chester W. Nimitz, USN, Commander-in-Chief, Pacific Fleet and Pacific Ocean Areas, Major General Delos C. Emmons, US Army and Rear Admiral Aubrey W. Fitch.

General Douglas MacArthur meeting Emperor Hirohito for the first time. The meeting was held at the US Embassy, Tokyo, 27 September, 1945.

Mamoru Shigemitsu, Japan's Minister of Foreign Affairs signs the *Instrument of Surrender*, assisted by Toshikazu Kase, Foreign Ministry Official and watched by Lieutenant General Richard K. Sutherland, USS *Missouri*, 2 September 1945.

Japanese representatives at the surrender ceremony on board USS *Missouri*, 2 September 1945.

Front row L-R Minister of Foreign Affairs Mamoru Shigemitsu, General Yoshijiro Umezu, Chief of the Army General Staff. Middle row L-R Major General Yatsuji Nagai, Army, Katsuo Okazaki, Foreign Ministry, Rear Admiral Tadatoshi Tomioka, Navy, Toshikazu Kase, Foreign Ministry and Lieutenant General Suichi Miyakazi, Army. Back row L-R Rear Admiral Ichiro Yokoyama, Navy, Saburo Ota, Foreign Ministry, Captain Katsuo Shiba, Navy and Colonel Kaziyi Sugita, Army.

ship damaged and retaining an unexploded bomb, eighteen Allied aircraft destroyed, and general damage inflicted upon Port Darwin's port facilities. The Imperial naval planes returned to their carriers as the land-based naval bombers flew back to Kendari, where they were joined at a later date by Admiral Nagumo and his four carriers. The carriers and land-based planes then supported the Emperor's troops in successfully completing the invasion of Java.

During the period 20/21 February, fifty-six Mitsubishi G3M2 converted transport planes flew from the naval air station at Kendari to Kupang on the island of Timor to drop 631 parachute troops of the Imperial marines of the Navy's 3rd Special Marine Corps. These troops, within a few days, captured the airport and crushed the Allied resistance, for the loss of twenty-four marines killed in battle.

At the same time, on 20 February, a four-engine Kawanishi flying-boat out on routine ocean patrol discovered US Task Force II, comprising the carrier USS *Lexington*, four heavy cruisers and nine destroyers on a course projected for Rabaul. An aerial engagement took place in which four US Grumman Wildcats shot down the patrolling flying-boat, which was replaced by a second, which itself was shot down; the shadowing of Task Force II was then taken up by an elusive third flying-boat, reporting the situation to the Rabaul Air Command headquarters of the Naval Air Service. The task force had now been shadowed to within 350 miles of Rabaul, with eighteen Imperial Navy torpedo-bombers seventy-five miles west of the US force at 1542. At 1620 the torpedo-bombers were attacked by a force of Grumman Wildcats, and in a running air battle fourteen naval torpedo-bombers were shot down by the Americans. Eventually four torpedo-bombers attacked the US carrier *Lexington* without result, and the engagement was broken off.

At 0900 on 23 February, a high-flying attack-bomber shadowed two American warships, which were later identified as the USS *Langley* seaplane carrier and USS *Seawitch*, with a cargo of US P40 fighter planes on board. The commander of the American force requested fighter cover, but none was available. At 1140 more aircraft were sighted from Java, including a squadron of nine twin-engine Mitsubishi G4M1 attack-bombers. USS *Langley*, commanded by Commander Robert P. McConnell, was blasted by five direct hits and two near-misses. As the seaplane tender blazed furiously, the wreck was strafed by six Mitsubishi A6M2 carrier fighters. Eventually, at 1332, the surviving crew abandoned ship and USS *Seawitch* successfully arrived at the Javanese port of Tjilatjap. With the invasion of the East Indies progressing so well in favour of the Imperial forces, on 25 February Admiral Nagumo with his four-carrier strike force, escorted by battleships and cruisers, sailed from Staring Bay in the Celebes. The fleet steered south of Java on a course to cut off the escape route of any Allied ships attempting to make for Port Darwin. The US destroyer *Pope* was attacked on 28 February by six Aichi D3A1 dive-bombers and six Nakajima B5N2 horizontal bombers from the carrier HIJMS *Ryujo*. The result was inevitable: the destroyer was sent to the bottom while escaping from the area of the Java Sea. In the period 28 February to 1 March, Admiral Nagumo's fleet sank a US destroyer, a British destroyer, a US gun boat and an Australian escort ship. In all these operations the Naval Air Service's 21st and 23rd Air Corps co-operated with the Army Air Corps' 3rd Joint Air Group in completing the successful invasions of Borneo, Sumatra, Java, the Celebes and the island of Timor.

Meanwhile, back in Japan the Mitsubishi Company had completed the first prototype J2M1 land-based fighter plane for final examination by the Navy. Though the first flight tests could proceed on schedule, the development work on the A6M3 had been difficult to complete.

Some 3,000 miles away to the south, in the Dutch East Indies, in a period between 1 and 7 March, the Imperial Army Air Corps' 3rd Joint Air Group, co-operating with the Naval Air Service's 21st and 23rd Air Corps, destroyed 300 Allied aircraft. These machines included 166 Martin bombers, some Curtis-Wright CW21 fighters and some Boeing B-17 four-engine bombers.

On 1 March, some reconnaissance planes from the Naval Air Service sighted USS *Pecos*, an oiler, steaming in the area of Christmas Island to the south-west of Java at 0945. *Pecos* was carrying survivors from USS *Langley*, previously attacked and sunk on 23 February. At 1145 *Pecos* was attacked by carrier-based bombers from HIJMS *Soryu*, and in a series of attacks the American warship was eventually sunk at 1448. In these attacks by aircraft from the Imperial Navy carrier *Soryu*, accompanied by battleships and escorting warships, the destroyer USS *Edsall* and a Dutch merchant ship were sunk. Four days later, on 5 March, the four-carrier force sent in 180 aircraft to bomb the Dutch Javanese port of Tjilatjap, sinking some twenty merchant ships. On 8 March, the port of Tjilatjap was occupied, and on the following day the Dutch forces in Java surrendered, while the carrier strike force proceeded to Staring Bay in the Celebes to re-form for the Indian Ocean foray.

The first prototype plane of the Mitsubishi J2M1 Raiden land-based fighter successfully completed initial circuits and bumps, when flown by test pilot Shima at the Kasumigaura Naval Air Station, thirty miles north-east of Tokyo. The machine displayed good stability and controllability characteristics, while the tail unit and control surfaces did not require any further modifications. However, complaints were made, particularly of the use of Plexiglas, which deformed the pilot's vision when landing. Some difficulty was experienced when an attempt was made to retract the landing-gear above a speed of 100 mph. Unhappily engineer Sadahiko Kato, who was in charge of the design of the retractable undercarriage system, contracted pneumonia and subsequently died. Nevertheless, engineering modifications were introduced to the undercarriage gear design by incorporating differential gears to the undercarriage transmission and improving the slip joints. Two prototypes were eventually produced, the J2M1 with a Mitsubishi Kasei 13 radial engine of 1,460 hp, and the J2M2 with a Mitsubishi Kasei 23 radial engine and armed with two 20 mm cannon and two 7.7 mm machine-guns, a total of 155 being made. The next variant was the J2M3, with four 20 mm cannon, and eventually a J2M4 high-altitude model powered by a turbo supercharged motor and carrying six 20 mm cannon, of which only two aircraft were manufactured. Subsequently in May 1944, a total of forty Model J2M5 fighters were produced, powered by the Mitsubishi 1,820 hp Kasei 26 radial engine at a maximum speed of 381 mph and firing twin 20 mm cannon. The original prototype J2M1 was flown to Suzuka Naval Air Station twenty-five miles south-west of Nagoya, where company test pilots Aratani and Mantani assisted Shima in further company flight tests.

The morning of 26 March saw the carrier strike force under Admiral Nagumo, without HIJMS *Kaga*, which had been sent back to Japan for repairs and replenishments, sail out of Staring Bay in the Celebes. It was accompanied by four battleships, one light cruiser, two heavy cruisers, eight destroyers and a fleet supply group consisting of six oilers escorted by three destroyers on course for the Indian Ocean.

At about this time during March 1942, the Nakajima Aircraft Company successfully flew the B6N1 reconnaissance and torpedo-bomber as a replacement for the B5N2. This new machine was powered by a Mamoru 11 radial engine developing 1,870 hp, and a total of 498 examples were built. Development continued, and the B6N2 Tenzan was produced, which used a Mitsubishi Kasei 25 radial motor, and eventually 770 machines of this type were manufactured. This aircraft was one of many used in the famous kamikaze attacks against the wooden-decked American aircraft carriers. By 1 April the Mitsubishi G4M2 bomber formations had been built up to a strength of 277 aircraft available for operational use.

Air war in the Indian Ocean

The Naval Air Strike Force was divided into two groups under Admiral Takeo Kurita, which consisted of the carrier HIJMS *Ryujo* with sixteen Nakajima B5N2 bombers and twenty-two Mitsubishi A6M2 carrier fighters. Escorted by four heavy cruisers and four destroyers, the strike force entered the Bay of Bengal. On 4 April the Kurita group conducted commerce raiding among the Allied ships bringing supplies to India. This was the only known occasion that Japanese warships conduct commerce raiding, as the B5N2 bombers sank twenty-three Allied merchant ships, representing 112,312 tons of shipping. These raids lasted from 4 to 11 April, before *Ryujo* left the Bay of Bengal.

On 4 April a Consolidated PBY Catalina flying-boat of RAF Coastal Command discovered the main carrier strike force under Admiral Nagumo as the fleet cruised to an attack position off the island of Ceylon. Regrettably the RAF plane was shot down, and therefore was unable to conduct shadowing operations, but the alarm had been given. The next morning, 5 April, Easter Sunday, a force of 125 Imperial Navy planes, comprising thirty-six A6M2 fighters, thirty-six D3A1 dive-bombers and fifty-three B5N2 bombers, attacked Colombo harbour and installations. Under radar guidance, forty-two RAF fighters were already airborne and waiting. However, as the RAF machines were older and slower than the Japanese aircraft, the outcome of the air battle was predictable. In the ensuing mêlée twenty-four RAF machines were shot down, but the naval planes under the command of Commander Fuchida reached their targets, attacking workshops, repair installations and shipping, and sinking one destroyer and a merchant ship.

Later an Imperial Navy plane out on routine reconnaissance patrol from the carrier strike force discovered two warships 300 miles south-west of Colombo. These warships were identified as the British cruisers HMS *Dorsetshire* and HMS *Cornwall*. At 1340, fifty-three Aichi D3A1 dive-bombers flew over the horizon, peeling off in flights to blast the cruisers. By 1359 both ships had been sunk, leaving a large slick of oil and 1,100 survivors floating in the water. No Japanese planes were lost in this particular action, which was led by Lieutenant-Commander Egusa HIJMN, veteran of the attack on the US warships at Pearl Harbor. However, the Colombo raid cost the Imperial Naval Air Service one A6M2 fighter and six Aichi D3A1 dive-bombers. Meanwhile the ships of the carrier strike force refuelled at sea from the support group of oilers.

Three days later an RAF Consolidated PBY Catalina flying-boat located a force of thirty-eight A6M2 fighters and ninety-one B5N2 bombers flying north in the direction of Ceylon under the command of Commander Fuchida HIJMN. Though advance warning had been given, only eleven RAF fighters were available to take off, and in the ensuing aerial attack nine of them were shot down at 0725 over Trimcomalee. The Imperial Navy bombers roared in and commenced their bombing runs, and port installations were smashed, warehouses exploded in flames and ships were sunk. However, the Royal Navy warships had gone sometime prior to the Japanese attacks, and only individual units were still at sea making for the open ocean.

Once again an Imperial Navy reconnaissance plane flew over a small group of British ships, which were identified as HMS *Hermes*, an old aircraft carrier, and her destroyer escorts. Immediately eighty-five D3A1 dive-bombers took off, escorted by nine A6M2 fighters, and attacked HMS *Hermes* at 1035. The aircraft carrier was mortally struck by twenty direct hits, and the great ship gradually heeled sideways as men slid over the side into the sea, and just twenty minutes later the ship had gone. Not only had HMS *Hermes* been sunk in these series of attacks by the Imperial carrier strike force, but also an Australian destroyer, a British corvette, a naval auxiliary and a merchant ship had been destroyed, for the loss of seventeen navy planes, lost with crews. At the conclusion of the operation the carrier strike force turned for the Straits of Malacca and was ordered back to the Home Islands by Admiral Nagumo.

The main target of the carrier strike force while operating in the Indian Ocean was the annihilation of the British Indian Ocean Fleet. However, the bulk of the fleet had left the Ceylon area before the Japanese had struck. By the use of technical superiority and radar surveillance, the British admiral had cheated the Imperial Navy planes by continually altering course and generally moving westwards, always just out of range of the carrier strike force. Eventually the British ships reached the safety of the east African ports. However, in four months of continual war the Imperial carrier strike force had sailed a third of the way around the world, lost no ships sustained no damage and had been responsible for sinking five Allied battleships, one aircraft carrier, two heavy cruisers, seven destroyers, several escort ships, merchant ships and naval auxiliaries, and had damaged three battleships.

While all the sweeping victories were taking place at sea, spearheaded by the Naval Air Service, the development of new aircraft types continued in the Home Islands. On 14 April an engineering conference took place at the naval aircraft research and development centre. At this meeting the Mitsubishi A7M Reppu was renamed the 17-Shi carrier fighter, since it had taken three years to reach operational status. Unfortunately, it had to be admitted that this situation had arisen owing to the poverty of industrial potential and an effort to keep aircraft types to a minimum, while maximising the nation's manpower to maximum efficiency. The engine finally selected for this aircraft was the Mitsubishi MK9a, developing 2,000 hp and incorporating a supercharger driven by a Fuldan hydraulic coupling, which enabled the maximum sea-level horsepower to be maintained to an altitude of 27,000 feet.

Policy decisions

At the end of April and the completion of the operations in south-east Asia, the Dutch East Indies and the Indian Ocean, decisions had to be taken concerning the next series of operations to establish an island defence perimeter. A naval conference was held on board the 63,000-ton super-battleship HIJMS *Yamato*, in the presence of Admiral Kusaka, Chief of Staff, and Admiral Nagumo, carrier strike force commander, and others. The meeting was briefed by Admiral Yamamoto with a situation report and information regarding future objectives. Admiral Kusaka felt future operations by the Kido Butai, or carrier strike force, would be foolhardy. The carriers had steamed 50,000 miles since the Pearl Harbor attack, the ships required renovating and the crews needed rest to recuperate. This accepted, Admiral Yamamoto outlined the next phase of the sea war. This was to establish an outer defence perimeter of islands as bases to detect shipping movements intending hostile operations against the Home Islands. The plan would require the occupation of Kiska, Midway Island, Wake Island and Port Moresby.

The task of transforming Admiral Yamamoto's idea into a war plan was undertaken by Captain Kameto Kuroshima HIJMN, and involved 200 ships, with a battlefield stretching over 2,000 miles from the far-off northerly Aleutian Islands to Midway Island in the Hawaiian Archipelago of the Central Pacific.

Operation MO

Operation MO was the plan for the invasion and occupation of Port Moresby in southern New Guinea, the fall of which would facilitate the conquest of New Guinea and place Australia in great peril. Planning proceeded accordingly, but was interrupted by a traumatic event with ominous signs for the future.

On 18 April 1942, nine twin-engine North American Mitchell bombers of the US Army Air Corps suddenly appeared over Tokyo and dropped bombs on the Shinbashi Station area of the city, three bombers attacked factories and oil tanks in the Yokohama vicinity, two other Mitchell bombers dropped bombs on factories in Nagoya and one attacked the Kawanishi aircraft plant at Kobe. Other targets bombed included steel plants, oil refineries and docks. Altogether the Japanese sustained fifty dead, 252 injured, ninety factory buildings gutted and a serious fire at a fuel farm, with six fuel tanks destroyed. Immediately confusion and terror struck the Imperial General Headquarters, but eventually staff officers calculated the estimated time of arrival of the aircraft carriers, which found the Imperial defences not ready. It appeared that the planes flew individually at treetop height, crossing the Nipponese coast at sixteen different places over a distance of 200 miles of coast-line. Some late arrivals were attacked by Naval Air Service fighters, with the result that the American bombers jettisoned their bombs, hitting a middle school, a hospital and a residential area.

It appeared that the sixteen Mitchell bombers were under the command of General James Doolittle, US Army Air Corps, flying from the carriers USS *Enterprise* and *Hornet* under the command of Vice-Admiral William S. Halsey, whose flagship, the cruiser USS *Nashville*, provided escort at a point 668 miles due east of Tokyo. Prior to the take-off the fleet had been discovered by the Imperial Navy picket boat HIJMS *Nitto Maru* number 23, which had been sent to the bottom in a barrage of shells twenty-three minutes later.

Because of the prospect of amphibious operations among the South Sea Islands, in which case aircraft would operate from beach-heads, float-planes had been developed. In February 1941, production began of the A6M2N, a float-plane fighter/reconnaissance aircraft, and from April 1942 until September 1943, over 327 aircraft of this type were manufactured. They were encountered in the Aleutian Islands off Alaska and off Guadalcanal in the role of reconnaissance and home-base defence. Powered by a 925 hp Nakajima Sakae 12 radial engine, the machine attained a top speed of 270 mph at an altitude of 14,110 feet.

The Allied losses due to offensive operations conducted by the Imperial Navy and spearheaded by the Naval Air Service, supported by units of the Army and the Army Air Corps, included 340,000 Allied prisoners of war, one million tons of Allied shipping sunk, including six battleships, one aircraft carrier, one seaplane carrier, seven cruisers and transports. They had captured 4,000 guns, 240 aircraft and 1,500 tanks, and destroyed 3,500 Allied aircraft, 1,400 claimed by the Army and 2,100 falling to the guns of the Naval Air Service.

The listed objectives of Operation MO included the occupation of Port Moresby and the establishment of a seaplane base at Tulagi and of a second seaplane base on the Louisiades, off the south-east tip of New Guinea.

The invasion fleet was organised under the command of Vice-Admiral Shigeyoshi Inouye at Rabaul with the 25th Air Flotilla. The carrier strike force was led by Vice-Admiral Takeo Takagi with two heavy cruisers and the 5th Carrier Division, consisting of HIJMS *Shokaku* and *Zuikaku* plus six destroyers and one oiler. The invasion force was commanded by Rear Admiral Aritomo Goto, escorted by a support group and a covering force made up of the carrier HIJMS *Shoho*, four heavy cruisers and one destroyer. The invasion force was formed by the Tulagi group as well as the Port Moresby group.

On 1 May 1942 the Japanese naval codes were scheduled for revision, but owing to the pressure of work the old codes were still in use. Unhappily for the Imperial naval planners, the vital new code was eventually broken by the US Naval Combat Intelligence Unit at Pearl Harbor under the command of Lieutenant-Commander J.J. Rochefort USN and 120 officers and men. This unit was housed in a windowless basement building within the Navy yard at Pearl Harbor, protected by steel gates and vault-like doors under constant guard. As a result of information deciphered by Lieutenant-Commander Rochefort and his unit, the American Naval Commander-in-Chief Admiral Nimitz knew something big had been organised.

On 2 May, the seaplane carrier HIJMS *Mizuho* was sunk by an American submarine as the various ships were assembling for the projected operation. The Port Moresby invasion force consisted of eleven transports, several auxiliaries, one light cruiser and six destroyers, sailing in company with the Tulagi invasion group, which contained one transport, two minelayers, several patrol craft, two minesweepers and two destroyers. The support group, covered by one seaplane carrier, two light cruisers and three gunboats, eventually set up the seaplane base on the Louisiades group of islands. The naval aircraft involved in the operation were twenty-one aircraft aboard HIJMS *Shoho* of the covering group and 126 aircraft aboard HIJMS *Shokaku* and *Zuikaku* of the 5th Carrier Division, which were used in preventing US ships intervening, as well as raiding airbases in northern Australia. At Rabaul was located the 25th Air Flotilla with 120 land-based twin-engine attack-bombers. By this date all was ready, and Rear Admiral Goto and Vice-Admiral Takagi boarded their respective heavy cruisers so as to direct air operations. Carrier strengths for the forthcoming battle were as follows:

	Fighters	Dive-bombers	Torpedo-bombers	Total
HIJMS *Shoho*	12 A6M2	-	9 B5N2	21
HIJMS *Shokaku*	21 A6M2	21D3A1	1 B5N2	43
HIJMS *Zuikaku*	21 A6M2	21 D3A1	21 B5N2	63
Total	**54 A6M2**	**42 D3A1**	**31 B5N2**	127

By 3 May, the naval forces had landed and captured Tulagi, twenty miles off Guadalcanal, to construct a seaplane base, while Rear Admiral Goto's covering group steamed south of New Georgia Island at the same time as Vice-Admiral Takagi with the two aircraft carriers cruised off the north of the Solomon Islands, beyond Allied aircraft range. The next day, 4 May,

fourteen transports, light cruisers, six destroyers, covered by the carrier HIJMS *Shoho*, with four heavy cruisers and a destroyer, advanced south to complete the next phase of Operation MO.

Air battle over the Coral Sea

On 4 May, Rear Admiral Fletcher USN, with his US Carrier Division, consisting of USS *Lexington* and USS *Yorktown*, attacked the Imperial forces' beach-head. In three waves totalling 105 missions, the American pilots dropped twenty-two torpedoes, twenty-six 1,000 lb bombs and fired 80,000 rounds of machine-gun ammunition. The result was disappointing, for one Japanese destroyer was beached, five seaplanes of the Naval Air Service were shot down, four landing-barges sunk and three minesweepers seriously damaged. Admiral Nimitz, the US C-in-C, disappointed with the result, determined that the flying crews required more target practice. However, Rear Admiral Fletcher retreated southwards and then turned westwards, calculating that the Imperial invasion force of Rear Admiral Goto should be off the eastern coastline of New Guinea. Admiral Inouye Shigeyoshi, accompanied by heavy cruisers, superior planes and pilots, cruised south in pursuit of Admiral Fletcher.

At daylight on 7 May, a reconnaissance plane of the Imperial Naval Air Service sighted an eastward-sailing convoy consisting of a US destroyer and the tanker USS *Neosho*, which had been refuelling Fletcher the previous day. Mistaking the two American ships for a cruiser and a carrier, the reconnaissance plane called in fifty-two carrier-based Imperial Naval Air Service bombers, consisting of Aichi D3A1 dive-bombers and B5N2 torpedo-bombers. Within minutes the US destroyer went to the bottom and the tanker became a blazing wreck.

Meanwhile Fletcher was informed that two carriers and four cruisers were steaming towards him from the north-west, but in reality these were the two light cruisers and gunboats of the Port Moresby invasion force escorts, which were attacked by ninety-three US carrier planes.

Admiral Inouye received a reconnaissance report that accurately located Fletcher's ships, but the Japanese admiral awaited a further report that a second US carrier force was located to the west. This force was the Australian squadron under the command of Rear Admiral J. Grace RAN, with cruisers and destroyer escorts but no carriers, whose mission was to locate the invasion transports. For Admiral Inouye, the intelligence reports became confusing, and he awaited confirmation before proceeding further in the action. The Americans were known to possess only five carriers, two being on their way to Pearl Harbor, and so four carrier sightings were mistaken identity on the part of the Imperial Naval Air Service.

Admiral Fletcher's planes discovered HIJMS *Shoho* with cruiser and destroyer escorts, which were attacked by ninety-two US dive-bombers and torpedo-bombers under the direction of Lieutenant-Commander William Hamilton, flight leader from USS *Lexington* and *Yorktown*. In just twenty-six minutes HIJMS *Shoho* was sunk with indifferent bombing by the US pilots; the carrier was very buoyant and then suddenly blew up eighty miles south of Woodlark Island. Three US Navy planes were shot down and one bomber crash-landed on the deck of *Shoho* to blow up.

During the afternoon Admiral Fletcher arrived at the passage through the Louisiades Isles, which the invasion force had to pass, but Admiral Inouye ordered the Port Moresby force to hold back, and sent forty-two D3A1 dive-bombers and B5N2 torpedo-bombers to attack the Royal Australian Naval Squadron in three waves. All the bombs were avoided by Rear Admiral Grace, who was promptly accidentally attacked by US Army bombers from Australia. A protest was lodged with General MacArthur, the US Commander-in-Chief, but subordinates foolishly suppressed the report.

By dusk the opposing forces had failed to locate each other, and Admiral Fletcher postponed operations till morning. Meanwhile Admiral Inouye sent twenty-seven D3A1 dive-bombers and B5N2 torpedo-bombers to make a night attack on Fletcher. The Aichi dive-bombers and Nakajima planes located Fletcher's carriers, but radar-directed Grumman carrier fighters engaged the attackers, and the Naval Air Service lost nine planes to three American machines shot down. An inconclusive attack was achieved, and on the return flight eighteen Imperial Navy planes lost their way, scattering in the dark. One naval bomber was shot down, and a second, being lost, attempted to land mistakenly on USS *Yorktown*. Eventually eleven of the eighteen bombers ran out of fuel and crash-landed on the sea to quickly sink, drowning their crews in the watery darkness.

At sun-up on 8 May, the carrier strike force sent a further attack group flying from the carriers HIJMS *Shokaku* and *Zuikaku* to attack the ships of Fletcher's fleet. *Yorktown* was heavily strafed by Aichi D3A1 dive-bombers and Nakajima B5N2 horizontal bombers, but by taking violent evasive action she avoided the onslaught till one bomb hit the ship, penetrating four decks and killing sixty-four men. The attack-bombers of the Naval Air Service turned their attention to *Lexington*, which was struck by two torpedoes and two bombs, creating a heavy list, which was corrected by counter-flooding. Fortunately the carrier maintained a speed of 25 knots, but later in the day, at 1247, a serious and heavy petrol leak was discovered. A raging fire broke out, and in a series of explosions *Lexington*, fully trimmed, sank to the bottom. More fortuitously, HIJMS *Zuikaku* found herself under heavy cloud cover in a tropical storm, rain lashing down from a cloud-burst as the carrier pitched in the storm and ploughed on in a swell of white spray and mounting seas. HIJMS *Shokaku* was less fortunate, being attacked by American carrier-borne planes and sustaining damage to her decks. The Imperial ships found they were able to outrun the US torpedoes, which at this stage of the war appear to have been faulty. Both sides now disengaged inconclusively to return to their respective bases, HIJMS *Shokaku* to Japan for a refit, and *Zuikaku* to take on board new planes, pilots and crews.

On 20 May, Admiral Yamamoto, C-in-C Combined Fleet, transmitted a series of very long messages, which were intercepted by the US Combat Intelligence unit at Pearl Harbor under Lieutenant-Commander Rochefort's command. Bits and pieces of these transmissions had to be pieced together, with over 15% of the messages missing. The information appeared to refer to a series of targets, including one with the precise target information missing. Both Lieutenant-Commander Rochefort and Admiral Nimitz suspected Midway Island. To obtain proof, a subterfuge was resorted to in which a false message was transmitted from USN radio stations to the effect that the water distillation plant had broken down at the US base on Midway. Two days later confirmation was received, when an Imperial Navy transmitter communicated the information to naval headquarters, Tokyo, that the distillation plant was inoperative. The worst fears of Admiral Nimitz were realised.

Six days later Imperial Navy submarines watching the approaches to Pearl Harbor reported the arrival of the aircraft carriers USS *Enterprise* and *Hornet* subsequent to the General Doolittle raid on targets in Japan. The crews repaired and replenished their carriers in readiness to sail in less than forty-eight hours. The next day, 27 May, the Imperial submarine patrol reported USS *Yorktown* as limping into Pearl, apparently after steaming 4,500 miles at 10 knots with a destroyer escort. Once in harbour, welders and fitters worked on the carrier non-stop for forty-eight hours to make good any battle damage sustained.

Meanwhile Admiral Yamamoto had sailed from Hashirajima anchorage off Hiroshima when Imperial radio monitors detected unusual radio traffic from US warships and the Pearl Harbor headquarters. Nimitz would be waiting! As the Yamamoto fleet progressed, US submarines shadowed the ships to the battle zone.

The Imperial submarine fleet was under the command of Vice-Admiral the Marquis Komatsu Teruhisa, whose picket submarines were eventually on station too late. The snooper submarines were unable to take a look at the Pearl Harbor berths as the scouting seaplanes could not be launched.

This failure to advise Admiral Yamamoto of the departure of the US carriers was crucial to the success or failure of the operation. As a result the US ships were in between the advancing Imperial warships steaming on Midway and the Imperial picket submarines now looking east for the American carriers, which had already cruised by westwards four days previously. On 28 and 30 May, USS *Enterprise, Hornet* and *Yorktown* cleared Pearl Harbor and Diamond Head for the open sea.

The strategy about to be embarked upon had been agreed between representatives of the Japanese Navy and Army after some misgivings on the part of the Army strategists, who wished to pursue a policy of fortifying the island possessions and consolidating their gains so rapidly achieved with long lines of communications. The Navy believed that it was necessary to invade Hawaii and Australasia so as to push out the outer perimeter of the island defensive chain. Eventually it was agreed to the Aleutian and Midway invasion during June, followed by the invasion of the oceanic islands in July of 1942.

At this time, May 1942, the Aichi B7A1 torpedo-bomber first flew, but production was delayed owing to difficulties with the Homare 11 radial engine, as well as the destructive forces of a recent earthquake in the region of Japan in which the company was located. This new torpedo-bomber had a wingspan of 47 feet 3 inches, and was powered by one Homare radial engine giving a power output of 1,875 hp, with a maximum speed of 337 mph at an altitude of 20,345 feet. The machine was armed with two 20 mm cannon and one movable 12.7 mm machine-gun with an offensive capability of 2,205 lb of high-explosive bombs or the equivalent weight of one torpedo. The Mitsubishi J2M1 Raiden land-based fighter was also under development at this time, with a modified Mitsubishi Kasei 23 radial engine giving 1,430 hp at take-off. The modification to the engine had included increasing the diameter of the engine supercharger impeller, as well as widening the area of the air intake scoops. Furthermore the engine was fitted with a methanol injection apparatus to increase the boosted power output, an arrangement that was subsequently fitted to the Mitsubishi Kasei 26 radial engine and powered the Raiden J2M5 model. Test flying for this new land-based fighter was undertaken by Katsuzo Shima, Mitsubishi's senior test pilot, a former petty officer 3rd class in the Imperial Naval Service. Later during 1942, Shima resigned from the Mitsubishi Aircraft Company as senior test pilot and joined Japan Airways Ltd, conducting aerial photographic survey missions over south-east Asia. Though the development work on the J2M1 was incomplete, the Navy changed the specification, which included replacement of the 20 mm armament with 13 mm machine-guns, which were standard at this time as fitted to the Mitsubishi A6M Zeke carrier fighter.

By the summer of 1942 the Imperial Naval Intelligence Service had received confirmed reports that the US Navy had developed a new carrier fighter – the Chance Vought F4U1 Corsair – reputedly powered by a 2,000 hp Pratt and Whitney double-row Wasp radial engine. This plane, armoured and heavily armed, could out-perform the Mitsubishi A6M2 Zero fighters of the Imperial Navy. The task now was to convince the Imperial Naval Air Service of the need to develop new power plants. The new engine was to be a Mitsubishi MK9B, which would be essential in the design of a new fighter plane capable of matching the performance of the new American carrier fighters. The engine developed 2,200 hp at sea level, reducing to 1,760 hp at an altitude of 19,700 feet and a further reduction of hp to 1,700 at a height of 26,200 feet. With this engine weighing 2,530 lb, the 17-Shi fighter design would have an all-up weight of 7,260 lb with a useful load of 2,640 lb. The calculated speed at an altitude of 20,200 feet was 425 mph, with a landing speed of 80 mph for an endurance time of thirty minutes at maximum power output and a cruising time of two hours. The projected armament was to be two Type 99 MK II 20 mm cannon with 200 shells for each gun, as well as four Type III 13 mm machine-guns, each with 400 rounds.

During June 1942, amendments to the 5th Expansion Programme were revealed, making provision for the construction of twenty-eight new aircraft carriers, the formation of 629 new squadrons of aircraft and the training of 45,000 flight crews, to be attained by the year 1947. Under this programme the Naval Air Service would have a total of 787 squadrons of aircraft, of which 541 would be operational, with 246 squadrons of training machines. This would have given a grand total of 19,083 aircraft. This expansion programme, of course, was quite beyond the industrial productive capacity of Japan at that time! Indeed, before the end of 1945, some three years later, when the Western Barbarians had peacefully landed on the sacred coast from the Yankee invasion fleets at the agreement of His Imperial Majesty the Emperor, the Imperial aircraft industry had delivered only some 50% of the original programme.

However, during the summer of 1942, a new class of aircraft carrier was launched – the HIJMS *Hiyo*, similar to HIJMS *Soryu* and *Shokaku*. The new carrier was fitted with an island structure surmounted by a bridge on the starboard of the vessel, with a large funnel, the upper section of which was inclined at 26 degrees to the horizontal.

By 3 June 1942, the ships and planes of the Imperial Navy had been assembled for the invasion of the Aleutian Islands and Midway. The forces of the Yamamoto fleet were made up of 189 ships and 685 aircraft, comprising:

1. 10 battleships
2. 12 aircraft carriers (two heavy, three medium, three light and four small seaplane carriers)
3. 685 aircraft:
   352 Mitsubishi A6M2 carrier fighters
   105 Aichi D3A1 dive-bombers
   162 Nakajima B5N2 torpedo-bombers
   56 reconnaissance aircraft
   10 Mitsubishi G4M2 attack-bombers
4. 24 cruisers
5. 70 destroyers
6. 15 submarines I class boats
7. 18 tankers
8. 40 auxiliary and transport vessels

Air attack on Dutch Harbor, Aleutian Islands

On 3 June 1942, the operations commenced with an attack on the US base at Dutch Harbor in the Aleutian Islands by twenty-two Nakajima B5N2 bombers escorted by twelve Mitsubishi A6M2 carrier fighters. Flying in from the carriers, the twenty-two bombers were unable to locate the US Army Air Force or US Navy planes on the airfield. Nor could any warships be detected in the harbour or the roads, so the raiders took their vengeance on the oil tanks, the radio station and the port facilities. Leaving Dutch Harbor wrecked, in smoking shambles and with a heavy cloud of smoke towering above the sea, the Imperial Naval Air Service bombers headed out to sea for the safety of the carriers. On the way back the flight commander signalled the position of a group of US destroyers, reporting course and speed. Once they were safely stowed below the flight decks deep in the bowels of the aircraft carriers, the weather closed in and flying ceased for the next three days. The opening move had been made, a feint attack, which had discovered the total absence of US warships and planes in those far northerly climes. For three days the carriers cruised in the fog, mist and rain as the ships pitched in between the white-topped seas and rolled into the troughs at precarious angles, keeping station with difficulty.

The air battle for Midway Island

By 3 June the Imperial naval forces were in position to execute the second phase of the strategic plan. Admiral Nagumo, with a four-carrier strike force, namely HIJMS *Kaga*, HIJMS *Akagi*, HIJMS *Soryu* and HIJMS *Hiryu*, steamed on course under the protection of a bad weather front with heavy rain-cloud cover, navigating from the north-west and hiding in the rain squalls to avoid detection. Unhappily the Imperial naval planners thought the US forces were 1,200 miles east, beyond the submarine picket boats. At the same time a fleet of escorted troop transports was advancing into the Midway area from the south-west in readiness to execute the invasion landings and subjugation of the US marine garrison.

At 0420, Admiral Nagumo ordered an airstrike against the Midway airstrip and airfield facilities, as well as an attack on the locations of the US Marine garrison. This strike force consisted of thirty-six Nakajima B5N2 bombers, thirty-six Aichi D3A1 dive-bombers and thirty-six Mitsubishi A6M2 carrier fighters. The planes were airborne as the carriers turned and steamed into the wind. Half an hour later at 0450 Imperial naval reconnaissance planes were launched to watch for the approach of the forces of the hitherto unknown location of the US fleet. But as a precautionary measure the Imperial Naval Air Staff had organised a force of seventy-two B5N2 bombers armed with torpedoes, as well as thirty-six A6M2 fighters on a standby basis, to strike at the US carrier fleet whenever its location could be established. Meanwhile, at 0553 the Imperial naval air armada was nearing its target when detection was established by US Navy radar surveillance units.

Shortly the Imperial planes swooped on an apparently unguarded island base as Aichi D3A1 dive-bombers peeled off, plummeting down in vertical lines from the sky, engines roaring, to release a hail of high-explosive bombs on runways, workshops, hangars, barracks and communication centres. This attack was augmented by the B5N2 bombers dropping bombs, which exploded in vast plumes of smoke and debris as the hospital, fuel tanks, the power station, mess halls, telephone exchange and fuel lines were reduced to a smouldering shambles.

The Americans had been waiting as Grumman F4F Wildcats and Brewster F2A3 Buffaloes underground control attempted to attack the Imperial bombers. The Mitsubishi A6M2 carrier fighters, flying escort, peeled off into a general mêlée of twisting, rolling and tail jumping. The American naval planes had climbed for altitude to pounce, but the Zero fighters were able to shoot twenty-two US machines out of the sky in a hail of 20 mm cannon fire, exploding in balls of fire and streaking down to the surface of the sea, pilots killed or maimed slumped over their controls. The attack on Midway had been a success, except that no US planes had been destroyed on the ground and the island firepower remained undestroyed. So the Imperial Navy flight commander, Lieutenant Tomonaga Joichi, signalled Admiral Nagumo aboard the flagship, 'There is need for a second strike.'

Meanwhile Admiral Nagumo's carrier strike force and escorting warships were cruising some two hundred miles away and came under heavy air attack by planes from the US Army Air Corps, US Navy and US Marine Corps. At the Admiral's command, the carrier strike force with the escorts increased speed and violently manoeuvred in great unwieldy circles of white boiling foam, churning up the sea, the ships pumping white-hot lead into the air in a curtain of fire surrounding the Imperial ships. A force of fifty-two bombers from Midway Island commenced a series of attacks at 0705, including Boeing B-17 Flying Fortress bombers coming at low level. During the next 3 hours 15 minutes, the A6M2 fighters defending the carrier strike force shot down forty-four planes to disappear into a watery grave. But still the Americans had been unable to score a single vital hit on any of the warships advancing on Midway.

At the same time as the Flying Fortresses commenced their bombing runs, the carrier planes from USS *Enterprise* under Rear Admiral Spruance and USS *Yorktown* under the command of Admiral Fletcher, located the Imperial fleet and started a series of attacks. At 0705 six Grumman Avengers attacked with torpedoes, and at 0711 four US Army B26 Martin Marauders attempted a torpedo attack. But at 0715 Admiral Nagumo ordered the de-arming of the 108 planes on standby and rearming with bombs. The hangar decks of the Japanese carriers became feverish with activity as heavy-calibre bombs were brought up the ammunition lifts from magazines deep in the bowels of the ships. Armourers, assisted by ground crews, changed the weapons, discarding the torpedoes to convenient corners of the deck to be lowered back into the magazine at a more opportune moment. By 0728 an Imperial reconnaissance plane reported a US force of ships to the north-east of the carrier strike force – identification unestablished! Fortuitously the three American carriers were 150 miles off and closing.

A hurried conference took place between Admiral Nagumo and his staff as to the next move in the attack, and it was decided to suspend the previous order to rearm the standby force of bombers until further clarification could be achieved concerning the whereabouts of the US carriers. By 0755 the US carrier fleet launched a third, fourth and fifth series of attacks on the fleet of Admiral Nagumo. A force of sixteen Douglas Dauntless dive-bombers made a glide-bomb attack on the Imperial carriers, but without success.

Meanwhile the Mitsubishi A6M2 carrier fighters patrolling the skies outside the immediate area of the carrier strike force were fighting a brilliant action. Fighter commanders marshalling the lightweight carrier fighters in the air attack frustrated bombing runs and shot down the US torpedo-planes and dive-bombers as they approached. The sea became littered with the wreckage of sinking American planes and the dead bodies of the shot-up and drowned crewmen floating aimlessly in the deep, white-capped, rolling swell.

The Air Staff officers aboard the Imperial carriers began to realise that a tactical error had been committed as the flight and hangar decks aboard the aircraft carriers became a scene of confusion and contradiction.

Then at 0814 a force of fifteen Boeing B-17 Flying Fortresses flew in to attack the ships, dropping a total of 127 bombs of 500 lb calibre, without achieving one direct hit as all the planes hurriedly flew off, pursued by A6M2 fighters.

If the Imperial forces had made a tactical error, the US attacks failed to achieve any success, being ill co-ordinated, with poor weapons and inferior equipment against a highly professional antagonist. Three minutes later, eleven US Vindicator bombers attacked the Imperial battleship HIJMS *Haruna*, failed to hit the ship and were driven off by heavy anti-aircraft fire. Then at 0820 an Imperial reconnaissance plane sighted a US aircraft carrier, and immediately Admiral Nagumo ordered the standby bombers to be rearmed with bombs – more feverish work to attain the requirements on the carriers' hangar decks.

Meanwhile the admiral,, with the carrier strike force and escorts, by judicious station keeping and manoeuvring, had survived attacks by nearly a hundred US aircraft without any ship being put out of action. However, the fighter defence force of Mitsubishi A6M2 lightweight carrier fighters, with superb flying and marksmanship, had defeated the American aerial intervention.

By 0915, the Midway attack force had returned and was being recovered. Five minutes later, planes from USS *Hornet* arrived over the carrier strike force, and all were shot down except one plane, which escaped. Then fourteen torpedo-bombers from USS *Enterprise* flew in to drop torpedoes, but during the operation were shot down by the A6M2 fighters. At 1015, a total of 108 strike aircraft of Admiral Nagumo's Midway 2nd Strike Force were lined up on the decks of the four carriers, armed with torpedoes for an attack on the US aircraft carriers, preparatory to take-off. Suddenly twelve US torpedo-bombers, Douglas Devastators, arrived overhead, and immediately the fighter defence screen came alive as seven planes fell to the guns of the Zero fighters, two were shot down by anti-aircraft fire and two retreated severely damaged.

By 1022, thirty-two Douglas Dauntless dive-bombers flew in from the south-west, based on USS *Enterprise*, and in the ensuing aerial battle with the Mitsubishi A6M2 fighters, eighteen American planes were sent crashing, crippled and in flames into the cold, deep ocean far below. But as luck would have it, seventeen US dive-bombers flew in from the south-east, based on USS *Yorktown*, and were promptly attacked by the Imperial fighters. The *Enterprise* dive-bombers immediately attacked HIJMS *Kaga* (36,800 tons) and HIJMS *Akagi* (36,000 tons), while the survivors from *Yorktown* fell on HIJMS *Soryu* (17,500 tons). In a tremendous crescendo of anti-aircraft gunfire, staccato rattle of cannon and machine-gun fire, the noise of diving planes rose to an ear-splitting roar as bombs were heard whistling down to explode

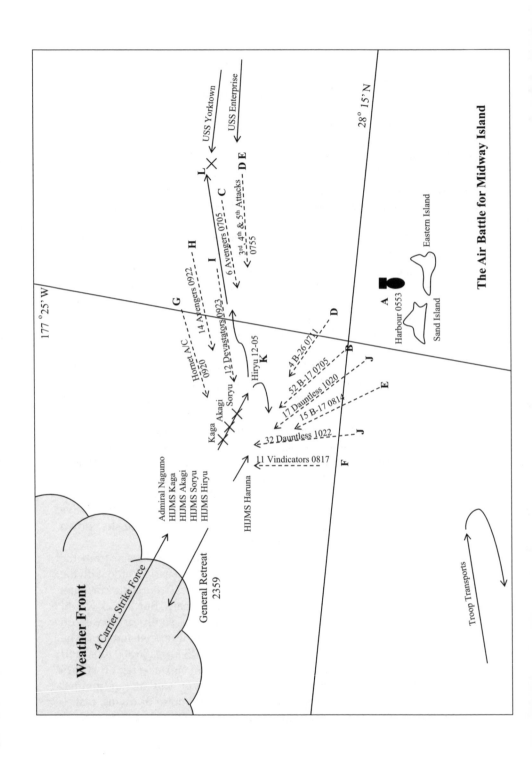

The Air Battle for Midway Island

on the decks of the Imperial carriers! *Kaga* received four direct hits, *Akagi* two hits and *Soryu* three direct hits, but *Hiryu* escaped the aerial onslaught to prepare and launch a reprisal raid on the US carriers at 1205. Nevertheless, during the previous 3¾ hours Admiral Nagumo had repelled ninety-three US bombers, destroying forty-four planes in the process, but in just six minutes the carrier strike force was crippled.

The Imperial Zero fighters were at low altitude to combat the attacks of the US naval torpedo-bombers. The sailors aboard the ships were concentrating on the operation of turning the carriers into the wind, preparatory to launching the 108 aircraft of Admiral Nagumo's 2nd Strike Force against the US fleet. Consequently the carrier decks were full of aircraft, lined up, refuelled and bombed up with torpedoes. A bombing attack against such vulnerable ships could not have been made at a more appropriate moment. But worse was to come! The American bombs had crashed through the flight decks to explode on the hangar decks. Here fuel lines were ignited and the de-armed torpedoes still lying around waiting to be stowed in the magazines in the bowels of the ships exploded into sheets of white-hot searing flame. The hangar decks now became uncontrollable, flaming, high-octane infernos, with flames spreading throughout the ships. Those below and in the engine-rooms became trapped as bulkheads popped and the flames licked into the magazines and around the aviation fuel tanks. All the engine-room staff were asphyxiated by the poisonous gases of the burning ammunition and fuel. Eventually the hulks heeled over and the order was given to abandon ship – at least to those who were still alive on the flight deck. In all, forty-two pilots had been lost in combat, with 140 more pilots killed in the flaming inferno of the carriers. Before the Midway operation, the Imperial Navy possessed ten carriers of 215,100 tons, but subsequent to this venture five carriers were afloat, totalling 96,100 tons.

The reprisal raid from HIJMS *Hiryu* had dispatched eighteen Aichi D3A1 dive-bombers against USS *Yorktown*, which had sustained three direct hits. A force of five B5N2 torpedo-bombers scored two hits in retaliation. However, HIJMS *Hiryu* was herself attacked and crippled, which finished all hopes of waging an offensive war, so that at midnight Admiral Yamamoto signalled the general retreat to all Imperial warships in the Midway area.

On the following day, at 1500, His Imperial Majesty the Emperor was officially informed of the débâcle and given details of the ships lost and damaged. The hulks of HIJMS *Hiryu* were sunk by the destroyers *Kazemugo* and *Yugumo*, while the remains of HIJMS *Akagi* were torpedoed by the destroyers HIJMS *Arashi*, *Hagikaze*, *Mikage* and *Nowake*. The aircraft carrier HIJMS *Soryu* finally went down at 1613 on 5 June 1942. The Midway operation was abandoned!

CHAPTER 16

War Among the Southern Islands

Though the Imperial Navy had sustained a very serious defeat, the war was far from won by the US Navy, and a series of actions followed for the prize of the South Sea Islands.

During June and July of 1942, while the outcome of the war in the South Pacific was being decided, the development of new aircraft by the Naval Air Service continued. Flight trials of the Mitsubishi J2M1 fighter took place under the command of Lieutenant-Commander Kiyoto Hanamoto of the Yokosuka Naval Air Depot and Lieutenant Takumi Hashi and Motonari Suo of the Naval Aircraft Research and Development. The results were disappointing, as the field of vision was inadequate, the pilot experiencing excessive warping of the canopy cover. For night operations the cockpit layout had to be modified to the same standards as those of the Mitsubishi A6M2 fighter. The pitch-change mechanism of the VDM-type propeller was unsatisfactory and would not operate properly. The rate of climb achieved by this prototype J2M1 was less than the planned technical specification. The maximum level flight of 350 mph at 18,000–19,000 feet was less than had been hoped for, while at 20,000 feet the machine only achieved 371 mph. As a result of these flight tests the Naval Air Service ordered that the Mitsubishi Kasei 14 radial engine be replaced by the Kasei 23 motor, which was fitted with a water methanol injection apparatus to increase the power output. To improve pilot vision, modifications were made to the cockpit canopy, and the armament was fixed at two 20 mm cannon with two 7.7 mm machine-guns.

Meanwhile, in July 1942, a squadron of Mitsubishi A6M2 carrier fighters were on routine patrol over the coast of Alaska when one of the planes experienced engine trouble, causing the machine to crash-land on soft terrain, catapulting along the ground and throwing the pilot from his cockpit. This machine was only partially damaged, although the pilot was dead with a broken neck and distorted features from the violence of his fall. At last the Allies were able to capture a Mitsubishi A6M2 carrier fighter in an undamaged condition so that from the cold, bleak, coastal water of Alaska this Zero could be taken to the United States for flight trials and evaluation.

On 6 July 1942, the Naval Air Service issued its specification for the 17-Shi carrier fighter to the Mitsubishi Aircraft Company. This machine was to be designated the A7M1 Reppu carrier fighter, and was to be a small single-seater, single engine monoplane, sufficiently small for carrier stowage and powered by an engine that had successfully completed tests by 1 April. The maximum speed was to be 397 mph at 20,000 feet, with a rate of climb from 0 to 20,000 feet of less than six minutes. Endurance was to be 2½ hours of flying time at 287 mph, with thirty minutes of combat. Take-off distance was to be 270 feet into a 49 mph wind, and landing back aboard the carrier at 77 mph. Armament required included two Type 99 Mk II cannon of 20 mm calibre, with provision for 200 shells per cannon, and two Type 3 machine-guns of 13 mm calibre with 400 rounds per gun. Extra equipment to be fitted into the new aircraft included a drift computer. At the same time the Navy abandoned the Nakajima J1N1C as a long-range fighter concept. Difficulties had been experienced during flight trials with the aileron controls, resulting in severe vibration due to buffeting at high angles of attack on looping or turning, but the machine was accepted as the Type 2 reconnaissance aircraft.

The day that the Naval Air Service issued its specification for the 17-Shi carrier fighter was also the day on which several hundred marines of the Imperial Navy, accompanied by 2,000 labourers, landed on the island of Guadalcanal and commenced the construction of an Imperial naval air station.

Despite the sinking of the Imperial Navy's carrier strike force at the Battle of Midway Island on 4 and 5 June, the Naval Air Service still possessed a potential to win the Great Pacific War, as an inventory of the time suggested:

Air Components of the United Fleet

A. The 2nd Fleet
  11th Air Wing. Seaplane carrier HIJMS *Chitose* and *Kamikawa-Maru*
B. The 3rd Fleet
  1st Air Wing. Carriers HIJMS *Shokaku, Zuikaku, Zuiho*
  2nd Air Wing. Carriers HIJMS *Ryujo, Junyo, Hosho, Daiyo, Unyo*
  Seaplane carriers. *Sanyo-Maru, Sanuki-Maru*
C. The 4th Fleet
  4th Base Defence Unit – land-based 21st Air Corps
  6th Base Defence Unit – land-based 19th Air Corps
  Seaplane carriers *Kunikawa-Maru, Kamui*

D. The 5th Fleet
    7th Base Defence Unit – land-based Chichi-jima Air Corps
    51st Base Defence Unit – land-based 5th Air Corps
    Seaplane carrier *Kimikawa-Maru*
E. The 8th Fleet
    Land-based – 2nd Air Corps
    Seaplane carrier – *Kiyokawa-Maru*
    11th Air Fleet
22nd Air Wing – Land-based Bihoro Air Corps, Genzan (Wonsan) Air
                 Corps
24th Air Wing – Land-based 1st Air Corps, 14th Air Corps, Chitose Air
                 Corps
25th Air Wing – Land-based 24th Air Corps, Yokohama Air Corps,
                 Tainan Air Corps
26th Air Wing – Land-based 6th Air Corps, Misawa Air Corps, Kisarazu
                 Air Corps

THE SOUTH-WESTERN DISTRICT FLEET

Originally formed to defend Malaya, the Philippine Islands, Indo-China and Indonesia in expectation of Allied counter-attacks.

A. 21st Air Wing. Land-based Kanoya Air Corps
    Toke (Tonkin) Air Corps
B. 23rd Air Wing. Land-based.   3rd Air Corps
    Takao (Koshun) Air Corps
C. 1st Southern Fleet. Land-based (Malaya & Indo-China) 40th Air Corps
    11th Base Defence Unit
D. 2nd Southern Fleet. Land-based (Indonesia)
    21st Base Defence Unit. 33rd Air Corps
    22nd Base Defence Unit. 34th Air Corps
    23rd Base Defence Unit. 35th Air Corps
    24th Base Defence Unit. 36th Air Corps
    4th Transport Air Unit
E. 3rd Southern Fleet. Land-based (Philippine Islands)
    31st Base Defence Unit. 31st Air Corps
    32nd Base Defence Unit. 32nd Air Corps
    1st Transport Air Unit. 2nd Air Corps
    2nd Transport Air Unit. Yokosuka Air Corps
    4th Transport Air Unit
    5th Transport Air Unit
    6th Transport Air Unit

The first air battle for Guadalcanal, 7–9 August 1942

Fighting continued in the south islands, resulting in the 1st US Marine Division capturing the Naval Air Service station on Guadalcanal. This happened after twenty-three US transports escaped destruction when, during 7–9 August 1942, the first action of Guadalcanal took place. Attacks were made by seventeen Mitsubishi A6M2 carrier fighters, nine Nakajima B5N2 torpedo bombers and twenty-seven Mitsubishi G4M2 twin-engine attack-bombers of the Rabaul-based 4th Misawa and the Tainan Air Corps, assisted by eight cruisers of the surface fleet. The outcome resulted in twenty-one US aircraft shot down for a loss of thirty-four machines of the Imperial Naval Service. However, the twenty-three US transports escaped and the US Marines on Guadalcanal were reinforced and revictualled.

The second air battle for Guadalcanal, 24/25 August 1942

The second air battle for Guadalcanal occurred over the period 24/25 August 1942. A force of planes from HIJMS *Zuikaku*, *Shokaku*, *Zuiho* and *Ryujo*, comprising Nakajima B5N2 torpedo-bombers, accompanied by Aichi D3A1 dive-bombers, escorted by Mitsubishi A6M2 carrier fighters, encountered the aircraft carriers USS *Saratoga* and *Enterprise en route* to Guadalcanal. The Imperial carrier bombers penetrated the US carrier defences, and in a series of running aerial battles with Grumman Wildcat fighters they successfully bombed USS *Enterprise*, scoring three direct hits and causing a serious number of casualties among the ship's company, largely because all American carriers had wooden flight decks, easily penetrated by well-aimed heavy-calibre bombs. During the course of this two-day air war the Imperial carrier bombers were joined by two Rabaul-based air corps, supporting the 2nd, the 6th and the Kisarazu Air Corps, all flying out of Rabaul Naval Air Station.

In a retaliatory air battle the carrier HIJMS *Ryujo* and the destroyer HIJMS *Mutsuki* were sunk by US planes from *Saratoga*, despite the valiant efforts of the A6M2 fighter pilots defending the Imperial squadron, but the most important advantage of this two-day air war was the successful landing of Imperial troops of the Army's 35th Brigade and 18th Division onto the beaches of Guadalcanal. The total number of men landed amounted to some 19,000.

In August 1942, the Kawanishi Company, celebrated for its flying-boat production, manufactured ninety-seven Kawanishi N1K1 Kyofu float fighter planes. This machine was designed for beach base operations, having a speed of 302 mph at an altitude of 18,680 feet and powered by a Mitsubishi Kinsei engine of 1,460 hp. With a wingspan of 39 feet 4½ inches, the fighter was armed with two 20 mm cannon and two 7.7 mm machine-guns. Later, in November 1942, this machine was produced as a land-based fighter under the designation Kawanishi N1K1-J Shiden.

Sometime in mid-September 1942, the submarine *I-25* surfaced off the west coast of the United States of America and launched a small Type-O reconnaissance seaplane. This machine dropped four 125 lb bombs while flying over the forest lands of Oregon in an attempt to start a major conflagration and disrupt the life and industry all along the west coast. The machine was successfully recovered and *I-25* continued on her patrol schedule.

The Naval Air Service now insisted during September in overriding the civilian and military objections to installing the Nakajima Homare engine in the new Mitsubishi A7M land-based fighter. The Mitsubishi Company was officially notified of this decision despite its dissatisfaction with this intention. However, the new plane was fitted with a bubble-type canopy, as well as a slender upper fuselage, to facilitate better vision for the pilot.

The chief designer of the Mitsubishi A7M1 land-based fighter advised the Navy that the reduced performance of the Nakajima Homare engine would necessitate reducing the plane's size, performance and equipment previously listed as being necessary for efficient operational usage. To obtain a superior performance, the use of flaps to achieve maximum manoeuvrability had not yet been proved. The drift computer had to be withdrawn and the number of rounds of ammunition were reduced, but the Navy was prepared to accept these recommendations.

In October 1942, the Naval Air Service was prepared to accept the J2M2 Raiden II as a service interceptor since Mitsubishi had commenced the flight testing. The new engine exhaust stacks were open and the engine exhaust gases gave the plane a certain jet propulsion effect, but though the plane was fitted with a four-bladed propeller the engine was rough, causing vibrations, and this could not be accepted for operational use. This vibration was eventually cured by increasing the propeller rigidity and increasing the resilience of the fore-and-aft direction of the engine-mounting rubber shock-absorbers. These engineering changes appeared to make the new fighter acceptable for production and eventual issue to operational squadrons.

At this time the opening phases of the Guadalcanal campaign had been undertaken, and for the Imperial Naval Air Service it had been necessary for the carrier fighters to fly 645 miles from Rabaul to the actual theatre of operations. Thus certain modifications were undertaken to give the Zero fighters increased range and extra combat time while over the front line. Extra fuel tanks were attached to each wing, increasing the tankage by twenty gallons, so that the cruising range was increased by one hour of flying-time. Modifications were made to the wings, restoring the wingspan to its original length, which improved the range and turning characteristics. Wing-folding equipment was introduced to facilitate ease of stowage while aboard aircraft carriers. Thus the Mitsubishi A6M3 Models 22 and 32 were born, and during the course of 1942, a total of 560 Model 22s were manufactured in Nagoya, in Japan. The new fighters were rushed to the South Pacific area to the naval air stations at Buin and Buka in New Guinea, as well as Guadalcanal. These machines had the prime responsibility of providing escort cover for naval ships from the northern bases and the Guadalcanal island area.

The third air battle for Guadalcanal, 26 October 1942

The third air battle for Guadalcanal island took place between 25 and 26 October 1942, off the Santa Cruz Islands, in the South Pacific, when HIJMS *Zuikaku*, *Zuiho* and *Junyo* aircraft carriers of the 1st and 2nd Air Wing of the carrier strike force were cruising under way to Guadalcanal. The Imperial naval squadron had made contact with the US carriers *Hornet* and *Enterprise* of the 16th and 17th task force and promptly launched three aerial attacks on the American ships. A total of fifty-nine Mitsubishi A6M3 carrier fighters, fifty-nine Nakajima B6N2 bombers and fifty-one Aichi D3A1 dive-bombers struck the American ships. Sweeping in across the ocean, the Imperial planes were attacked by US Grumman Wildcat fighters, and during the course of three separate attack missions seventy-four American planes were destroyed in a series of combats, twisting and turning, diving, zooming, with victims streaking seawards in flaming plumes to crash in an explosion of watery foam. But the Imperial naval bombers got through the US naval defences and struck mortal blows at *Hornet*, which subsequently sank, as did the US destroyer *Porter*. USS *Enterprise* was severely damaged, but remained the only operational US carrier in the Pacific.

But the Imperial Naval Air Service had to pay a price, for sixty-nine Nipponese aircraft were lost, with two Imperial carriers and one cruiser damaged and one cruiser sunk. A total of five out of six Imperial air companies had been lost, or 40% of the pilot strength of the 1st and 2nd Air Wings of the carrier strike force had gone. The Imperial naval forces withdrew. However, USS *Porter* was sunk later.

Late in 1942, HIJMS *Ryujo* (Dragon Phoenix) joined the carrier strike force. She had been converted in December 1941 from a former submarine tender, HIJMS *Taigei*, originally completed in 1934. Unfortunately, on 12 November, aircraft from USS *Enterprise* in a surprise raid sank the battleships HIJMS *Hiei* and HIJMS *Kirishima*, as well as one cruiser and three destroyers. In the course of this action the US task force lost two cruisers –*Atlanta* and *Juneau* – as well as seven destroyers. The lesson of this battle lay in the fact that only forty-nine Rabaul-based naval aircraft were available to defend the fleet, and more naval air squadrons should have been activated. By mid-November 1942, the Imperial Naval Intelligence Service obtained clear proof that the Americans had the operational usage of the twin-engine Lockheed P-38 Lightning fighter for missions in the South Pacific area.

As a result of the P-38 aerial activities and their strategic implications, during December of 1942, the 6th Joint Air Group, 12th Air Group, Hiyoshi Air Group and the 76th Direct Command Squadron of the Imperial Army Air Corps arrived in Rabaul. The duty of these air units was the defence of Rabaul and the New Guinea theatre of operations. Meanwhile the Naval Air Service became responsible for air operations in the Solomon Island area.

At this stage of the development of new machines for naval operations, a policy split appeared between the Mitsubishi design staff under the direction of Dr Horikoshi and the Naval Air Service concerning the new land-based fighter, the Mitsubishi A7M Reppu. The Mitsubishi engineers wanted to go for high-speed performance, but the Yokosuka Experimental Air Corps only required high-speed performance without the loss of manoeuvrability. But the Naval Aircraft Research and Development Command opposed these views, insisting upon manoeuvrability equal to that of the A6M3 carrier fighter, and considered the use of automatic combat flaps for sharp turning-radii at low speeds. Furthermore the NARDC required the installation of the Nakajima Homare engine NK9H, which had completed flight tests. The Mitsubishi concept had envisaged an aircraft that had to be small, high powered and with some equipment sacrifice to compensate for the heavier motor. Engine comparison data were submitted to the Naval Air Service for consideration, which indicated that the Mitsubishi engine possessed a higher horse power performance than the Nakajima power plant. The engineers at Nagoya waited impatiently for the Navy's reply.

During the course of 1942 the Naval Air Service, in association with the engineers of the Mitsubishi Aircraft Company, continually developed the G4M series of bombers. This was necessary owing to a lack of a comparable design to the American Boeing B-17 Flying Fortress and the B-29 Superfortress four-engine bombers. The lack of a long-range four-engine bomber was a serious omission from the Naval Air Service's inventory of offensive aircraft, which might have turned the scales in the Japanese favour in the air war over the South Pacific. This was particularly so in view of the fact that Admiral Yamamoto had conceived the policy of long-range bomber operations.

After the Mitsubishi G4M1 became operational, the G4M2 Model II series was introduced, and as a result of operational combat experience the Model 21 eventually replaced the earlier machine. The differences from the Model II were revised glazed flush panels in the waist, replacing side blisters. The tail cone was cut out for ease of movement for the 20 mm tail cannon. The power plants now installed were two Mitsubishi MK4E Kasei 15 motors, for high-altitude rating. A carbon dioxide fire extinguisher system was installed as standard, while the wing tanks had layers of rubber placed beneath each installation, and similar rubber strips with foam were located to the lower section of the fuselage fuel tanks.

The G4M2 Model 22 rapidly followed, resulting from the combat lessons of the air war in the South Pacific while engaged with American aircraft, and featuring a revised armament layout. This comprised the installation of one 7.7 mm machine-gun in the nose and one located either side of the nose, as well as two 7.7 mm machine-guns in side-beam hatches, one 20 mm cannon firing from the tail position and one from the mid-upper turret gunner's position. The power plants fitted were two Mitsubishi MK4P Kasei 21 engines rates at 1,800 hp each. A third G4M2 was tested with electrically operated bomb doors. From production aircraft number 65, all G4M2 Model 22 bombers were fitted with bulging bomb doors, and continued in production with a run of 274 machines.

The machine following was the G4M2, a Model 22a, which differed from the Model 22 in that the machine-guns mounted in the side-beam hatches were replaced by 20 mm cannon. A production batch of fifty of these machines was produced. It is interesting to record that these heavy 20 mm cannon were not mounted in a power-operated gun turret, but were manually operated by means of moving the pistol grip underneath the weapons. Movement of the pistol grip activated a series of electric motors, which altered the alignment of the weapon in elevation and azimuth in accordance with the gunner's desired aiming position.

The G4M2 Model 24J was similar to the G4M2a, but was modified to carry the Navy's suicide attacker, the Ohka (Cherry Blossom) Model II piloted bomb. With the bomb doors removed, the Ohka was carried under the belly of the mother plane and released within a short distance of the target ship. The Ohka was rocket propelled by three solid-fuelled rocket motors, which enabled the suicide craft to plunge into the target ship at very high speed, exploding the 1,760 lb warhead on impact. By 31 October 1942, the 1st Kokutai, later redesignated the 752nd Kokutai, was the only unit of the Imperial Naval Air Service to operate all the various types of the G4M2 bomber series throughout the war. As compared to the production statistics of aircraft manufactured for the more sophisticated Western air forces, bomber production of the G4M2 seemed pitifully small. Production of the G4M2 1942 was as follows:

January–March	82 planes
April–June	93 planes
July–September	84 planes
October–December	112 planes
Total	**371 planes**
Plus previous total production	194 planes
Grand Total	**565 planes**

Average production per week, January–December = 7.137

In December 1942, anticipating the forthcoming air battles in the Solomon Islands and New Guinea, the Air Command of the Imperial Naval Air Service embarked upon certain reorganisations. The Navy erected the South-East Fleet Command with headquarters situated in Rabaul, with the following operational units:

1. The 4th Fleet – 902 and the 952nd land-based Air Corps
2. The 11th Fleet – 22nd Air Wing, 252nd, 701st, 755th Air Corps
3. The 24th Air Wing – 201st, 552nd, 703rd, 752nd Air Corps
All land-based:
1. The 25th Air Wing – 251st, 801st, 702nd Air Corps
2. The 26th Air Wing – 705th, 582nd, 204th Air Corps, 802nd Air Corps

With the finalisation of the organisation of the South-East Fleet Command, the Navy turned its attention to the aircraft carrier arm of the carrier strike force. By 1942, the following aircraft carriers had been successfully built, and included:

- HIJMS *Juno*
- HIJMS *Hiyo*
- HIJMS *Unyo*
- HIJMS *Chuyo* converted from the NYK liner Nitta-Maru of 20,000 tons
- HIJMS *Ryujo* converted from submarine tender HIJMS *Daigai* of 15,000 tons

All these carriers were placed under the control of the Combined Fleet, and with the exception of HIJMS *Shokaku* and HIJMS *Zuikaku* were small carriers.

Operational bases used by the Mitsubishi G4M2 attack-bombers during 1942

Unit	Date	Bases	Attached to Unit
1st Kokutai		Tainan, Davao, Kendari,	21st Koku Sentai
		Ambon, Truk, Rabaul	
	Fr. 1 Apr	Rabaul, Taroa, Mili	24th Koku Sentai
	31 Oct	Redesignated 752nd Kokutai	
4th Kokutai	Fr. 10 Feb	Truk, Vunakanau, Rabaul, Lae	24th Koku Sentai
	10 Apr	Vunakanau, Lae, Surumi, Truk	24th Koku Sentai
		Kisarazu	
	31 Oct	Redesignated 702 Kokutai	
701st Kokutai	1 Nov	Redesignated from Mihoro	22nd Koku Sentai
		Kokutai Tinian, Roi, Vunakanau	
702nd Kokutai	1 Nov	Redesignated from 4th Kokutai	25th Koku Sentai
		Kisarazu	
703rd Kokutai	1 Nov	Redesignated from Chitose	24th Koku Sentai
		Kokutai Rabaul, Wake, Roi	
Misawa Kokutai	1 Mar	Kisarazu, Misawa, Saipan	26th Koku Sentai
		Vunakanau	
705th Kokutai	1 Nov	Redesignated from Misawa	26th Koku Sentai
		Kokutai, Vunakanau	
Kisarazu Kokutai	April	Kisarazu, Marcus Island, Vunakanau, Yokohama Naval District	
707th Kokutai	1 Nov	Redesignated from Kisarazu Kokutai Vunakanau	26th Koku Sentai
	30 Nov	Deactivated	
Kanoya Kokutai	until	Thudaumot, Davao, Kendari,	21st Koku Sentai
	1 Oct	Gelvembang, Kisarazu, Sabang Kcwieng	
	Fr. 1 Oct	Kewung, Sabang	21st Koku Sentai

Unit	Date	Notes	Sentai
752nd Kokutai	1 Nov	Redesignated from 1st Kokutai Kisarazu, Wake, Mili	24th Koku Sentai
Kanoya Kokutai	Until Oct	Thudaumot, Davao, Kendari, Gelvembang, Kisarazu & Sabang	21st Koku Sentai
751st Kokutai	Fr. 1 Oct	Redesignated Kavieng, Sabang	21st Koku Sentai
752nd Kokutai	1 Nov	Redesignated Fr. 1st Kokutai Kisarazu, Wake, Mili	24th Koku Sentai
Takao Kokutai	Oct	Takao, Jolo, Kendari, Koepang Rabaul	
753rd Kokutai	Fr. 1 Oct	Redesignated Takao Kokutai Kendari, Koepang	23rd Koku Sentai
755th Kokutai	1 Nov	Redesignated Wonsan Kokutai Kisarazu	22nd Koku Sentai

Operational bases by units equipped with G4M attack-bombers

Yokohama Naval District	21st Koku Sentai	22nd Koku Sentai	23rd Koku Sentai	24th Koku Sentai	25th Koku Sentai	26th Koku Sentai
	Ambon					
	Davao					
	Gelvembag					
			Jolo			
	Kavieng					
	Kendari		Kendari			
Kisarazu	Kisarazu			Kisarazu	Kisarazu	Kisarazu
			Koepang	Lae		
Marcus Island						
				Mili		
						Misawa
	Rabaul		Rabaul			
	Roi	Roi		Roi		
						Saipan
					Surumi	
		Tainan				
			Takao			

				Taroa		
	Thudaumot					
		Tainan				
	Truk			Truk		
	Sabang					
Vunakanau		Vunakanau		Vunakanau		Vunakanau
				Wake Island		

Units attached to the Koku Sentai

21st Koku Sentai	22nd Koku Sentai	23rd Koku Sentai	24th Koku Sentai	25th Koku Sentai	26th Koku Sentai
751	701	-	702	702	705
752	755	-	703	-	707
-	-	752	-	-	-

In the course of 1942 a number of technical developments took place among the aircraft and ships built for the Naval Air Service. The Nakajima Aircraft Company designed the JIN as a long-range reconnaissance aircraft, designated Type 2 land-based reconnaissance aircraft. Later this machine was developed as a twin-engine night-fighter known as the Gekko, or Moonlight, of which 477 examples were manufactured and delivered to operational squadrons.

Between January and November 1942, HIJMS *Chuyo* and HIJMS *Unyo* were converted to aircraft carriers from liners by the Kure Navy Yard. Though larger, faster and capable of accommodating more aircraft than Allied merchant ship conversions, they lacked necessary aircraft-handling equipment. These two aircraft carriers were used mainly as aircraft ferries and as training carriers.

Conversion to aircraft carrier specification of HIJMS *Chiyoda* commenced in December 1942, and that of her sister ship HIJMS *Chitose* a year later. Originally both ships had been designated as seaplane tenders and ordered under the 2nd Fleet Replenishment Law, being completed in 1937 and 1936 respectively. During 1941, HIJMS *Chitose* had her stern cut open and two large steel doors fitted. The seaplane storage decks were pared down to the waterline aft and part of the seaplane storage space was fitted out for launching twelve midget submarines in seventeen minutes. In July 1941, HIJMS *Chiyoda* was converted to the same specification, and eventually it was decided to reconvert the vessels to aircraft carriers, commencing in December 1942. The conversion rebuild took approximately twelve months to complete in the case of both ships, and they formed part of the 3rd Carrier Division upon completion. Later, in July 1944, the anti-aircraft armament was increased to forty-eight 25 mm anti-aircraft cannon. Eventually both ships formed part of Admiral Ozawa's decoy force and were sunk at the Battle of Leyte Gulf on 25 October 1944.

Details of the *Chiyoda* Class

Displacement	11,190 tons
Dimensions	621½ × 68¼ × 24½ ft
Machinery	4 shaft geared turbines, 44,000 shp
	Diesels 12,800 hp = 29 knots
Armament	8 × 5-in. DP AA guns, 30 × 25 mm cannon
Aircraft	30
Built	Kure Navy Yard
	Chitose 29.11.36, *Chiyoda* 19.11.37
Converted	*Chitose* January 1943, *Chiyoda* December 1942
Fate	Sunk at Battle of Leyte Gulf, 25.10.44

The conversion of the former luxury liner *Argentina Maru* commenced in December 1942 and was completed in the November of the following year as the aircraft carrier HIJMS *Kaiyo* by the Mitsubishi Nagasaki Ship Yard. The diesel engines were replaced by destroyer-type steam turbines, the smallest type of merchant ship conversion. The ship was first used as a ferry conveying aircraft to the southern islands, but after the Battle of the Philippine Sea, *Kaiyo* was used as a training aircraft carrier. Later, in July 1944, the anti-aircraft armament was increased to forty-four 25 mm anti-aircraft cannon. But in August of 1945, HIJMS *Kaiyo* was torpedoed in Beppu Bay by USN torpedo-bombers.

Details of HIJMS *Kaiyo*

Displacement	13,600 tons
Dimensions	545½ × 69 × 26½ ft
Machinery	2 shaft geared turbines, 52,000 shp = 23¾ knots
Armament	8 × 5 inch AA, 24 × 25 mm AA
Aircraft	24
Complement	829
Built	Mitsubishi Nagasaki Yard, 09.12.38
Fate	Torpedoed at Beppu Bay, 10.8.45

Meanwhile HIJMS *Shokaku* and HIJMS *Zuikaku* had been transferred from the 5th Air Fleet to the 1st Air Fleet.

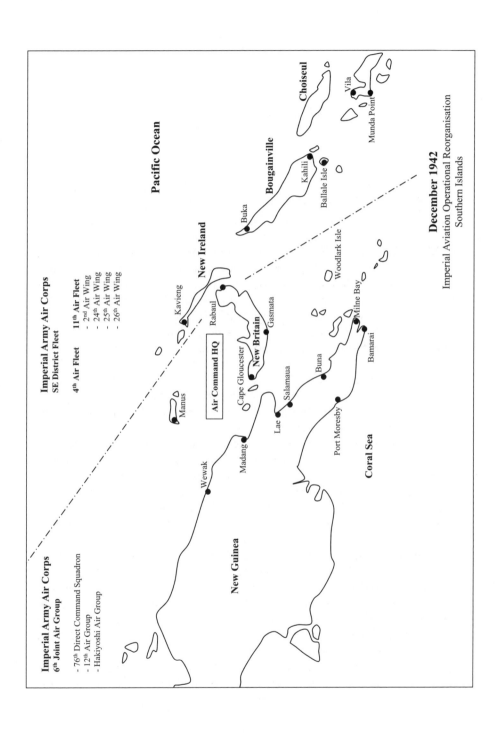

Imperial Army Air Corps
6th Joint Air Group

- 76th Direct Command Squadron
- 12th Air Group
- Hakiyoshi Air Group

Imperial Army Air Corps
SE District Fleet

4th Air Fleet

11th Air Fleet
- 2nd Air Wing
- 24th Air Wing
- 25th Air Wing
- 26th Air Wing

Air Command HQ

December 1942
Imperial Aviation Operational Reorganisation
Southern Islands

New Guinea

Coral Sea

Pacific Ocean

New Ireland

New Britain

Bougainville

Choiseul

Wewak

Madang

Manus

Lae

Salamaua

Buna

Port Moresby

Bamarai

Milne Bay

Woodlark Isle

Cape Gloucester

Gasmata

Rabaul

Kavieng

Buka

Kahili

Ballale Isle

Vila

Munda Point

CHAPTER 17

Flying High

During the course of the year 1943, aircraft production by the Japanese aviation industry had reached its peak.

The responsibility for bomber production was mainly in the hands of the Mitsubishi Aircraft Company of Nagoya, where the G3M2 and G4M2 series of attack-bombers were being manufactured for the Imperial Naval Air Service. The origin of the parallel production of these two types for the Navy lay in a sharp division of opinion in policy that existed in naval circles at that time, resulting in a total of 1,048 aircraft of the G3M type being manufactured by December 1943. At the same time the total production of the G4M series in the same period of time totalled 1,225 machines. This figure was made up of 148 machines built from January to March, 168 from April to June, 171 between July and September and 173 in the period October to December. This was a total of 660 aircraft of this type built for the year 1943; it was the principal bombing aircraft used for offensive operations by the Naval Air Service, and it was kept under constant updating by its manufacturer.

The Mitsubishi G4M2 Model 22b was developed from the Model 22a, but was equipped with increased armament, comprising four 20 mm cannon, which were belt-fed Type 99 Model 2 weapons, replacing drum-fed Model 1 aerial cannon. Only a few of this type were built. Hitherto the manufacture of the aircraft had been held up owing to production delays in building the Mitsubishi Type 21 Kasei engines, but quantity production commenced during July 1943. The Model 24 of this machine had the same quantity of armament as the Model 22, but was powered by two 1,825 hp Mitsubishi MK41 Kasei 25 engines. This was followed by the Model 24a, which possessed the same armament as the Model 22a and was powered by the same type of motor as was fitted to the Model 24, but only fifteen machines of this variant were manufactured. The next model was the Model 24b, of which 171 examples were built, and they differed from previous examples in that they retained the same armament as the Model 22b and were powered by the engines of the Model 24. In the Model 24c, which was the major production variant of the G4M2 attack-bomber, the nose armament was revised and replaced by a single 13 mm heavy machine-gun. The next variant was the Model 25, which was equipped with two Mitsubishi MK47-B Kasei 25B engines producing 1,825 hp, which were modified to test the motors at higher altitudes. This step was followed by the Model 26 in which the Kasei 25B engines were modified to take turbo superchargers. The Model 27 was modified as an engine test-bed, and had installed as a modification bulging bomb-bay doors. Eventually the new Mitsubishi G4M3 attack-bomber was designed and introduced into squadron service. The design work was led by engineers Takahashi and Kuroiwa, who planned a new single-spar wing plan housing self-sealing fuel tanks, with a reduction of the tankage capacity from the previous 4,490 litres. The fuselage length was shortened, which resulted in a redistribution of the weight layout, upsetting the centre of gravity of the machine. The tail turret was modified so as to increase the field of fire of the tail-end 20 mm cannon. Armour protection was provided for all crew members, and the plane was characterised by a dihedral on the tail surfaces.

An interesting aviation development was the Kawanishi N1K1 Kyofu, or Strong Wind, aircraft, introduced into squadron service in 1943 as a single-float-plane fighter aircraft for beach-head operations. Development of this machine commenced in 1940, and by 1942 it had been successfully flight tested. With a span of 39 feet 4½ inches and an overall length of 34 feet 9 inches, the aircraft was powered by a Mitsubishi double-row radial Kasei engine rated at 1,460 hp, which gave a maximum speed of 299 mph at an altitude of 18,700 feet, with a service ceiling of 34,645 feet. It was a conventional fighter plane with a single float and two outrigger floats, propelled by contra-rotating propellers. As a float-plane this machine was designated Type 2 sea fighter, of which 327 examples were manufactured. Armament comprised of two 20 mm cannon and two 7.7 mm machine-guns.

Meanwhile, Kawanishi has developed the N1K2-J Shiden land-plane from the N1K1 Kyofu float-plane fighter. The Shiden design commenced during November 1942, and by July 1943 four prototypes were being flight tested. As a result of the successful flying development programme, quantity production commenced during December 1943. This type was met by Allied Air Forces in most theatres of the Pacific Ocean in the course of the early part of 1944. In the course of time the Kawanishi Aircraft Company and the associated company Himeji manufactured 1,007 machines of this type. Eventually three different standards of armament were installed in the course of engineering changes and production developments. They comprised the N1K1-J with two 7.7 mm machine-guns and four 20 mm cannon, the N1K1-Ja with ventral bomb-rack attachments under the fuselage and four 20 mm cannon, and the N1K1-Jb, which had rocket booster motors attached to the wings and was armed with four 20 mm cannon as well. A later design was the N1K2 Shiden Kai, which had a low configuration as opposed to the previous machine of the same type with a mid-wing layout. The Shiden Kai possessed two thirds of the components of the previous N1K1 model, which resulted in a significant weight saving of some 500 lb. The first prototype batch was ready to be wheeled out of the hangars on 31 December 1943. This Kawanishi fighter type was to be built by seven shadow factories and the parent construction plant, but as a result of systematic bombing by United States Air Force Boeing B-29 heavy bombers, only a quantity of 428 were eventually delivered to operational squadrons. This Shiden land-based fighter was subsequently powered by a Nakajima Homare 21 radial engine developing 1,990 hp for a maximum speed of 369 mph, at a service ceiling of 18,370 feet. With a loaded weight of 9,039 lb, a range of 1,066 miles was achieved at cruising speed. The span measured 39 feet 4 inches, with a length of 30 feet 8 inches and a height of 13 feet. Offensive armament comprised four 20 mm cannon, two 7.7 mm machine-guns and two 550 lb bombs for ground attack operations.

In late 1943 the directors of the Mitsubishi Aircraft Company took a policy decision that further development of the A6M5 carrier-based fighter could continue, owing to an already too small technical and flying staff already committed to designing a new fighter plane. Mitsubishi did not posses sufficient technical personnel to run a further aircraft development programme. However, aircraft production could not stop but continued, although manufacturing schedules failed to supply machines, now under acute shortage at the various fronts, in sufficient numbers to match the American aircraft industry's fabulous output. Certain modifications were possible, and included the installation of belt-fed ammunition for the 20 mm cannon, increasing the available rounds from 100 to 125 cannon shells per gun. Heavier-gauge metal skin was used to cover the wing spars, and under test the new version of this plane exceeded an indicated air speed of between 410 and 460 mph. In all, Mitsubishi manufactured 757 aircraft of the A6M5 Model 52 variant. The invasion of the island of Guadalcanal in the South Pacific by US forces was now at its peak, and despite a high production and increased performance of the new Zero fighters, a serious shortage of naval planes existed. Several weeks later the Imperial Naval Air Service was forced to abandon the Imperial naval air stations at Rabaul.

The Mitsubishi J2M1 or 14-Shi land-based fighter design was not yet ready, being far behind on schedule. The United States planes were now able to outpace the earlier Zero models, but it was thought that by simplifying production methods, reducing wingspan, eliminating wing folding mechanisms and installing individual exhaust stacks to the motors a better performance could be achieved. On 16 June 1943, Lieutenant-Commander Hoashi HIJMN took off from Suzuka airfield to undertake vibration tests with a new propeller with resistance blades. At seventy feet the Mitsubishi J2M2 went into a nose-dive, crashed into a barn, slid into a barley field, burst into flames as the fuel tanks blew up and killed the pilot. This was a serious setback for the flight-testing programme. In the succeeding September a Mitsubishi test pilot, Eisaku Shibayama, took off in plane number 10 of the J2M2 Model II type. Upon retraction of the undercarriage the plane commenced to nose-dive, the pilot unable to return the control column to the climbing position. The undercarriage was lowered and the pilot recovered his control of the machine to land under normal circumstances. Upon close inspection it was discovered that the curved shock-strut of the tailwheel was located against the elevator torque tube lever after retraction. This jammed the control column in the dive position – a serious design fault, which had not been observed by the ground mechanics while inspecting the machine prior to take-off. In October 1943, the 14-Shi interceptor, the Mitsubishi J2M1 Raiden, was adopted by the Navy, but it was several months before quantity production was to make large numbers available over various battlefields. As a consequence the Mitsubishi A6M2 was the only fighter type to defend the skies above Rabaul before the Naval Air Service finally withdrew its units. Later armament revisions were to take place, with 20 mm cannon replacing the 7.7 mm machine-guns installed on earlier variants of the J2M1 and J2M2 Model II fighters. The Mitsubishi J2M3 Model 21a possessed four Type 99 Model 2 20 mm cannon as well as minor equipment changes, including the strengthening of the wings to retain torsional rigidity, so allowing a dive at an indicated air speed of 460 mph. In December 1943, the Imperial Naval Air Service accepted the Mitsubishi J2M2 Model II into service operation, with the first production models being delivered to Toyohashi Naval Air Station to the 381st Air Corps for training and familiarisation purposes.

By October 1943, the final delivery of the Mitsubishi A7M 17-Shi fighter was now uncertain. It had emerged from an idea as a Mitsubishi design originated by Dr Horikoshi, the chief designer. Late in 1942 and early 1943 the performance of US fighters was surpassing the performance of the Zero with the same range and manoeuvrability. Serious delays had occurred in engineering design work, and in the prototype construction schedule in the original flight-testing programme, and work on the A7M type had been stopped at various times, by the intervention of an earthquake and not least by the heavy bombing of the American Boeing B-29s. To rectify the acute aviation situation, the Imperial government established the Ministry of War Supply in November 1943, in which a division of responsibility between the Imperial Army Air Corps and the Imperial Naval Air Service was made, with units responsible for the development of specific planes. Hitherto the Naval Bureau of Aeronautics had controlled prototype departments of the aviation industry, but this responsibility was vested in the Naval Aeronautical Research and Development Corps, and production organising was now a separate delegated affair.

In the course of 1943 the Nakajima Aviation Company produced the C6N1 Saiun, specifically as a ship-borne reconnaissance plane, of which 498 were subsequently manufactured. During 1943 a total of twenty-three prototypes were built in various variants, with the C6N1-B as a torpedo-bomber, the C6N2 equipped with a Homare 24 engine and the C6N3 with a supercharged 2,000 hp Homare NK9L-L engine, but most important was the C6N1-S, a night-fighter version armed with two 20 mm cannon. In August 1943 the Nakajima J1N1-S night-fighter appeared, originally made as a reconnaissance plane and later adapted, with 479 planes being manufactured in this series. The J1N1-C and the J1N1-S were used by the Imperial Naval Air Service over the Solomon Islands from the spring to the autumn of 1943. The introduction of the Nakajima J1N1-S into the night skies over the Solomon Islands heralded the adoption of new night-defence tactics as conceived by Commander Yasuma Kozono HIJMN. These consisted of the use of twin 20 mm cannon aboard the aircraft amidships, fixed at an angle of 30 degrees forward and upwards, as well as twin 20 mm cannon at an angle of 30 degrees fixed forwards and downwards. Hitherto the US Liberator bombers had flown nightly some 2,000 miles to attack the Imperial bases and island installations, which stretched out like a string of pearls across the South Pacific. Consequently anti-aircraft gunfire had been the only means of defending Imperial naval locations, until the introduction of the Nakajima J1N1-S heavy fighter into the area. The original idea for the re-angled cannon was first conceived in Great Britain in 1925, when the Westland Aircraft Company of Somerset built two Westbury twin-engine, three-seater biplane fighters, armed with a 37 mm cannon manufactured by the Coventry Ordnance Company, which was fixed to fire forwards and upwards over the upper wing. Due to a serious lack of funding for aviation development, the British government did not continue the development of this revolutionary aviation idea. However, the project was resurrected in 1931 when Westland successfully designed the Westland COW Gun Fighter. This machine was a single-seater, low-wing monoplane equipped with a single 37 mm cannon fixed to fire forwards and upwards, and flanking the side of the cockpit. Once again the British government abandoned the idea due to a shortage of money for defence purposes and to supply the needs of other operational requirements. But in 1943, within the capable hands of the Imperial Naval Air Service, the original British idea was highly successful when installed on the Nakajima J1N1-S night-fighters.

An aircraft design initiated by the First Naval Air Technical Arsenal, Yokosuka Naval Air Depot, commenced in 1940: this was the Yokosuka PY torpedo-bomber. Eventually production was undertaken by the Nakajima Aircraft Company, and subsequently 1,002 machines of this design were manufactured. A night-fighter version of this twin-engine aircraft was developed under the designation P1Y2-S Kyokko, which was manufactured by the Kawanishi Aircraft Company. Equipped with two Mitsubishi Kasei 25 engines, each rated at 1,850 hp, and with aircraft interception radar, the plane was armed with three 20 mm cannon. A production run of only ninety-seven night-fighters was completed. The next version was the P1Y3-S, which differed from previous versions in that Nakajima Homare engines were installed for development purposes.

The original P1Y1 Ginga (Galaxy) achieved a maximum speed of some 345 mph when undertaking service trials in 1943 and with a range of 3,300 miles the Imperial Naval Air Service was delighted. So much so, in fact, that the discontinuation of the Mitsubishi G4M2 attack-bomber was seriously considered at the time. The machine possessed a span of 65 feet 7½ inches, a length of 49 feet 2½ inches and a height of 14 feet 1¼ inches, and the all-up weight loaded registered 23,148 lb. Armed with one 20 mm cannon and movable 13.5 mm machine-gun, the plane possessed an offensive load of 1,760 lb of bombs or was capable of carrying one 1,875 lb torpedo. The production and servicing of this machine by Imperial Naval Air Service units in the field was very difficult. The new machine was plagued with troubles, since the aircraft had been designed with a complicated structure. As a consequence, on entry to operational status this new twin-engine bomber had a bad reputation with its crews.

At this time the Aichi Aircraft Company was manufacturing the 12-cylinder German aircraft engine known as the Daimler-Benz DB liquid-cooled inverted V DB-601. Prior to 1931 this company made the Lorraine W type water-cooled engine of 400 hp and 450 hp, which influenced subsequent engine developments. The Kawasaki Company was in production at the time with the German aero engines BMW6, BMW6a and the Daimler-Benz DB-601 inverted V. The early Mitsubishi aircraft engines built under licence included the British Armstrong Siddeley Jaguar, the Hispano-Suiza 12x and 12y motors and early Junkers types, while Nakajima had built Lorraine engines and the British Bristol Jupiter aeroplane motors. Meanwhile the Mitsubishi A6M2 carrier-based fighter was now facing the American Lockheed P-38 Lightning twin-engine fighter. Below an altitude of 10,000 feet the Japanese fighter had the edge on performance, but at a height of 15,000 feet the P-38 had a more pronounced performance, which increased with altitude.

With the lessons from the air ocean battles still fresh in mind, the 25 mm cannon anti-aircraft armament was increased in number on all the operational aircraft carriers, as follows:

Carrier	Increase
HIJMS *Ryujo*	to twenty-four 25 mm AA cannon
HIJMS *Taiyo, Chuyo, Unyo*	to twenty-four 25 mm AA cannon
HIJMS *Hiyo, Junyo*	to forty 25 mm AA cannon
HIJMS *Zuiho*	to forty-eight 25 mm AA cannon

This increase in anti-aircraft armament was additional to the installation of rocket mortars on the foredeck of aircraft carriers as part of the AA defence weaponry. Aircraft carriers were not the only boats to be fitted with increased anti-aircraft armament since many large seaplane carrier submarines had been constructed. These boats included seventeen in which one bridge-mounted 13 mm machine-gun was replaced by twin 25 mm cannon.

The *I-12* was a headquarters submarine, but installed with less-powerful electric/diesel motors of 4,700/1,200 hp as compared to the *I-9*, which had an output of 12,400/2,400 hp. For the *I-12* submarine the reduction in electric/diesel hp produced speeds of six knots surfaced and two knots submerged, with sea ranges of 2,200 miles surfaced at 16 knots and 70 miles at 3 knots submerged. The submarine *I-54* had less-powerful engines producing diesel/electric hp of 4,700/1,200 as compared to the more powerful *I-15* or *I-40* class, whose electric/diesel motors generated 12,400/2,000 hp respectively. The operational range of the *I-54* submarine was 21,000 sea miles at 16 knots surfaced and 105 sea miles submerged at 3 knots. The seaplane hangar was located forward of the conning tower, and the catapult ran the full length of the deck forward to the bows, while the Type 22 radar aerial was located on a mounting above the seaplane hangar. Specification of the submarine *I-54* included:

Displacement	2,140 tons
Aircraft	One
Dimensions	356½ × 30½ × 17 ft
Complement	100 men
Machinery	2 shaft diesel/electric 4,700/1,200 hp
Armament	1 × 5.5 inch gun, 2 × 25 mm AA cannon, 6 × 21 in. torpedo tubes and 19 torpedoes
Built	*I-54* Yokosuka Navy Yard, completed 1943
	I-56 Yokosuka Navy Yard, completed 1943
	I-58 Yokosuka Navy Yard, completed 1944

On 24 January 1943, when the US fleet was assembling for the Marshall Islands assault, His Highness Commander Prince Takamatsu HIJMN went to the Imperial Palace, Tokyo, to petition His Imperial Majesty the Emperor Hirohito. The nature of the petition was to place before His Imperial Majesty a new aerial strategy as outlined by the aviation enthusiasts of the Naval General Staff. It was accepted that aircraft wastage was currently due to the endless bickering between the Imperial Army Air Corps and the Imperial Naval Air Service over the question of the allotment of aviation resources. It was therefore respectfully suggested that the Emperor might consider the creation of a separate air force along the lines of the more sophisticated air forces of the Barbarian nations. Now Admiral Yamamoto had built up the Naval Air Service and its policy of long-range bombing and carrier attack formations against the wishes of many senior naval officers who required victories by big-gun sea actions, now totally out of date. Prince Takamatsu, a naval commander of the Naval General Staff, suggested turning the full naval budget to aircraft construction, considering that 10,000 aircraft, less than five months' output, would ultimately be required. The Emperor could not reject these suggestions out of hand, and finally agreed to study the proposals, which unless very delicately handled could lead to a very serious crises of confidence between the Army and the Navy, particularly at a time when a difficult and serious war was being fought against the most powerful Western industrialised nation.

The air battle off Rennell Island, 29 January 1943

On 29 January 1943, Imperial headquarters, Tokyo, was informed that aircraft of the 701st and 705th Air Corps had made contact with the US naval carrier task force off Rennell Island. A series of air battles followed in which the Imperial Navy lost one aircraft carrier, two battleships, four cruisers, eleven destroyers and six submarines, as against the US Navy, which sustained two carriers, eight cruisers, fourteen destroyers and one submarine sunk. The Imperial Navy lost some 300 planes, which was virtually matched by American losses.

Back in Tokyo, Emperor Hirohito summoned the Lord Privy Seal, Marquis Kido, for a military briefing. The Lord Privy Seal advised the Emperor against the acceptance of the tyrannical instructions issued by the military services for the prosecution of the war, and in particular against the persons of the Army and Navy chiefs. The air navy idea, however commendable, was too late and could not assure Japan of victory. This Kido understood, but it was difficult to remove political obstacles. These difficulties arose with the peace faction in finding witnesses to excuse the Imperial decisions. The Lord Privy Seal had an ally in the Prime

Minister, General Tojo, who it was thought would take responsibility for dismissing the service chiefs, who might agree to resign for 'health reasons'. On 21 February 1943, the Tokyo daily newspaper *Mainichi* published an article in which it was stated that 'wars cannot be won with bamboo spears'. This was immediately seized upon for investigation by the 'Thought Police'. The article continued with the editorial supporting the government, pointing out that more aircraft were required, as well as improved uses for these machines in the various operations. At last the public had been given a hint at the strategic differences of opinion that had arisen.

Seven days later, on 28 February 1943, the Emperor called for 7,000 troops to reinforce the New Guinea enclaves from the naval base at Rabaul. On 2 March, US planes located the Imperial convoy of escorted troop transports and attacked.

Air battle of the Bismarck Sea, 2–4 March 1943
The Imperial convoy was protected by forty Mitsubishi A6M3 carrier fighters, which were engaged by a force of US Lockheed P-38 Lightning twin-engine fighters. A series of running air battles was fought from an altitude of 30,000 feet down to 10,000 feet. These tactics were embarked upon to engage and occupy the Imperial Zeros so that a US bomber force could fly in at low level against the ships. A force of US B-25 and A-20 bombers roared in at high speed low over the water and 'skip' bombed the Imperial troop transports, using five-second delayed-action fuses, which subsequently exploded on or near the Japanese ships. The Imperial troops and ships' crews took to the boats and any life-rafts that were available. Then the US bombers and P-38 fighters strafed the helpless survivors in the sea with machine-guns and cannon blazing till the surface of the sea foamed in red froth. Well over 4,000 troops were killed as they floated in the sea. Upon the news being transmitted to Imperial headquarters, Tokyo, the Emperor communicated to Admiral Yamamoto stationed on Truk Atoll in the Caroline Islands the Imperial displeasure at the results of this disastrous action.

For Admiral Yamamoto the Imperial displeasure sounded the realisation that Allied air power in the New Guinea area and the Solomon Islands had to be destroyed, otherwise defeat would be a stage nearer. So, on 25 March 1943, Operation I, GO was planned with the object of destroying the Allied air forces in the area. The admiral transferred his HQ from Truk to Rabaul and moved in 300 Zero pilots for the air offensive.

Air superiority battle for New Guinea
A total of four air strikes were launched against US bases, facilities and

airfields in the New Guinea and Guadalcanal area, using formations of 100 and 200 Imperial naval aeroplanes with the hope of outnumbering the American opposition, for the US planes possessed longer range and could retreat to return from their bases at will. US ships and airfields were shot up, including the sinking of six Allied ships as well as the shooting-down of twenty-five US planes for the Imperial loss of forty-two machines. In the fourth airstrike 174 naval planes took part in an attack on the city of Port Moresby, now used by General MacArthur as Southern Pacific Headquarters.

Meanwhile the Emperor had seen fit to upbraid General Sugiyama, Chief of the Army Staff, over the total failure of the reinforcing mission, which ended so ingloriously in the Bismarck Sea.

The operation had commenced on 7 April with the attacks on US bases and facilities in Guadalcanal by 224 fighters and bombers from the Imperial Naval Air Service. On the following day, persistent air attacks had been launched in the New Guinea area against facilities at Oro Bay and Milne Bay. Ultimately, by 12 April, a big aerial offensive had been launched against Port Moresby and General MacArthur's headquarters. On analysis of the bombing results, Admiral Yamamoto saw the air strategy as failing. He was determined to make a tour of inspection of the forward bases used by the Imperial forces.

The running of the fox to earth
The US Navy radio stations on 13 April picked up a radio transmission originating from Admiral Yamamoto's headquarters in Rabaul. It was known as message number JN25, the Imperial fleet code number, which was transmitted in a low-security fleet code. This message outlined an inspection itinerary to be undertaken by Admiral Yamamoto on 18 April. The first stop for inspection was to be the island of Ballale, off southern Guadalcanal, to see the troops under the command of General Maruyama subsequent to the division's ordeal on Guadalcanal. But the general warned the admiral of the dangers of an aerial ambush after he had narrowly escaped a similar situation from US fighter planes over Bougainville. The commander of the 11th Naval Air Fleet also warned Admiral Yamamoto, as did Commander Watanabe HIJMN, Air Staff officer, who eventually wrote out a coded message for transmission to forward units. He was assured by the coding officer that the code was unbreakable as from 1 April 1943.

Moments after the message was transmitted at 1755 on 13 April by Imperial naval radio, the message was intercepted by the US naval radio listening station at Dutch Harbor in the Aleutian Islands and by the naval radio listening station at Wahiawa on the island of Oahu in the Hawaiian

Islands group, by the US Naval Combat Intelligence headquarters, Pearl Harbor, 'the men in the cellar', the Midway code-breakers under the command of Commander Edwin Layton USN.

Eventually the 'men in the cellar' had decoded the message into a plain-language Japanese script, which was translated by Lieutenant-Colonel Alva Lasswell USMC, the Marine Corps' language officer, with coded place-names being recognised from previous decoded messages.

Decoded transcripts of the message were received in Washington by the Deputy Chief of Naval Intelligence, Captain Ellis Zacharias USN, who communicated the information to President F.D. Roosevelt.

Meanwhile, at Pearl Harbor, the Fleet Intelligence Officer, Commander Layton USN, had an interview with Admiral Nimitz at which a discussion took place as to the feasibility of assassinating Admiral Yamamoto! If Admiral Yamamoto was assassinated did any suitable replacement exist? No! Would any retaliation take place against a similar high-ranking US naval officer? No! The eventual decision for this operation, stated Admiral Nimitz, would have to be taken by the President himself.

That night, 13 April, in the Imperial officers' mess at Rabaul, Admiral Yamamoto held a drinking party attended by all base unit officers, in which everyone drank too heavily. During the course of the proceedings the admiral penned letters to all dead and alive members of the Imperial Naval Academy Class of 1904. All the other officers present were forced to sign the various letters. Subsequent to this strange behaviour, fellow naval officers of the admiral began to become concerned at the situation, which grew in intensity over the next four days.

By 15 April discussions continued as to the feasibility and morality of an operation that would result in the killing of the Commander-in-Chief of the Imperial Navy, a man who had no equal as to the understanding of the strategy and tactics of the war. The US Secretary of the Navy, Colonel Knox, required a second opinion on the whole affair, referring the matter to the Navy Advocate General's department regarding the legality of the operation. Senior churchmen were consulted concerning the Christianity of the operation, and Air Force General 'Hap' Arnold was asked about the feasibility of the operation. The Deputy Chief of Naval Intelligence, Captain Zacharias, collected documents on assassinations, kidnapping and intimidation for evidence to prove that such acts were ethical. At long last the US Secretary of the Navy, Colonel Knox, agreed to the implementation of the scheme. Planned as Operation Vengeance, the idea was about to be put into the operational stage. By the evening of 15 April, orders were issued in Washington granting permission for the operation, and the best P-38 Lockheed Lightning fighter squadron on Guadalcanal was fully briefed. Special drop tanks for the planes were brought in from

New Guinea after authorisation had been radioed by Admiral Nimitz at Pearl Harbor to Admiral Halsey, in whose area the action would take place.

On 17 April Admiral Yamamoto lunched with Lieutenant-General Imamura Hitoshi, the commander of the Rabaul Imperial Army Group, at which the general recounted how he had narrowly missed an aerial ambush while flying recently over Bougainville. The general warned the admiral of the present dangers, but Yamamoto promised to use the facilities of a six-Zero fighter escort group whenever flying.

That afternoon Admiral Yamamoto was with a very good friend of his – Rear Admiral Joshima Takaji, who had flown into Rabaul from Bougainville. The rear admiral considered the inspection tour sheer madness, and pleaded with his old friend to give up the idea. This Admiral Yamamoto refused to do, but by this date President Roosevelt had given the US Navy permission to proceed with the operation. At the time the Imperial admiral remembered that he had accepted responsibility for the 'Doolittle' violation of Imperial air space, for the Midway naval defeat and for the disastrous results of the Guadalcanal air offensive. Nevertheless he was only too well aware of the superiority of the performance of the Lockheed P-38 Lightning fighter over that of the Mitsubishi A6M3 carrier fighter. Furthermore, in January 1943 the Imperial Intelligence Service had acquired knowledge of the fact that President Roosevelt and Prime Minister Churchill had decided at their Casablanca meeting to initiate the policy of unconditional surrender. Moreover, the A6M3 replacement types, the Shiden and Raiden fighter planes, were still on the drawing-board. Operation Vengeance was about to commence.

Operation Vengeance, 18 April 1943

Sunday 18 April 1943 dawned clear and humid, the appointed day for the inspection of the southern Imperial enclaves. The pre-arranged itinerary had been scheduled for departure from Rabaul at 0800, arriving at Shortland Island at 1040; departure from Shortland at 1145, to arrive Buin, Bougainville, for luncheon at 1310. At the end of the afternoon a scheduled departure was to be made at 1800, so as to arrive back at Rabaul during the evening at 1940. The message previously transmitted had been signed by Vice-Admiral Samejima Tomoshige, the local Air Officer Commanding, previously posted to the Rabaul command some five months previously by the Emperor to indicate his concern at the progress of the air war. This officer had been a former naval aide-de-camp to the Emperor himself.

The two staff bomber planes had been specially prepared and crews briefed for the forthcoming inspection flight, as well as the escorting fighter group, consisting of five Mitsubishi A6M3 carrier fighters. Eventually Admiral Yamamoto appeared, not wearing his number one white admiral's uniform, but his jungle-green fatigue uniform, with his sword of office. The farewell party was already assembled. To Vice-Admiral Kusaka, the commanding officer of the Rabaul Naval Air Station, Admiral Yamamoto handed two scrolls containing hand-written poems, composed by the Emperor Meiji, which he had written out himself on scrolls of paper to be handed to the commander of the 8th Fleet. Having completed the farewells, the two parties boarded their respective staff bombers, with engines turning they taxied around the perimeter track and one by one they raced down the runway to rise majestically into the clear blue sky. As each plane in turn roared past the farewell party, each officer bowed respectfully to the Commander-in-Chief as his plane flew out over the jungle and over the sea. The two Mitsubishi G4M3 bombers took off at 0600, Tokyo time. In the first plane was Admiral Yamamoto, the admiral's secretary, the fleet medical officer and an air staff officer. In the second bomber sat Vice-Admiral Ugaki Matome and several staff officers, including the fleet paymaster and the admiral's chief of staff.

The two staff bombers and a five-plane fighter escort took a bearing and headed south at an altitude of 5,000 feet, flying over the deep blue ocean bathed in a shimmering heat haze. Looming on the horizon was a deep green smudge, edged in a golden brown, which became larger as the island of Bougainville gently slid under the noses of the seven-plane formation.

Over on Henderson Field, eighteen Lockheed P-38 Lightning twin-boom fighters under the command of Major John Mitchell USAF awaited orders for take-off. Of this squadron four planes were earmarked to attack the staff bombers to achieve the killing of Admiral Yamamoto and his staff officers, while the remaining fourteen Lightnings would engage the fighter escort or any other Imperial naval fighters that might intervene. On take-off one of the four killer Lightnings piled up on the runway, becoming a total wreck, but the remaining seventeen flew off successfully. Major Mitchell's squadron flew low, below the effective detection height of the Imperial naval radars, the pilots flying by horizon and avoiding the sun's blinding reflections from the pulsating wave-tops as best they could. The pinpoint interception zone was some 450 miles off from the base on Henderson Field. At 0933, the Lockheed P-38s climbed for altitude, rising up from the west, when suddenly the voice of Captain Douglas Canning USAF was heard over the radio-telephone, warning, 'Bogeys 11 o'clock high!' Having reached combat altitude of 2,000 feet, the killer P-38s dived on the Imperial Naval Air Group and in a confused mêlée lasting just five minutes blasted in cannon shell and machine-gun fire at the two staff bombers and five defending fighters. Immediately the admiral's G4M3 bomber burst into flames, emitting thick black smoke, and commenced to spiral earthwards to crash in the jungle rain forests. The second G4M3 bomber careered crazily around the sky, equally in flames and billowing black smoke, to crash into the ocean below, with no survivors. The Lightnings, having achieved their mission, returned to Henderson Field. The most brilliant mind in the Imperial Navy had been obliterated, with very serious consequences for the Naval Air Service. It was to take three weeks before the Japanese government was to announce the death on active service of the Naval Commander-in-Chief. Imperial ground troops were sent to search for the bodies of the dead officers in the thick green jungle. When they arrived at the wreckage of the G4M3 bomber they found the remains of Admiral Yamamoto still sitting in his seat clutching his sword – in an attitude of complete composure. The cremated remains of the admiral were sent back to Tokyo, where he was given a state funeral.

Japan now faced a very serious shortage of raw materials and machine tools, which resulted from wasting of resources in the period of December 1942 to June 1943. A conservation programme was initiated to reduce the use of light alloys, the government announcing that the aircraft industry would use 30% less aluminium per plane manufactured, and the elimination altogether of the use of magnesium. A greater quantity of wood was now being used, which increased the all-up weight of aircraft, and thin sheet-steel was being used to replace light alloys. Scrap-metal wastage was reduced by rearranging fabrication, and an increase in the use of precision casting and synthetic resins was ordered.

Aircraft and engine designation policy

Until mid-1943 the designation of Imperial naval planes had been based upon the Japanese calendar, in which comparable dates were 660 years ahead of Western calendars. Thus 1936 was 1936 + 660 = Year 2596, and known as the year 96. Thus the Mitsubishi OB96 was the heavy bomber designed for the year 1936. It was decided to change the system, using a combination of letters and numbers. This combination would consist of a letter representing type function, a number indicating the name of such types accepted into general naval service, a letter indicating the name of the manufacturing company and the model of that type manufactured, and in some instances a further letter or number if modifications had been introduced in production. Thus the Mitsubishi G4M2 attack-bomber indicated

G = Heavy bomber type

4 = 4th bomber plane design entering Naval Air Service

M = Mitsubishi Aircraft Company
2 = Second version of the original design

Aircraft type code:
    A = Carrier-borne fighter
    B = Carrier attack-bomber
    C = Carrier reconnaissance plane
    D = Carrier dive-bomber
    E = Seaplane reconnaissance
    F = Seaplane observation plane
    G = Land-based attack-bomber
    H = Flying-boat
    J = Land-based fighter
    K = Training plane
    L = Transport aircraft
    M = Research aircraft
    N = Seaplane fighter
    P = Land-based bomber
    Q = Anti-submarine patrol plane
    R = Land-based reconnaissance plane
    S = Night-fighter
    Y = Glider

Manufacturer's code letter:
    A = Aichi Aircraft Company
    D = Showa Aeroplane Engineering Company
    G = Tokyo Gasu Denki Naval Arsenal
    H = Hiro Naval Arsenal
    K = Kawanishi Aviation Company
    M = Mitsubishi Aircraft Company
    N = Nakajima Aircraft Company
    P = Nippon Aeroplane Company
    S = Sasebo Navy Air Depot
    W = Kyushu (Watanabe Iron Works Ltd)
    Y = Yokosuka Naval Air Depot

Eventually the type number system was replaced or added to by names:
    1. Carrier-based fighters & seaplane fighter – Winds
    2. Interceptor fighters & land-based fighters, single engine – Lightning
    3. Land-based fighters, two or more engines – Thunder
    4. Night-fighters all types – Light
    5. Single-engine bombers – Stars
    6. Bombers, two or more engines – Constellations
    7. Attack-bombers, including torpedo-bombers – Mountains

8. Anti-submarine, patrol aircraft, flying-boats & seaplanes – Seas

9. Reconnaissance aircraft of all types – Clouds

I. Manufacturer's code first letter:
   A = Aichi Watch & Electric Machinery Company Ltd
   K = Tokyo Gasu Denki
   I = Tokyo Ishikawajima Shipping Company, Aircraft Department
   K = Kawanishi Aircraft Eng. Company Ltd
   M = Mitsubishi Heavy Industries Ltd
   N = Nakajima Aircraft Company Ltd
   Y = Yokosuka Naval Aircraft Depot

II. Engine type second letter:
   D = Diesel
   F = Liquid-cooled
   K = Air-cooled

III. Class of engine third symbol

IV. Variant of engine fourth letter

Therefore, Mk4D = Mitsubishi, air-cooled, 4th class, 4th variant.

A further designation system was added, known as the Ha system:
   Ha 1 = Air cooled incline engine
   Ha 2 = Air cooled radial single row engine
   Ha 3 = Air cooled radial double row engine 14 cylinders
   Ha 4 = Air cooled radial double row engine 18 cylinders
   Ha 5 = Air cooled radial multi row engine 22 cylinders
   Ha 6 = Liquid cooled engine 12 cylinders
   Ha 7 = Liquid cooled engine more than 12 cylinders
   Ha 8 = Diesel engine 2 cycle
   Ha 9 = Engine for special use e.g. target aircraft

Further to this Ha system was added a number combination representing the bore and stroke of the reciprocating piston in mm within the cylinders:
   1. 140/150 mm
   2. 150/170 mm
   3. 140/150 mm
   4. 146/160 mm
   5. 130/150 mm
   6. 130/160 mm

Opposition and challenge in the skies hitherto dominated by the Mitsubishi Zero fighters came on 1 September 1943, during the Battle for Markus Island. Here the US Navy used the Grumman F6F3 Hellcat fighter for the first time. The Grumman Company of New York had commenced feasibility studies in the spring of 1941, subsequent to the successes achieved by the previous navy fighter, the Grumman Wildcat. In-depth design was followed with a prototype maiden flight on 26 June 1942, and eventually the new fighter plane succeeded the F4F fighter on the production lines in January 1943. Between January 1943 and November 1945 over 12,272 aircraft of this type were manufactured. The Grumman F6 F3 Hellcat fighter specification included:

Engine	1 × 2,000 hp Pratt & Witney
Span	42 ft 10 in.
Length	33 ft 6½ in.
Height	13 ft 0 in.
Weight empty	9,042 lb
Weight loaded	11,381 lb
Maximum speed	376 mph at 17,300 ft altitude
Service ceiling	38,400 ft
Range	1,090 miles
Armament	6 Browning machine-guns of ½ in. calibre
Offensive load	2 × 1,000 lb bombs or rocket projectiles to an equivalent load weight

With the aerial opposition mounting against the Imperial Naval Air Service in both quantity and quality, additional arming of the ships of the carrier strike force took place. During October 1943, HIJMS *Amagi*, a similar design to *Hiryu*, with a bridge on the starboard side, was completed, and both she and her sister ship of the class, HIJMS *Unryu*, were constructed with four power-operated 25 mm cannon turrets on the starboard side aft of the funnel. Furthermore, *Amagi*, *Unryu* and *Katsuragi* were fitted with four triple 25 mm cannon turrets placed at port and starboard near the stern, as well as one turret in front of the bridge. In addition twenty-five anti-aircraft cannon were located about the ships, as were 168 5-inch rocket launchers, placed in banks of three on the bow.

HIJMS *Aso* and *Katsuragi* were powered with destroyer turbines, giving the ships a maximum speed of two knots less than the rest of the class, which were powered by cruiser-type steam turbines. The remaining aircraft carriers of the class were never completed due to surrender on the part of the Nipponese government and because the materials supply situation for ship building virtually broke down at the end. The HIJMS *Aso* and *Katsuragi* specification included:

Displacement	17,150 tons
Dimensions	741½ × 72 × 25¾ ft
Machinery	4 shaft geared turbines producing 152,000 shp = 34 knots

Armament		Twelve five inch AA, 51 25 mm AA guns
Aircraft		65 planes
Class:	*Amagi*	Lost 24.7.1945
	64 aircraft *Aso*	Constructive loss January 1945
	53 aircraft *Ikoma*	Constructive loss January 1945
	64 aircraft *Kasagi*	Scrapped April 1945
	64 aircraft *Katsuragi*	Scrapped 1947
	Unryu	Lost 19.12.1944 in East China Sea

Besides opposition in the air, the Imperial Navy was beset with a critical sinking of the Imperial merchant fleet, which by September 1943 had lost to the action of US Navy submarines a total of 172,082 mercantile tons. By 15 November 1943, Admiral Koshiro Oikawa had been appointed commander of the Grand Escort Command, controlling four escort carriers as well as the 901st Naval Air Group established for anti-submarine warfare. Unfortunately the airmen of this group were hitherto untrained in this type of combat. At this time the need for carrier escort ships was becoming critical, and as a result HIJMS *Ibuki* was towed to Sasebo Navy Yard for conversion to an aircraft carrier from her original design as a high-speed oiler. In this capacity she had never been commissioned, although when the hull was originally laid down on 24 April 1942 it was anticipated she would be launched as the carrier HIJMS *Suzuya* from the Kure Navy Yard. Her specification included:

Displacement	12,500 tons
Dimensions	658 × 70 × 21 ft
Machinery	4 shaft geared steam turbines, 72,000 shp = 29 knots
Armament	4 × 3.9 in. AA, 48 × 25 mm AA, 30 depth charges
Aircraft	27
Built	Kure Navy Yard, 25.5.1943
Fate	Scrapped 1947

In the course of November 1943, the Nipponese government under Prime Minister General Tojo established a Ministry of War Supply due to the crisis of the raw material supply situation and the ever-increasing demands of the armed forces for more weapons and machines. The new ministry was concerned with aircraft mass production, the manufacture of special aerial weapons, raw material acquisitions and production of auxiliary equipment. To achieve these ends the production staff of the Bureau of Aeronautics of each service were transferred to the new ministry to form a bureau of air weapons. Unfortunately the ministry concerned itself only with mass production and saw little to concern itself with in the way of new prototypes, or of engineering changes to be introduced in production as a result of a recommended battlefield modification, or alternatively as a suggestion on the part of the work of some service development group, but mass production had to continue. This aim was established in order to attempt to combat the enormous output of the American war industry. On 4 December 1943, the aircraft carrier HIJMS *Chuyo* was sunk by the US submarine *Sailfish* 250 miles south-east of Tokyo Bay, off the coast of Honshu.

Towards the end of 1943, an Imperial Navy peace study group had been organised under the direction of Vice-Admiral Nagumo, and in its concluding report it stated that the Nipponese Empire had lost the war. A similar study group organised by the Imperial Army concurred, and both groups stated that Japan would lose all the territorial gains achieved since 1880. It was recommended that in some way the Imperial homeland had to be spared from the ravages of the war, and that it was vitally important to ensure that the Imperial dynasty should be left on the throne.

CHAPTER 18

The Fall of the Cherry Blossom

Sometime during January 1944, air firing tests were being undertaken over Toyohashi airfield by aircraft number 30 of the Mitsubishi J2M2 Model II type. The fighter plane swooped over the aerodrome and sharply turned onto the towed target streamer from above and behind. Promptly Petty Officer Michito Yamanouchi fired his twin cannon and machine-gun installation for a two- or three-second burst. To the horror of the ground party, the fighter plane broke up in mid-air and eventually crashed, spewing wreckage over a wide area. The investigation committee had discovered, on inspecting the remains of the wreckage, cracks in the engine attachment lugs, which had caused excessive vibrations in the structure of the aircraft. Furthermore, these vibrations had caused a secondary disintegration in the fuselage, resulting in a complete collapse of the structure. Immediately engineering changes were introduced to strengthen the engine attachments and other points within the aircraft, which had caused such a disastrous accident. As a consequence, no further trouble of this type was to be experienced.

In January 1944, production of the Mitsubishi G4M1 Model 21 ceased after 1,200 bombers of this type had been manufactured by the 3rd Airframe Works located at Oe-Machi and the 7th Airframe Works at Okayama of the Mitsubishi aircraft complex. Of the production total, which had started in October 1940, 661 planes had been the Model II variant manufactured in parallel to the G4M2 bomber. The Model 22 and Model 24 of this later bomber type were produced to a total of 1,154 machines.

A projected anti-submarine bomber, designated Type G4M3a Model 34a, was planned, but the Model 36 development was never implemented prior to the Nipponese surrender. However, the Type G4M3 series was equipped with turbo supercharged Kasei 25b RU engines, as first tested on the Mitsubishi G4M2c Model 26 attack-bomber, when engine teething troubles were experienced. The Kasei 25 series of engines were installed on fifty-eight bombers of the G4M3 Model 34 variant, but only on two of the

G4M3 Model 35 versions. Now with the air war passing into more sophisticated phases of operation, fighter defence tactics had changed. Forward- and upward-firing armament was introduced on a variety of Naval Air Service aircraft, including the P1Y1-S, the P1Y2-S, of which ninety-seven were built, the D4Y2- S, the C6N3, the Mitsubishi A6M5 and the Nakajima J1N1-S. At the time a total of seven weapons of different calibres were in use, ranging from the 7.7 mm rifle-calibre machine-gun up

Development of the Aerial Machine-Gun

Name of machine-gun	Calibre (mm)	Projectile wt (oz)	Mg wt	Muzzle velocity (ft/sec)	Rounds per gun per min
Type 97 F	7.7	0.40	26.4	24.60	1,000
Type 1 M	7.92	0.41	15.2	25.90	1,000
Type 2 M	13.0	1.20	37.5	24.60	900
Type 3 F & M	13.2	-	66.1	25.90	750/800
Type 5 F	30.0	12.30	154.0	24.60	50
Prototype F	40.0 (inc. case)	46.0	-	23.00	200

By early 1944 the United States air forces had achieved superiority both numerically and in performance against the Imperial Naval Air Service's Mitsubishi A6M2 carrier fighter planes. The US Navy had introduced the Grumman F6F Hellcat carrier fighter, whose superior performance had made the air war critical. As a result the Naval Air Service had endeavoured to introduce hasty modifications to fight in the desperate air battles against the Lockheed P-38 Lightning fighters equipped with twin engines of 1,425 hp, as well as the Corsair F4U4 Navy fighters using a 2,000 hp motor. To senior air officers of the Imperial Naval Air Service the Nipponese Empire could win or lose on the basis of the outcome of the air war, and to many the balance was heavily weighted against success.

In February 1944, representatives of the Naval Air Service and top aeronautical engineers of the Mitsubishi Aircraft Company held a conference at the Naval Air Research and Development Centre headquarters. The purpose of this meeting was to study the feasibility of modifying the Mitsubishi A7M2 fighter plane to carry heavier weapons, to achieve a higher maximum speed and to increase the rate of climb to a greater altitude. As a result of the conference deliberations, the A7M3 fighter aircraft was born and adopted at the meeting. This machine was intended to intercept the US Boeing B-29 heavy bombers, which were protected by heavy armament, raiding from very high latitudes and propelled by engines equipped with exhaust-driven turbo superchargers. At the time the majority of Nipponese fighter pilots were junior men, the senior pilots having been lost in the air battles for Rabaul in the South Pacific. As a result of the junior pilots manning the majority of the fighter squadrons, a torrent of complaints were received regarding, among other points, the plane's poor visibility. Mitsubishi countered by proposing to modify the upper fuselage and fit a bubble canopy, as was the case with the Zero fighters. The Naval Air Service refused to accept the modification, fearing possible production delays at the factories, and though the cockpit canopy was enlarged there was little improvement in pilot visibility. Trouble was now experienced with the latest model of the Mitsubishi A6M6c Model 53c. Self-sealing fuel tanks were now fitted as standard equipment, but the Sakai 31 motor, while incorporating a water methanol injection apparatus, was decidedly underpowered, and upon investigation it was discovered that the methanol metering device had failed. Furthermore, for ground crew maintenance the servicing of this new motor was a nightmare, so that only one experimental model of this machine was manufactured. A former Naval Air Service officer and Mitsubishi test pilot – Katsuzo Shima of Japan Airways – joined the Domei Press Service Corporation, flying liaison and transport missions from Japan to south-east Asia. However, by March of 1944 five prototypes of the Nakajima J5N1 twin-engine fighter aircraft, designated Tenrai, had completed test flights armed with the Type 5 aerial cannon of 30 mm calibre.

During April 1944, a further 291 bombers of the Mitsubishi G4M series had been produced and were available for active duty, but the pressing need at this juncture was to ensure the supply and technical superiority of the manufacture of fighter planes. The Mitsubishi A6M5b Model 52b, a joint Mitsubishi/Naval Air Service project, had been completed with an armament. which now included one 7.7 mm machine-gun as well as one 12.7 mm machine-gun, both fuselage mounted. Windscreen armoured glass was now standard, as well as the equipping of wings and fuselage with automatic fire extinguishers. The new machine went into production in April 1944, and before this model was replaced a total of 470 aircraft of the type were built, ready for the commencement of Operation A-GO. During the Battle of the Philippine Sea on 19 June 1944, the air battles accounted for a loss of 300 Imperial Naval Air Service machines out of a total of 400 flying from Nipponese aircraft carriers. Unhappily the Mitsubishi A6M5b had been unable to utilise its improved advantages. However, the Mitsubishi A7M Reppu fighter plane was rolled out of a hangar during the course of April, and the prototype proved to be a sleek low-wing monoplane. On 22 April the number one prototype underwent

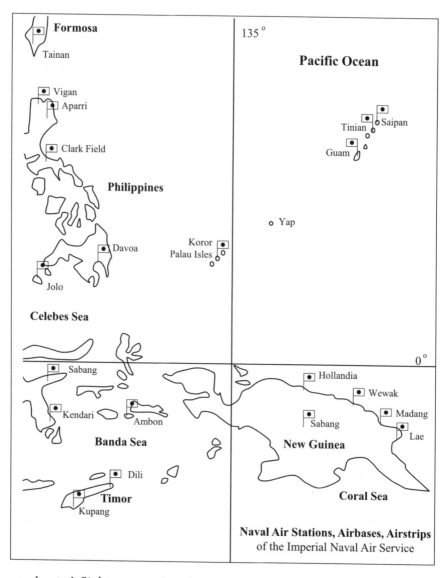

Naval Air Stations, Airbases, Airstrips
of the Imperial Naval Air Service

accelerated flight tests at Suzuka airfield with satisfactory results, while the number two prototype was flight tested by the Imperial Naval Air Service at Yokosuka Airbase.

On 1 May the number three prototype was completed and ready for flight testing. At a meeting between the company and senior members of the Naval Air Service, Lieutenant-Commander Kofukuda HIJMN reported on the plane's good performance with a variety of weapons, as

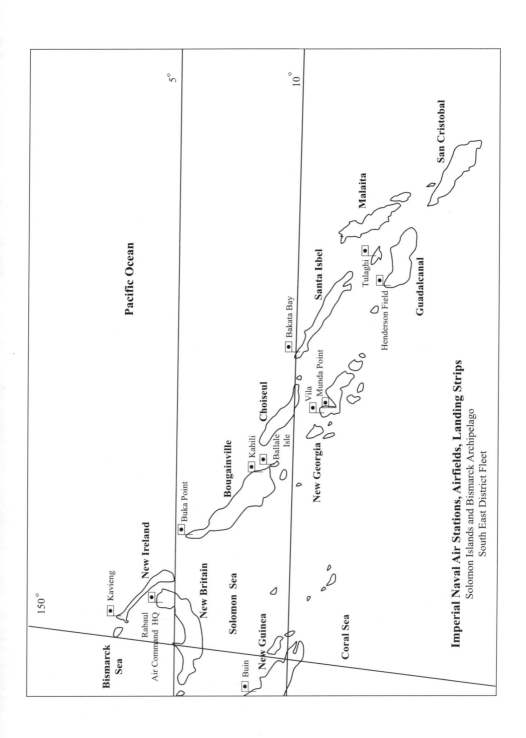

Imperial Naval Air Stations, Airfields, Landing Strips
Solomon Islands and Bismarck Archipelago
South East District Fleet

well as the ease of handling, particularly for the inexperienced service pilot. On the following day the second prototype machine flew to the Misawa Naval Air Station in Aomori, and while making a forced landing it was severely damaged. This was a dispersal airfield for the Air Proving Division of the Yokosuka Experimental Air Corps, but while flying again, a forced landing had to be made at the Matsushima Naval Air Station, where the aircraft was totally wrecked owing to a faulty fuel gauge. The importance of successfully completing flight tests and going into full-scale production of the new A7M2 could not be overestimated. Indeed, ever since approximately December 1943, every effort had been made by the Engineering Production Works and by the Engineering Research and Development Units, which had worked around the clock to see the new fighters successfully launched into active operational service. Unfortunately, repeated aerial attacks by high-flying US Boeing B-29 bombers had damaged Suzuka airfield, with the result that the prototype Mitsubishi A7M2 fighter aircraft had been severely damaged. On 6 May a Mitsubishi A7M2 Reppu, piloted by Eisaku Shibayama, had developed trouble with the undercarriage gear, but after landing, with modifications having been carried out, the problem was successfully rectified. On the fourteenth test flight, this time piloted by Mitsubishi's test pilot Ogawa, stability and controllability flight tests were successfully completed so that the prototype aircraft could be handed over to the Naval Air Service. By 25 May prototype planes four to seven had been completed and were available for participation in the flight test programme. Between 31 May and 2 June 1944, three major test flights were undertaken by Lieutenant-Commander Toshio Shiga HIJMN of the Naval Air Research and Development Centre, who affirmed that the aerobatic controllability under 276 mph was excellent, and in fact the new machine was considered as manoeuvrable as the Zero fighter had been. Four days later, on 6 June, Lieutenant-Commander Mitsugi Kofukuda, also from the Naval Air Research and Development Centre, flew prototypes numbers one and two, fully endorsing the findings on 2 June of his fellow pilot. However, the prior calculated performance figures worked out by the Mitsubishi chief designer were completely vindicated when the machine was equipped with the Homare engine. Achieving a maximum speed of some 345 mph, with a rated engine of 1,300 hp and climbing to an altitude of 20,000 feet in 10 minutes, a new test programme was inaugurated to increase the power output. The chosen project leader was Lieutenant-Commander Shima HIJMN, who unfortunately fell ill, and so he was replaced by Lieutenant-Commander Kofukuda.

In the course of May 1944, while the flight test programme was in process for the new carrier-borne fighter, the A7M2, a parallel development programme had been undertaken with the land-based fighter plane, the Mitsubishi J2M5. Flight tests had proved successful and the new fighter was ordered into immediate production. The Model 33 had achieved a maximum speed of 375 mph at an altitude of 22,300 feet, while at an altitude of 21,600 feet the maximum speed increased to 381.6 mph. The rate of climb was such that in 6 minutes 20 seconds 19,680 feet was attained, and in a time of 9 minutes 45 seconds an altitude of 26,240 feet could be reached. Though the J2M5 Model 33 was highly successful in combating the Boeing B-29 Superfortress, a criticism of poor pilot visibility and short-range duration was levelled against the design, and as a result only thirty to forty machines of this type were manufactured for the Naval Air Service.

With the prospect of the air/ocean war increasing in intensity, two aircraft carriers were modified in June 1944 to increase the anti-aircraft defences of the ships. The carrier HIJMS *Ryujo* had increased armament installed, including 168 5-inch rocket launchers, sixty-one 25 mm anti-aircraft cannon, six depth charges and an extension of the flight deck, which now extended over the bows. These modifications were made at the Kure Naval Dockyard, where the aircraft carrier was still berthed at the time of the surrender. Similarly the two sister ships, HIJMS *Unyo* and *Taiyo*, had sixty-four 25 mm anti-aircraft cannon as well as ten of 13 mm calibre installed aboard the ships. Meanwhile the Mitsubishi A7M3-J Model 34 prototype was under development, the schedule calling for prototype completion within the next four months. The design had become bogged down owing to the Naval Air Service continually changing the specification, resulting in further modification to the original design.

The air battle for the Marianas, 19–20 June 1944
On the air stations in the chain of Mariana Islands the Imperial Naval Air Service had amassed some 500 aircraft for what was considered to be a decisive sea battle of the war. In addition, with five Nipponese aircraft carriers a further 473 aircraft were available for offensive operations. But prior to the commencement of this most important engagement, carriers of the US fleet had undertaken intruder operations with the specific task of destroying as many Imperial aircraft of the Naval Air Service as was possible. Nevertheless, Admiral Ozawa HIJMN could land his aircraft on the island of Guam, where refuelling operations could take place.

By 19 June long-range flying-boats of the Imperial Naval Air Service had detected and shadowed the advancing US carrier task force under the command of Admiral Nimitz.

At approximately 1000, the first wave of Nipponese attackers had made contact with the US carriers *Lexington* and *Essex* at a distance of some seventy-two miles. Immediately the Imperial planes were engaged in fierce air battles with US Grumman Hellcat fighters, and with the aid of carrier radar-controlled interception the Naval Air Service lost forty-one planes. The second wave of Imperial Naval Air Service attackers was similarly engaged by US Grumman Hellcats under carrier radar-controlled interception at approximately sixty miles from the US task force. A third and fourth Imperial strike force endeavoured to bomb the US carriers and escorts, but to no avail and with great loss of machines and crews. On the same day the Nipponese carrier HIJMS *Taiho* was torpedoed and sunk by the US submarine *Albacore*, as was HIJMS *Shokaku*, torpedoed and sunk by the US submarine *Cavalla*. On the following day torpedo-bombers and dive-bombers from the US carrier *Belleau Wood* stalked the carrier HIJMS *Hiyo* and sent her to the bottom amid the water plumes of exploding bombs. When the final survey was made and Admiral Ozawa's battle report was radioed to Imperial headquarters in Tokyo, the staggering loss sustained included three aircraft carriers from a carrier strike force of five ships. The total number of aircraft of the Naval Air Service lost amounted to some 400 machines, from a force of over 500 aircraft. Admiral Nimitz had lost a hundred planes, but all his ships had remained intact.

This had been a decisive sea battle and had underlined the predictions of Lieutenant-Commander Prince Takamatsu. Upon his informing His Imperial Majesty on 22 June of the disastrous loss, the Emperor could not be anything but highly incensed, and he ordered that the carriers would no longer be required, as well as demanding the resignation of the Navy Minister and Naval Chief of Staff, Admiral Shimada HIJMN, who would take the blame for the defeat.

The previous day, 21 June 1944, Captain Jo Eiichiro HIJMN forwarded a cablegram from the aircraft carrier HIJMS *Chiyoda* on the high seas in which he outlined a suggestion for the introduction of kamikaze tactics to be used for attacking the various American naval task forces at sea. Vice-Admiral Onishi received the message and spent some time in considering the suggestion and discussing the matter in naval circles.

However, for the Battle of the Philippine Sea, the Zero fighters on active service had been converted for use in dive-bombing operations. Modifications had been made to facilitate the carrying of a 500 lb bomb and the aircraft was used to do this during the course of Operation A-GO. Unfortunately, due to a technical error, the bomb-release mechanism failed to operate and the externally slung bomb caused an increase in the air drag with a consequential uplift in the overall weight of the machine. A severe reduction in the operational range had resulted, causing unacceptable losses due to 'ditching' while overflying at sea. Nevertheless a new version of the Mitsubishi carrier fighter was developed, with a reinforced tailplane, a reliable bomb-release mechanism and provision for the installation of two external wing fuel tanks. This model was not to be available for operational use until approximately May 1945.

With the ever-increasing aerial activity of US naval aviation, it had become necessary to embark on a programme of strengthening the anti-aircraft defence capability of the Nipponese naval vessels. As a result, during the course of July 1944, the aircraft carrier *Junyo*, of the Hiyo class, had her 25 mm anti-aircraft guns increased to a total of seventy-six, with the additional installation of 168 5-inch rocket launchers. At the same time, the carrier *Zuiho*, of the Shoho class, had a total anti-aircraft armament of sixty-eight 25 mm automatic weapons, as well as 168 5-inch rocket launchers built onto an extended bow during the course of a refit. The carrier *Zuikaku* was similarly rearmed, with ninety-six 25 mm anti-aircraft guns and 168 5-inch rocket launchers mounted on the bow. Unfortunately, despite the rearming of *Zuiho* and *Zuikaku*, both aircraft carriers were to be sunk later at the Battle of Leyte Gulf on 25 October 1944.

On 23 July 1944, the Naval Air Service issued priority orders for the introduction of new models to replace the existing Zero carrier fighters. Suggestions and recommendations included the installation of two 13 mm machine-guns outboard of the landing gear, and the fitting of one 7.7 mm and one 13 mm machine-gun to fire through the whirling propeller disc, with one 20 mm cannon in each wing. Armoured windscreen glass was to be provided, with the addition of armoured plate behind the pilot's accommodation and a 31-gallon fuel tank located at the rear of the pilot's compartment. A maximum diving speed of 460 mph was projected, provided the wing skin had an increase in thickness. The result of all these modifications and revision in armament installations was a further increase of weight by as much as 660 lb. However, if the Sakae 21 engine was to be retained, despite the use of a water methanol injection apparatus, the plane would lose valuable performance. Thus operational units calling for increased fuel tank protection and increased performance considered the Mitsubishi A6M5c as capable of matching the US Navy's Grumman F6F Hellcat.

The intended A6M5c replacement was to be the A7M Reppu fighter plane, designed to meet the needs of a carrier-borne fighter and land-based interceptor. It was powered by a Mitsubishi MK9A engine with exhaust-driven turbo superchargers and a designed maximum speed of 393 mph at an altitude of 32,800 feet, attained by a rate of climb in 18 minutes 15 seconds. The designed armament was four 30 mm cannon and overloaded with a further two cannon elevated at 30 degrees aft of the fuselage. The Mitsubishi design engineers were bedevilled by a constant flow of design changes demanded by the Imperial Naval Air Service, which eventually made it quite impossible for the company to meet the design deadline date of March 1945. Nevertheless, an engineering conference was convened for 30 July 1944, held at the Naval Air Research and Development Command headquarters, where it was unanimously agreed to discontinue the use of the Nakajima Homare engine. The Mitsubishi representatives earnestly endeavoured to persuade the Navy to accept the Mitsubishi MK9B engine by examining an analysis of the technical data. But despite the arguments in favour, the Naval Bureau of Aeronautics and Ministry of War Supply disapproved of the recommendation, though nevertheless some officers of the Naval Air Research and Development Command thought otherwise. So a compromise agreement was arrived at which provided for the installation of the MK9B engine in prototype planes numbers five and six, but installation costs would have to be borne by the Mitsubishi Company, as concluded at the insistence of Vice-Admiral Misao Wada HIJMN, commanding officer of the NARDC and Dr Jiro Horikoshi, chief designer of the Mitsubishi Aircraft Company. In the meantime the Kawanishi N1K2-J Shiden fighter plane was reaching operational units. This machine had a span of 39 feet 3 inches and an overall length of 30 feet 8 inches, and was powered by a Homare 21 radial engine of 1,990 hp, giving a maximum speed of 369 mph at an altitude of 18,370 feet. Armed with four 20 mm cannon and an offensive load of two 550 lb bombs, too few planes were being produced for the front-line units, and it could not be used as a carrier-based interceptor, so that operational squadrons persistently called for a new Zero replacement.

Meanwhile, Yokosuka had produced the D4Y, which could be equipped with either an in-line motor or a radial engine. The latter-powered machine was for shore-based operations, while the former engine was used on carrier-based aircraft. The D4Y1 Model II was powered by a 1,185 hp Aichi Atsuta 21 in-line motor, which was a replica of the imported German Daimler-Benz DB601. The D4Y2 Model 32 was fitted with a 1,400 hp Atsuta radial motor, which gave a maximum speed of 360 mph and necessitated an increase in the area of the tail. The D4Y3 Model 33 was a radial-engine version, of which 2,319 aircraft were built and introduced as a replacement for the Aichi D3A1. The Yokosuka D4Y3's specification was:

Engine	Mitsubishi Kinsei 62 radial of 1,560 hp
Span	37 ft 8³/₄ in.
Length	33 ft 4 in.
Height	10 ft 9¹/₈ in.
Weight empty	5,514 lb
Weight loaded	8,270 lb
Crew	Pilot and observer
Maximum speed	350 mph at 19,360 ft
Service ceiling	34,450 ft
Range	944 miles
Armament	1 movable 7.9 mm machine-gun
	2 fixed 7.9 mm machine-guns
Offensive load	Bombs up to 1,650 lb weight

During the summer of 1944, the engineer designers from the Mitsubishi Aircraft Company examined an American Wright Cyclone 18R-3350-23 engine taken from a Boeing B-29 bomber shot down over the Nipponese island of Kyushu. It was discovered that the vibrations produced by the reciprocating pistons were eliminated by the use of dynamic vibration damper counterweights. This concept and other innovations were used by engineer Matsudaira on a modified Homare 21 engine along similar lines to the arrangement used on the Wright Cyclone engine. With the modified Homare 21 engine installed in the airframe of a Nakajima C6N1 carrier reconnaissance plane, and a thin-bladed propeller fitted, the vibrations were discovered to have been eliminated, and an increase in the maximum speed of 17 mph was obtained. Meanwhile the Mitsubishi engineers were incorporating other ideas discovered in the Cyclone engine. The Mitsubishi J2M4 Model 32 fighter had the most spacious fuselage of all the Imperial naval fighters, and the newly acquired technology was incorporated in this machine by installing a turbo supercharger for very-high-altitude bombing attacks, which made fighter defence interception very difficult. As a consequence the Imperial Naval Aircraft Research and Development Command had modified one version of the Mitsubishi J2M4 and experienced excessive buffeting on flight testing. But the Mitsubishi Company had modified a further machine with successful results by 4 August 1944. Nevertheless, a further blow was struck at the heart of the Imperial carrier strike force when the aircraft carrier HIJMS *Taiyo* was torpedoed by the American submarine USS *Rasher* on 18 August 1944 in a position north-west of the island of Luzon, and was a total loss. The ever-worsening maritime situation was now demanding excessive counter-measures.

During the month of August 1944, Naval Ensign Mitsuo Ohta, a transport pilot for the Navy's 405th Transport Kokutai, proposed to his commanding officer the design of a simple rocket-propelled aircraft for kamikaze, or Divine Wind, operations. These attacks would be suicide missions principally against Allied aircraft carriers and capital ships. A small aircraft, driven by solid fuel propellants, would be attached to the bomb bay of a Mitsubishi G4M2 naval attack-bomber and released when in close proximity to Allied naval vessels to dive down under full power to crash onto a ship's deck and explode. Initial design work was undertaken by Ensign Ohta, assisted by personnel from the Aeronautical Research Institute of the University of Tokyo. Detailed designing was led by a team of engineers from Dai-Ichi Kaigun Koku Gijitsusho, the 1st Imperial Naval Air Technical Arsenal at Yokosuka. Eventually a prototype was constructed and was designated Yokosuka MXY-7 Model II naval suicide attacker Ohka, or Cherry Blossom. This was a small-size monoplane powered by three Type 4 MKI Model 20 rockets located in the tail position. In the nose was a warhead containing 1,200 kg, or 2,646 lb, of high explosives, with a pilot's cockpit located behind the wing trailing edge. Using this weapon against American aircraft carriers was considered by the Imperial Naval Air Service to ensure instant success. It was known that the flight decks of all American aircraft carriers were of wooden construction. Such a projectile, with over a ton of explosives crashing at high speed, would cause immeasurable destruction on any deck so poorly constructed, penetrating the below-decks locations. However, British aircraft carriers were built with metal decks and were not so easily put out of action.

Early in September 1944, Eitaro Sano, an engineer from the Mitsubishi Aircraft Company, was sent on a liaison visit to the Naval Aircraft Research and Development Command to co-operate with the Imperial Navy's aeronautical engineers and to assist in the design work now proceeding on the Mitsubishi A6M5c Model 52c. About one month prior to the American invasion of Leyte, the first A6M5 was ready for flight testing. Equipped with a Sakae 31A engine with water methanol injection apparatus, the motor failed on power output, and as these carrier fighters were not built with self-sealing fuel tanks, the outcome was very much as the Mitsubishi Aircraft Company had predicted, and so a production run of only ninety-three machines was completed.

The American 'steamroller' tactics were now advancing the US military juggernaut nearer the Home Islands, as sledgehammer blows were wielded to crack the Nipponese island defensive chain. By 6 September 1944, Vice-Admiral Mitscher USN was subjecting Palau, 550 miles east of Mindanao, to aerial bombardment. The carrier planes from Mitscher's Task Force 38 were attacking Mindanao itself three days later.

On 15 September, US Marines made spectacular landings on Morotai in the Spice Islands and on Peleliu in the Palau Island chain. Imperial Headquarters, Tokyo, received the signals informing the Imperial General Staff of the latest military situation with some dismay. However, Emperor Hirohito, in consultation with senior officers of the armed forces, issued General Staff Order No. SHO. This was a plan entitled 'Operation Victory' for a campaign of eventual victory before the culmination of the final defeat of Japan. The Imperial Air Navy was to preserve itself for the defence of the Home Islands of Japan, where most units were in constant training. The remaining naval aviation units, comprising some 2,000 planes in all, were to operate from naval air stations in Taiwan and Luzon, attacking primarily the American and British aircraft carriers. Upon occupation of further Imperial territory, the remaining naval air units would be permitted to retreat to naval air stations still operational, but near the Home Islands. On the following day, the aircraft carrier HIJMS *Unyo* was torpedoed and sunk by the submarine USS *Barb* in the South China Sea. By 21 September, Vice-Admiral Mitscher's task force was forty miles off the east coast of Luzon, and carrier planes were bombarding the Manila area. The airbases at the former American airfields of Clark and Nicholas Field were ploughed up in massive bombing attacks. During the desperate air battles the Imperial air forces lost 200 machines, for the fifteen American planes shot down. Despite every attempt, planes from the Imperial Naval Air Service were unable to break through the defensive aerial cordon surrounding US Task Force 38, due to massive fighter defensive flights.

On 10 October 1944 the Imperial Naval Air Service and the Imperial Army Air Corps fought air battles to the last plane over the skies of the Philippine Islands and were wiped out by the aircraft from Task Force 38, which was now turned northwards, steaming up to Okinawa. At the same time, Imperial naval air stations were bombed on Okinawa and along the Ryukyu Island chain to Taiwan, by aircraft from Admiral Halsey's US task force, which was escorted by nine heavy cruisers and eight light cruisers. During the ensuing air battles the Imperial Naval Air Service lost heavily, but more importantly the naval air stations were heavily damaged, which would otherwise have been used for counter-offensive action by Imperial naval planes against American installations in the Philippine Islands. To Task Force 38 the Nipponese lost over a hundred aircraft, four cargo ships, a submarine tender and four torpedo-boats, during which the American planes made 1,396 sorties.

The next day, 11 October, the naval air stations and facilities in Luzon were attacked by Task Force 38, which had altered course southwards especially to support an aerial bombardment by some of the 500 US Boeing

B-29 bombers flying in from bases in south-east China. The Imperial Naval Air Service facilities were completely decommissioned.

On 12 October, Task Force 38 launched fighter planes from four separate carrier groups, supported by the B-29 bombing groups flying from China, against 1,000 Imperial naval planes of the 6th Base Imperial Naval Air Force, Formosa, under the command of Vice-Admiral Shigeru Fukudome. In the course of the following two days the Naval Air Service lost 50% of its effective force, with innumerable installations damaged beyond repair. As the ebb of battle flowed back and forth, the Americans lost a hundred aircraft, damage to the carrier USS *Franklin* and two cruisers sunk. But the US forces were able to amass a total of 1,600 aircraft from forty-seven American carriers, as well as 500 B-29 bombers based in south-east China. In the meantime, US forces had consolidated the conquest of beach-heads in the Philippine Islands campaign, with aircraft located in bases in China, Tinian, Morotai and Peleliu.

Meanwhile, on 13 October, the Mitsubishi A7M2 Reppu was flown from Nagoya Harbor airfield adjoining the Mitsubishi Aircraft Works, Nagoya, to the naval air station at Suzuka. The machine had been completed early in October and was powered by a Mitsubishi MK9A Model 1 engine, the pilot reporting a faster take-off and landing speed, with a faster rate of climb to altitude than had been calculated hitherto. This machine had been produced with very serious setbacks due to the heavy aerial attacks by B-29 bombers during the course of the autumn and eventually the winter months. As a result the aircraft industry was in chaos. Nevertheless, the Navy continued to press for increased production and increased performances. The Kawanishi aircraft for the Naval Air Service were ineffective interceptors, with a poor climb rate to operational altitude, so that the Navy pressed for refinements and modifications to the Mitsubishi J2M Model 23 and 33 fighter planes. This particular type of Imperial naval fighter plane was the only machine that was an effective defence against the US B-29 bombers, for the J2M fighters were fast-climbing, high-altitude interceptors. At this stage an organised aircraft industry disposal programme was introduced, but so disruptive had been the aerial attacks that the new Kasei 26 aircraft engine production programme could not be increased, with a consequent limited manufacture of the new Mitsubishi J2M fighter planes. As ever, in any international fighter development programme, the manufacturing departments were bedevilled by constant engineering changes not calculated to increase productive output. At the request of the Imperial Naval Air Service, modifications were constantly requested. The turbo supercharger placed under the pilot's seat required considerable fuselage alterations. A new fuel tank had to be installed and structural changes were necessary so that obliquely fixed cannon could be installed behind the pilot. Changes in the undercarriage landing-gear increased the overall weight, necessitating further alterations of design. The wings had to be redesigned and made anew, so as to accommodate the 30 mm cannon, while the rudder and tail were repositioned as a consequence upon wind tunnel spin testing. Yet despite all this rework, the fact remained that the J2M fighter was the best slayer of the B-29 scourge.

The Naval General Staff, Imperial headquarters, Tokyo, recognised that an air defeat had been inflicted on the Naval Air Service in the air battles over Taiwan, so that by 17 October, the Emperor dispatched Vice-Admiral Takijiro Onishi, Prince Takamatsu's go-between, to the Philippines. Here he was to take charge of the small number of remaining planes of the Imperial Naval Air Service and form a kamikaze air corps at the Imperial Air Command Centre, Mabalacat, north of Manila, with the 5th Base Naval Air Force, Luzon. With fewer than a hundred serviceable planes of all types, it was suggested that Zero fighters should be bombed-up with 250 kg bombs and that each plane should crash-land into the vulnerable flight decks of the American carriers in a support operation to the ships of Admiral Kurita HIJMN, who would raid Leyte Gulf in an endeavour to cripple the US fleet. By 0800 of 24 October, US carrier reconnaissance planes had discovered Admiral Kurita's fleet of Imperial warships, and despite his radioed message to the Air Command Centre of the 5th Base Naval Air Force for fighter defensive cover, only twelve serviceable A6M planes were available. In any case, the Imperial ships were too far off for the single-engine fighters, which did not possess sufficient range. At the same time, 180 Nipponese naval planes were attacking Admiral Halsey's US 3rd Fleet, stretched out from mid-Luzon to Leyte Gulf. In the resulting air battles the Imperial Naval Air Service lost 179 aircraft to the US Grumman Hellcat fighters. The light carrier USS *Princeton* was severely damaged by a 550 lb high-explosive bomb, and subsequently sank. By now the US 3rd Fleet under Admiral Halsey and the carrier Task Force 38 commanded by Vice-Admiral Mitscher had formed a combined force. In the course of the Battle of Leyte Gulf on 25 October 1944, the Imperial aircraft carriers HIJMS *Zuiho*, *Zuikaku*, *Chiyoda* and *Chitose* were sunk by US carrier-based planes of the US 3rd Fleet. During the day of 25 October, kamikaze attacks were mounted by bombed-up Mitsubishi A6M fighters, under the command of Vice-Admiral Onishi Takijiro of the 5th Base Naval Air Force. This was the first occasion that kamikaze operations were officially sanctioned. As a result, one US carrier was sunk and four severely damaged.

Meanwhile developments were taking place for the use of the Ohka piloted missile. The Mitsubishi Aircraft Company had introduced a production modification to the G4M2a naval attack-bomber while on the assembly line, known as the G4M2e Model 24J, equipped to carry the piloted missile while attached to the parent aircraft. Later the G4M2e was specially produced with the bomb-bay doors permanently removed and special shackles installed so that the Ohka bomb protruded beneath the belly of the parent bomber during flight. At the same time the A6M7 carrier fighter was produced as a dive-bomber from the A6M5 model, and the A6M8 carrier-borne fighter was equipped with a 1,500 hp Mitsubishi Kinsei 62 engine, of which 3,879 were manufactured by the Mitsubishi Aircraft Company, 6,215 by Nakajima Aviation and 327 as Rufe float-plane fighters, production eventually totalling some 10,611 aircraft.

Interestingly enough, during the course of October 1944, the first four-engine bomber prototype made its maiden flight. This was the Nakajima G8N1 Renzan (Mountain Range), powered by four Nakajima Homare 21 radial engines.

In the course of November 1944, great efforts were made to develop the Ohka piloted bomb. Initially unmanned flights were made with a prototype at the airfield of Kashima, followed by unpowered glider flights and eventually powered flights, all the tests having been conducted by aerial launching from a specially converted Mitsubishi G4M2a Model 24 naval attack-bomber. The Ohka Model II was the production version, of which 155 aircraft of this type were manufactured by the 1st Naval Air Technical Arsenal, with 600 machines being assembled by Dai-Ichi Kaigun Kokusho, or 1st Naval Air Arsenal, Kasumigaura, supported by sub-contractors Nippon Hikoki KK and Fuji Hikoki KK. The Ohka KI was an unpowered training version of the Model 11 but was fitted with retractable landing-skids and had water ballasting in place of the high-explosive warhead; only forty-five examples of this version were produced.

The projected Ohka Model 21 was a smaller version of the Model 11 to be carried by the Yokosuka P1Y1 Ginga twin-engine Model 33 high-speed bomber. The weight of the warhead of this version of the Ohka was reduced to 600 kg, or 1,323 lb, of explosives, but none of this version were manufactured. The Model 22 was similar to the previous version, but the rocket motors were replaced by a Tsu-11 jet engine based upon the Italian Campini jet engine. Fifty of these projectiles were built by Dai-Ichi KK Gijitsusho, with a planned follow-on production by Aichi Koku KK, Fuji Hikoki KK, Miguro Hikoki KK and Murakami Hikoki KK, but no large-scale production ever resulted. The next projected variant was the Ohka Model 33, initially planned to be carried by the four-engine Nakajima G8N1 Renzan heavy bomber. The Model 33 was to be powered by an NE-

200 turbojet engine, carrying an 800 kg warhead, but this scheme never came to fruition. The Model 43A followed on, being an enlarged version of the Model 33, but equipped with folding wings for catapult launching from a surfaced submarine, but none were built. Similar to the previous version was the Model 43B, designed for catapult launching from within caves on the shores of the Home Islands, but none appear to have been manufactured. The Model 43 K-I Wakazakura (Young Cherry) was a two-seater training version, powered by one Type 4 MK I Model 20 rocket engine for limited power-handling experience, of which only two appear to have been made. Finally the Model 53 was projected for glider towing aloft by a G4M2 attack-bomber.

While all the above developments were going on, the Navy issued a modification order for the A6M8c Model 64 carrier fighter. The plane was to be equipped with an automatic fire extinguisher, an increased fuel capacity for a maximum of thirty minutes' full-power combat time, plus 2½ hours cruising time. The two 13 mm fuselage guns would be removed to minimise the weight factor. However, despite these and other modifications, the prototype was seriously delayed due to the heavy air raid damage to Mitsubishi's Suzuka factory by the B-29 attacks and because plant dispersal had caused confusion as well as disorganisation, resulting in costly delays. As a result the Model 64 was not to appear until March 1945.

By mid-November 1944 ten test flights had been completed on the prototype Mitsubishi A7M2 Reppu over a period of six weeks. The climb to 20,000 feet took just six minutes, with a maximum speed of 387.5 mph at 19,000 feet altitude. These test figures corroborated the calculated figures of performance analysis dated 15 July, and presented at an engineering conference at the end of that month. Two days later, on 17 November, the aircraft carrier HIJMS *Shinyo* was torpedoed by the American submarine USS *Spadefish* in the south of the Yellow Sea.

Tactics for the use of the new Ohka bombs had been worked out and tried. The Mitsubishi G4M2e attack-bombers would fly at an altitude of 16,000–18,000 feet, carrying the Ohka bombs suspended from the bomb bay of each attack-bomber by shackles. Flying with a strong fighter escort to within twenty miles of an Allied fleet, the Ohka bombs would be released and the attack-bombers would return to base. The Ohka pilots meanwhile could either glide down and then fire the three rockets simultaneously, to crash-land at high speed onto the decks of the American carriers, or the projectiles' range could be extended by firing each of the three rockets in turn and then crashing into one of the Allied ships. Unfortunately the heavily laden G4M2e attack-bombers were easy prey to the American Grumman Hellcat and Chance Vought Corsair carrier fighters, which engaged their prey between fifty and a hundred miles ahead of the Allied fleets. Consequently, many of the Ohka bombs were lost due to fuel and range shortage, and crashed into the ocean far below. On 29 November, the aircraft carrier HIJMS *Shinano* was sunk by the US submarine USS *Archerfish* just 150 miles south of Tokyo. This ship was carrying Ohka piloted bombs to the forward areas, with the 721st and 722nd Kokutai having previously been activated at Hyakurigahara and Konoike respectively. The carrier was originally laid down as a Yamato-class battleship, but subsequent to the Battle of Midway in 1942, design changes were made and she was completed as an aircraft carrier in November 1942. Fitted with an armoured flight deck and open hangar (HIJMS *Taiyo* was the only other carrier in the Imperial Navy so designed), it was planned to use the ship as a mobile airbase. So equipped, the carrier had few aircraft, but large stocks of aircraft stores and maintenance equipment, as well as Ohka bombs. Other Ohka bombs were to be discovered in underground shelters on the island of Okinawa, when the US Marines stormed the beaches and fortifications at a later date.

On 6 December 1944 arrangements were being made for an engineering conference to take place at Yokosuka between the engineers of the Mitsubishi Aviation Company and representatives of the Imperial Naval Air Service to study the proposals for A7M2 Reppu and A7M3-J fighter planes. Next day, 7 December, a severe earthquake struck the Tokai district of Nagoya and totally wrecked beyond repair the airframe works of the Mitsubishi Combine, so that production was temporarily suspended. Immediately operations commenced to pull machine tools and production equipment from the wreckage, with a view to recommencing aircraft construction elsewhere. Meanwhile the American Boeing B-29 heavy bombers were rampaging and bombing Japanese cities with devastation, US troops were ashore in the Philippine Islands, and Formosa and Okinawa were also under continual aerial bombardment. In the Home Islands war production was beginning to fall; nevertheless on 12 December B-29 heavy bombers attacked the Daiko Engineering Works of the Mitsubishi Combine located at Nagoya. However, three days later, by 15 December, the Imperial Navy and representatives of the Mitsubishi Aircraft Company had agreed details of specification and layout for the projected Mitsubishi A7M3 fighter planes. On 18 December the B-29 heavy bombers returned to batter in a pulverising attack the renovated and reorganised Oe Airframe works of the Mitsubishi Combine located in another area of Nagoya. In a particularly ferocious attack the raiding bombers smashed machines, machine tools and production lines, killing hundreds of workers, supervisors, production managers and aviation engineers, so that the production facilities at the large factories were in a complete state of confusion. Next day, 19 December, the Imperial aircraft carrier HIJMS *Unryu* was torpedoed and sunk by the American submarine USS *Redfish* in the East China Sea.

By the end of December 1944 it had become apparent to the engineering department of the Mitsubishi Aircraft Company that the A7M2 Model 22 fighter plane appeared more promising than the development of the A7M3 high-altitude interceptor, which was powered by a Mitsubishi MK9c engine, equipped with a mechanically driven three-speed supercharger.

With the escalation of the American bombing campaign, war production in Japan entered a very erratic period. The production of the Mitsubishi J2M2 fighter became chaotic. It was obvious that the military establishment was unable to cope with the war situation, with the Navy ordering full production of the J2M3 in late 1943 to replace the A6M3 fighter. The scheduled production programme for 1944 fixed manufacture at 3,600 aircraft, but at the year end 500 per month were being turned out. The plan was to phase out the Zero fighter, replacing production with the new J2M2, but technical difficulties arose and it was decided to continue with the A6M3 due to the J2M2 production falling to thirty machines per month. As a result the Navy decided to transfer manufacture of the new plane to the Suzuka plant, leaving the Nagoya facility for Zero production. A number of J2M3 Model 21s were produced at the Koza Naval Arsenal, assisted by the Nippon Kentetsu, in the autumn of 1943, as planned by the Naval Bureau of Aeronautics. Commencing in October 1944 top priority was given for the development of the A7M2 and A7M3. During this period Mitsubishi manufactured 476 aircraft of the J2M3 Model 21 variant, thirty-five of the Model 33 type and a very limited quantity at the Koza Naval Arsenal. The production of the naval attack-bomber, the Mitsubishi G4M series, was more uniform. During January–March 1944, 152 were manufactured, 219 during April–June, 265 during July–September and 277 during October–December. The grand total of this type produced to December 1944 amounted to 2,138 bombers.

War production, in particular aircraft manufacture, was becoming chaotic and disorganised, as well as the moves to find suitable replacements for the aircraft carriers sunk as a result of actions fought on the high seas. The aircraft carrier HIJMS *Hosho* had the 5.5-inch GP guns removed and the flight deck extended fully fore and aft, with the anti-aircraft armament reduced to six guns. In December 1944, two oil tankers were being constructed at the Kawasaki Shipyard at Kobe, but both vessels were commandeered by the Navy and hastily converted into aircraft carrier escorts. A flight deck was built over the bridge and engine-room structure, and the resulting covered space was used as a hangar, with an elevator rising to the flight deck. The funnel was sited horizontally and outward to starboard aft. HIJMS *Shimane Maru* was sunk at Kobe during an air raid on 24 July 1945, and *Otakisan Maru* was mined in Kobe Harbour on 25 August 1945. Each vessel had a capacity for internal stowage of twelve aircraft, displacing 14,500 tons, measured 526 × 65½ × 29¾ ft and was powered by one geared turbine developing 8,600 hp, which gave a speed of 18½ knots. They were each armed with two 4.7-inch GP guns, fifty-two 25 mm anti-aircraft cannon and possessing a magazine containing sixteen depth charges for anti-submarine operations.

The modified 1942 building programme had envisaged the construction of what was to be the world's largest class of submarines capable of carrying two, later three, small bombing monoplanes with a view to attacking such targets as the Panama Canal installations. The design had combined the duties of a headquarters attack and scouting submarine, which duties were hitherto undertaken by the previous A, B and C designs. The original design was for a boat displacing 4,550 tons and carrying two bombing aircraft. Later displacement was increased to 5,223 tons, with an enlarged hangar to accommodate three aircraft. Each plane was specified to carry either four torpedoes or three 800 kg bombs, or alternatively twelve 250 kg high-explosive bombs. The conning tower was to be from the *I-13* submarine design, with a catapult that extended to the bows. Propelled by four sets of diesel engines, assisted by a primitive snorkel apparatus, a radius of action of 30,000 miles at 16 knots could be achieved. With battery and electric motor propulsion, a range of sixty miles could be attained. Measuring 400¼ × 39⅓ ft × 23 ft, each was propelled by two shaft diesel-electric propulsion of 7,700 hp/2,400 hp. *I-400* was built by the Kure Navy Yard, and her sister ships *I-401* and *I-402* were constructed by the Sasebo Navy Yards. All three boats surrendered to the US Navy and were scuttled by the Americans in 1946. Meanwhile some sort of production of the Mitsubishi A7M2 fighter had commenced, despite aerial attacks on the factories and the devastating effects of an earthquake.

CHAPTER 19

The Final Sortie

During the course of January 1945, the American Boeing B-29 heavy bombers attacked the Mitsubishi Daiko Engineering Works where the Mitsubishi radial engine MK9a was produced. In a particularly heavy air attack, the engine plant was badly hit, and consequently delivery of the motors for the A7M2 fighter production programme fell to a trickle, so that the service introduction of this carrier fighter had to be postponed. Now every effort was made to recommence production at the Oe Airframe Works, Nagoya, as well as the Mitsubishi Nankai factory situated in the southern district of the city of Osaka. At a much later date Vice-Admiral Rikizo Tada, chief of the Navy Air Research and Development Centre, was to tell US Navy officers that the main reason for Japan's defeat lay in the inability to overcome delays in replacing the Mitsubishi A6M3 Zero carrier fighter. Due to much earlier successes with this machine, the Imperial Navy had been lulled into a false sense of security until the coming of the Grumman Hellcat F6F fighter plane aboard US carriers. The Mitsubishi A7M2 Reppu follow-on design had been beset with innumerable design difficulties, material shortages and production engineering problems, so that manufacture had almost failed to commence.

However, the news was sadly received when on 3 January 1945 Katsuzo Shima, a former chief test pilot of the Mitsubishi Aircraft Company, was shot down over the south of Kaohsiung, Formosa, while flying an L2D transport plane. This plane was the Japanese licensed-production version of the American Douglas DC3 transport aircraft. Unfortunately, in the course of a running encounter with US carrier-based Grumman F6F Hellcat fighters, Shima was killed at the controls of his transport and shot down in flames.

At about this time an operational decision was taken by the Naval General Staff for Operation Ketsu-Go was drawn up by Imperial headquarters, Tokyo. This was a defence plan, which organised the assembly of some 10,000 aircraft, mainly training planes, for the final defence of the homeland. A total of two-thirds of this figure were allocated to the aerial defence of the southern island of Kyushu, while the remaining machines would be used in the defence of the Tokyo Bay area.

On 17 January 1945, the Imperial Navy lost a further escort aircraft carrier when *Yamashio Maru* was sunk by aircraft of the US 5th Fleet in the roads of Yokohama Bay.

By 3 February 1945, the engineering mock-up of the Mitsubishi A7M3 carrier fighter was available for inspection at Mitsubishi's development centre. Detailed blue prints were available for distribution to production departments of the company's various 'shadow factories'. Air officers of the Imperial Navy and the chief engineers of the Mitsubishi Aircraft Company undertook the viewing. Unhappily, the factory dispersal programme was to interrupt the retooling for the manufacture of the airframe components, and the productive programme was to make a very disjointed and erratic start. The next day, 4 February, Lieutenant-Commander Kofukuda HIJMN flew prototype number two A7M2 to the naval air station, Yokosuka, after the Navy had accepted the new plane. As a special effort, he had completed the flight test programme in just two months by the simple process of driving himself very hard. However, if personal effort was to win the war (and many were prepared to give a maximum effort), nothing could stop the continual offensive of the long-range heavy bombers and carrier bombers from US naval task forces off the shores of the Home Islands. By day and night waves of American bombers smashed cities, devastating both urban areas and factories. In many factories, production came to a complete halt. Aero-engine production almost ceased: indeed the Daiko Engine Works was reduced to twisted girders, so that production of the Mitsubishi A7M2 fighter plane ceased altogether. Meanwhile the US bomber attacks continued, and to counteract the American aerial attacks the Imperial Navy proposed to equip the Mitsubishi J2M5a Model 33a with four 20 mm cannon. With the proposed introduction of the J2M6a Model 31a, the armament was revised. While four 20 mm cannon were to be installed, two of these weapons were Model 1 Type 99 cannon and the remaining weapons were to be Model 2 Type 99 cannon; regrettably, the war was to finish before production of the J2M5a fighter commenced.

The bomber aircraft of the Imperial Navy launched an attack on the US carriers on 21 March 1945, using Mitsubishi G4M2e naval attack-bombers equipped with the new Ohka flying-bomb. When they were some sixty miles from the American ships, Grumman F6F Hellcat fighters attacked the Imperial naval planes, and during the course of the air battle the G4M2e attack-bombers were forced to release the Ohka bombs, which eventually fell into the sea, losing the lives of their suicide pilots and being unable to reach the assigned targets. Having jettisoned their offensive loads, the naval attack-bombers fought for their lives but were overwhelmed, and all eighteen bombers were shot down.

During the present aerial fighting, Pilot 1st Class Takeo Tanimizu was recognised as an air ace, with six recorded kills using a Mitsubishi A6M5 Model 52c carrier fighter from Kagoshima Naval Air Station while attached to the 303rd Fighter Squadron of the 203rd Naval Air Corps. Meanwhile the carrier conversion of the ship *Ibuki* ceased in March 1945; though 80% completed, work stopped due to the necessity of dealing with more urgent construction and repair jobs. Eventually this ship was surrendered and later scrapped in 1947.

On 1 April 1945, squadrons of Mitsubishi G4M2 attack-bombers equipped with Ohka flying-bombs took advantage of dawn, dusk and moonlight conditions to make aerial attacks on units of the Allied fleet invading the island of Okinawa. As a result of these attacks by the 721st and 722nd Kokutai, the battleship USS *West Virginia* was seriously damaged. For use in the attacks on the Allied invasion fleet, some 516 Mitsubishi G4M2 attack-bombers were available. Six days later a preliminary study meeting took place at the Yokosuka Experimental Air Corps headquarters between officers of the Imperial Naval Air Service and the various engineers of the aircraft industry with the purpose of discussing the 20-Shi fighter project and programme. From 12 April the use of the Ohka flying-bomb was discontinued owing to poor results being achieved after innumerable aerial attacks and despite the sinking of the US destroyer *Mannert L. Abele,* so that conventional bombing tactics would be made by the Navy's attack-bombers on Allied ships and positions.

During April 1945 the flight testing had taken place of the Mitsubishi J2M5 Model 33 land-based fighter plane, which in the course of an experimental run had disintegrated in mid-air. This plane had been manufactured by the Koza Naval Arsenal and was a modification from the J2M6 Model 31. With this serious failure in the testing programme, it was decided to replace this machine with the J2 M4 Model 32, which proved very satisfactory, attaining a maximum speed of 362 mph at 30,200 feet altitude. This machine had a turbo supercharger installed, which proved faulty and was eventually replaced by a mechanical supercharger. However, the aircraft was of complex construction, which proved difficult and complicated to maintain. The engine cowl was enlarged to take a Kinsei 21 motor, and a bigger manifold was designed to increase the high-altitude performance. Meanwhile developments continued with the J2M7 Model 23 and a J2M7 Model 23a, being continual modifications of the J2M3 Model 21 and J2M3a Model 21a in which the Mitsubishi Kasei 26 engine was installed and fitted with an improved oil cooler system. At the same time, the new Mitsubishi A6M8c Model 64 flew for the first time, in which the prototype finally reached 355 mph at 19,700 feet in 6 minutes 50 seconds, and thus halted the trend of a declining performance. The Navy and the Mitsubishi engineers were very pleased.

During the period 1–23 May 1945, the Naval Air Service and aeronautical industry representatives held a series of conferences on the designing and productive planning for the 20-Shi fighter project. These conversations were held at the First Navy Technical Research and Development Command. Those present included representatives from the Nakajima Aircraft Company, Kawanishi Aviation and the Mitsubishi Aircraft Company. It was agreed that the requirements of the new machine should include the highest of performance, using a Homare-type engine with a two-stage, three-speed supercharger. The new fighters A7M2 and A7M3 required the improved Mitsubishi MK9c engine, while the carrier or escort fighters subsequently to be designed required a high-altitude performance. However, it was necessary to accelerate the development programme of the Nakajima NK11-20 and Mitsubishi MK9c engines if they were to be of any use in the fighter development programme. On 25 May 1945, the Navy accepted the first prototype Mitsubishi A6M8c Model 64 following the results of successful flight tests. This machine was service tested at the Misawa Naval Air Station situated in Aomari Prefecture by the Yokosuka Experimental Air Corps Proving Division. Its acceptance by the Naval Air Service resulted in production contracts being placed for a total of 6,500 machines, but the war finished before quantity production came off the assembly lines.

By the end of May 1945, neither the Imperial Navy nor the Mitsubishi Aircraft Company had anticipated the fast-flowing development of aviation technology. The wartime requirements of the numbers of engineering staff had never been calculated, and as a consequence a serious shortage of qualified engineers, technicians, draughtsmen and skilled craftsmen had arisen. Accelerated engineering development programmes had been introduced for the construction of the Mitsubishi J2M1 land-based fighter and the A7M1 carrier-based fighter to replace the ageing A6M8 interceptor. It would be no exaggeration to record that Japan suffered from a second-rate industrial capacity, which was not likely to match up to the industrial potential of her adversary, the United States of America. With too few experienced qualified engineers and poor natural resources within the country, supporting industry was by any standards backward. With rampant administrative inefficiency, incompetent planning and leadership, too many aircraft projects were being embarked upon by too few qualified engineers.

On 10 June 1945 the final design requirements for the Mitsubishi A7M2 carrier fighter were agreed at a meeting held at the Naval Aircraft Research and Development Centre. The Mitsubishi MK9c engine would be installed and the armament increased to a total of six cannon of 20 mm calibre. Aft of the pilot were to be located rubber-lined, self-sealing fuel tanks, with armour plating for pilot protection and bullet-proof glass for the cockpit canopy. However, the wing folding mechanism was now eliminated from the specification. Some nine days later, the final conference took place between the Navy and the aircraft industry, when it was agreed that a joint development agreement would be drawn up between Mitsubishi and Nakajima, using as a basis the A7M3J Model 34 airframe for the future prototype. The Navy required the new plane to be fully operational by the end of 1946. Meanwhile, the two industrial contractors became worried over the supply of completed engines, anticipating insufficient motors for the airframes completed.

The prototype design and construction of the A7M Reppu was delayed owing to the same design team being responsible for modifications to the Zero fighter, as well as the J2M plane and the design of the A7M. As a result twenty engineers and draughtsmen were detached by the Navy from the dockyards at Nagasaki, Kobe, Yokohama and the Hitachi Aircraft Company to assist the Mitsubishi engineering design office. However, Mitsubishi maintained a special prototype assembly factory where sub-assemblies were put together. The procedure was for the blueprints to be prepared from which parts would be manufactured, and the final blueprints would then be rushed to the factory assembly shop. The first subsequent prototypes were completed for flying tests by 19 July, when the third prototype was flown to Misawa Naval Air Station for the commencement of flight tests.

In June 1945, while feverish preparations were being made at the Mitsubishi Aircraft Company for the eventual production of the A7M fighter plane, the Kyushu Aircraft Company had equipped experimentally the J7W1 Shinden canard-type fighter plane with Type 5 cannon of 30 mm calibre. In the following month the Mitsubishi Company produced the Shusui J8M1 rocket-powered interceptor, armed with 30 mm cannon, and also manufactured four German Messerschmitt Bf 163 fighter planes, which were armed with 30 mm cannon. As an experiment a 40 mm weapon was ordered into production. During July 1945, the Tsukuba Naval Air Corps was organised at Oita Naval Air Station as an advanced training unit for squadrons earmarked for defence of the Home Islands. About this time two naval air aces were mentioned in public – Saburo Sakai with sixty-four victories and Hiroyoshi Nishizawa who claimed 102 kills. Meanwhile the J2M6 Model 31 had been modified with a new cockpit canopy two inches higher and three inches wider and with the top portion of the engine cowling trimmed to improve pilot visibility, but regrettably only one model was built. However, the development of the A7M2 fighter fared no better when Allied heavy and carrier-based bombers wrecked the Suzuka production plant. As a result prototypes numbers one and five were completely wrecked, while prototypes six and seven were slightly damaged but quite repairable. Thus, with the undamaged prototype number four, and accompanied by numbers six and seven, they successfully flew to Matsumoto Naval Air Station. But bombing damage to the production factories was serious, and the Nagoya factory had completed the number one prototype and was rushing completion of the other planes when components became very scarce, so that the number two plane was never completed. In Osaka the damage was sufficiently serious to cause a halt in the production of the southern district factories. In another particularly heavy aerial attack by carrier bombers of the US 3rd Fleet on the dockyard and roads in Kure Harbour, the warships *Aso*, *Ikoma*, *Amagi* and *Shimane Maru* were sunk at their moorings on 24 July.

During August 1945, the second machine of the J7W1 Shinden type was completed by the Kyushu Aircraft Company. These two canard interceptors had a six-bladed contra-rotating propeller, with twin fins mounted on mini-planes, and had achieved a top speed of 467 mph. Though much earlier mass production of this type of aircraft might well have stemmed the American steamroller tactics long before the US forces were within sight of the Home Islands, on 6 August 1945 a single US Boeing B-29 Superfortress heavy bomber attacked Hiroshima and dropped an atomic bomb over the city.

Existence of the atomic bomb had been discovered by the 'Spanish Network' of Imperial Naval Intelligence Service by agents operating in the USA and duly reported to Naval Intelligence headquarters, Tokyo, during the early part of the war. However, the selected target, while of military significance and perfectly legitimate – it was the headquarters of the Combined Naval Fleet, a former headquarters of Admiral Yamamoto and contained innumerable war production targets – was struck by an airburst bomb, which did not cause secondary radiation. The target contained domestic habitations constructed in wood and papier mâché as an anti-earthquake measure. These buildings were consumed in the explosion, killing and maiming many of the inhabitants, who were obviously unprotected to withstand such an assault. Buildings in concrete and steel withstood the force of the exploding bomb, protecting those within. Had all the buildings in the city been constructed in concrete and steel, with appropriate air raid shelters for their occupants, the disaster would have been minimised into insignificance. It was some three days or so before Imperial headquarters, Tokyo, could bring itself to reveal to the Nipponese population the depth of the disaster that had befallen them. At about the same time, Misawa Naval Air Station had been attacked, and the third prototype of the Mitsubishi A7M2 fighter was so severely wrecked that the machine would never fly again. At the same time, HIJMS *Kaiyo* was sunk in Beppu Bay.

On 15 August 1945, His Imperial Majesty the Emperor Hirohito issued the surrender order to all units of the Imperial forces, whether located in the Home Islands or still within the provinces of the Imperial Empire. On the same day, at Atsugi Naval Air Station, Captain Yasuna Kozono HIJMN addressed the pilots of the 302nd Air Group while standing on a raised platform at the end of the runway. In his speech he made the point that surrender equated the end of national essence, and called upon the pilots to join him in one last operation to destroy the enemy. The pilots greeted the speech with shouts of 'Banzai!' Similar scenes were re-enacted at other naval air stations. At Oita Airbase in north-east Kyushu, Admiral Matome Ugaki, a former chief of staff to Admiral Yamamoto, and now commanding officer of the kamikaze units, remembered his 'old chief's' death with some responsibility, and similarly wished to die fighting. Many naval air personnel took off in one final operational sortie, never to return, shot down and lost among the white-capped rollers, embraced and engulfed in the green depths of the northern Pacific Ocean in the planes that they had so lovingly tended and fought in the name of His Imperial Majesty the Emperor and to the sacredness of the Home Islands – gone!

In considering the aircraft delivered to the Imperial Naval Air Service by the Mitsubishi Aircraft Company from October 1940 to September 1945, the following machines were delivered: just two prototype 12-Shi fighters, thirty Model G6M1s, 1,200 Model G4M1s, 1,154 Model G4M2s and sixty Model G4M3 attack-bombers. Unfortunately, company production for various reasons lagged 14% behind government delivery schedules. During the year bomber production by Mitsubishi had run as follows: 164 attack-bombers from January to March 1945, 105 from April to June and seven between July and September. Meanwhile, on 16 August, the mock-up of the A7M2 fighter was completed at Matsumoto Naval Air Station. Though the first production fighter was completed, it was nevertheless the end of the Reppu; but for the troubles that had been experienced in its development, mass production could have commenced by mid-1944, even before the Mariana Islands had been captured by US forces and used as operational bases by the Boeing B-29 Superfortress bombers.

It was on 19 August that General Kawabe Torashiro, Army Vice-Chief of Staff, flew in a Mitsubishi G4M3 bomber to the American-held island of Okinawa and thence on to the Philippine Islands, to Nicholas Field just south of Manila. Here an eight-hour conference took place at General MacArthur's US headquarters in which it was stated that the first US transport aircraft would arrive at Atsugi Naval Air Station, Japan, on 23

August, just eighty hours away. Unfortunately, General Torashiro had to point out that Atsugi was out of commission due to ruts and holes that required repairing. The American General Sutherland agreed to a five-day delay to permit renovations to take place.

On 21 August a labour racketeer friend of Prince Takamatsu, the brother of the Emperor, marshalled his trucks and construction men to repair Atsugi Airbase so as to enable the American transport aircraft to land. By the morning of 28 August, forty-five US Curtiss C47 transport aircraft snaked their way over the bronze statue of Amida Buddha at Kamakura, and after circling the airfield landed in rapid succession to taxi to the far end of the runway. Immediately US Marines swarmed out of the enormous bellies of the transport planes, arms at the ready, somewhat nervously hesitant – no shooting took place – all was quiet – and then relaxed. Unloading commenced immediately, and within half an hour the Imperial Naval Air Service men, having admired the American planes, were assisting in the unloading operations. Nevertheless the Barbarians had landed. The sacredness of the Imperial Home Islands had been desecrated.

APPENDIX 1

An Analysis of Imperial Military Aviation Organisation

Imperial Naval Air Service, 1927–30

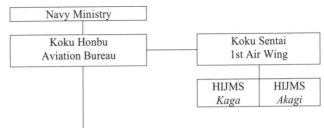

Kasumigaura Kokutai Air Corps	Yokosuka Kokutai Air Corps	Tateyama Kokutai Air Corps	Kure Kokutai Air Corps	Sasebo Kokutai Air Corps	Ohmura Kokutai Air Corps	
-	0.5	0.5	-	-	1.0	Fighter squadrons
1.5	0.5	-	-	-	-	Reconnaissance land-based squadrons
-	0.5	1.0	-	-	1.0	Land-based attack squadrons
1.5	0.5	1.0	0.5	1.0	-	Reconnaissance sea squadrons
-	0.5	1.0	-	0.5	-	Flying-boat squadrons
1.5	-	-	-	-	-	Land-based training squadrons
1.5	-	-	-	-	-	Sea training squadrons
1.0	-	-	-	-	-	Research squadrons
7	2.5	3.5	0.5	1.5	2.0	= 17 squadrons
6						Balloon squadrons
4						Airship squadrons

Imperial Army Air Corps, 1927–30

```
                    ┌─────────────────────┐
                    │   Army Ministry     │
                    └─────────────────────┘
                    ┌─────────────────────┐
                    │    Koku Honbu       │
                    │  Aviation Bureau    │
                    └─────────────────────┘
```

1st Hiko Rentai	2nd Hiko Rentai	3rd Hiko Rentai	4th Hiko Rentai	5th Hiko Rentai	6th Hiko Rentai	7th Hiko Rentai	8th Hiko Rentai	
4		3	2		1		1	Fighter Chutai. EA 12, Nieuport 29 or 24 a/c
	2		2	4	2		1	Reconnaissance Chutai. EA Type 88 or Salmson 2A2
						2		Light bomber Chutai. EA 9, Type 87 a/c
						2		Heavy bomber Chutai. EA 6, Type 87 or Farman F60 a/c
Kasumigaura	Kasumigaura	Yokaichi	Tachiari	Tachiari	Pyongyang Korea	Hamamatsu	Pingtung Formosa	LOCATIONS
3rd Div	3rd Div	16th Div	12th Div	Imperial Guard Div	20th Div	3rd Div	Formosa Command	Attached ground formations
48	18	36	42	36	30	30	21	= 261 a/c

For:	Combat	11 squadrons	132 a/c
	Reconnaissance	11 squadrons	99 a/c
	Light Bombers	2 squadrons	18 a/c
	Heavy Bombers	2 squadrons	12 a/c
		26 squadrons	261 a/c

Hiko Rentai = Air Wing
Chutai = Air Squadron

Imperial Army Air Corps – Kanto Command, Manchuria, 1932

Manchurian & Shanghai Incidents 1931–36

Kanto Command

Hiko Tai

10th Hiko Daitai	11th Hiko Daitai	12th Hiko Daitai	
1, 2, 3 27 a/c			Reconnaissance squadrons. Type 88
	1, 2, 3, 4 36 a/c		Fighter squadrons. Type 91
		1 9 a/c	Heavy bomber squadrons. Type 87
		1 9 a/c	Light bomber squadrons. Type 87

Imperial Naval Air Service – Shanghai, January 1932

	Naval 1st Air Wing		

Total A/C	Aircraft Carrier HIJMS *Kaga*	Aircraft Carrier HIJMS *Hosho*	Seaplane Carrier HIJMS *Notoro*	
26	16	10		Fighters. Type 3
41	32	9		Torpedo-bombers. Type 13
8			8	Sea reconnaissance planes. Type 14 or 90 ii

Cumulative 75

Imperial Army Air Corps – Kanto Command, Manchuria Reorganisation, 1936

Kanto Command
Joint Air Group

10th Air Wing	11th Air Wing	12th Air Wing	15th Air Wing	16th Air Wing
1	1	1	1	1
2	2	2	2	2
3	3	3	3	1
	4	4		2
Light bomber squadrons	Fighter squadrons	Heavy bomber squadrons	Reconnaissance squadrons	2 light bomber & 2 fighter squadrons
		Locations		
Tsitsihar	Kharbin	Kungchuling	Changchun	Mutanchiang

Fighter	Chutai =	6
Reconnaissance	Chutai =	3
Light bomber	Chutai =	5
Heavy bomber	Chutai =	4
Grand Total	=	18

Imperial Army Air Corps, 1936

Koku Heidan
Air Command
AOC: Lieutenant-General
Yoshitoshi Tokugawa

1st Hikodan

1st Air Group	Squadrons	No.	Type	Location
1st Air Wing	1 2 3 4	4	Fighter Chutai	Kasumigaura
2nd Air Wing	1 2	2	Reconnaissance Chutai	Kasumigaura
3rd Air Wing	1 2 3	3	Reconnaissance Chutai	Yokaichi
7th Air Wing	1 2 1 2	2	Light bomber Chutai	Hamamatsu
		2	Heavy bomber Chutai	Hamamatsu
13th Air Wing	1 2 3	3	Fighter Chutai	Kikugawa

2nd Hikodan

2nd Air Group	Squadrons	No.	Type	Location
6th Air Wing	1 1 2	1	Fighter Chutai	Pyongyang, Korea
		2	Light bomber Chutai	Pyongyang
9th Air Wing	1 2 1 2	2	Fighter Chutai	Hoeryong
		2	Light bomber Chutai	Hoeryong

3rd Hikodan

3rd Air Group	Squadrons	No.	Type	Location
8th Air Wing	1 1	1	Fighter Chutai	Pingtung
		1	Light bomber Chutai	Pingtung
14th Air Wing	1 2	2	Heavy bomber Chutai	Chali
4th Air Wing	1 2 1 2	2	Fighter Chutai	Tachiari
		2	Reconnaissance Chutai	Tachiari
5th Air Wing	1 2 1 2	2	Fighter Chutai	Tachikawa
		2	Reconnaissance Chutai	Tachikawa

Fighter Chutai   = 15
Reconnaissance Chutai = 9
Light bomber Chutai   = 7
Heavy bomber Chutai   = 4
Grand Total   = 35

Imperial Naval Air Service Air Corps Expansion, 1927–41

	Kasumigaura Air Corps	Yokosuka Air Corps	Tateyama Air Corps	Kure Air Corps	Sasebo Air Corps	Ohmura Air Corps	Maizura Air Corps	Kanoya Air Corps	Kisarazu Air Corps	Chinhe Air Corps	Ohminato Air Corps	Saeki Air Corps	Yokohama Air Corps
1927–1930													
1931													
1932													
1933													
1934													
1935													
1936													
1937													
1938													
1939													
1940													
1941													

Imperial Army Air Corps Squadron Expansion, 1927–41

	No. of Squadrons					
Type	1937	1938	1939	1940	1941	1942
Reconnaissance	12	13	14	19	24	29
Fighter	21	24	29	35	40	42
Light bomber	12	16	28	34	37	49
Heavy bomber	8	16	22	26	30	30
Super heavy bomber	0	1	1	1	2	2
Total	53	70	94	115	133	152

Deployment 1942

Japan	58 squadrons
Korea	16 squadrons
Formosa	11 squadrons
Manchuria	57 squadrons

Imperial Naval Air Service, 1937

Squadrons	Naval General Staff – Tokyo												
	Yokosuka Air Corps	Kasumigaura Air Corps	Tateyama Air Corps	Yokohama Air Corps	Ohminato Air Corps	Kure Air Corps	Sasebo Air Corps	Ohmura Air Corps	Saeki Air Corps	Kanoya Air Corps	Chinhe Air Corps	Maizuru Air Corps	Kisarazu Air Corps
Fighter	1	0.5	1	-	-	-	-	1.5	0.5	1	-	-	-
Dive-bomber	0.5	-	0.5	-	0.5	-	-	0.5	1	-	-	-	-
Torpedo-bomber	1.5	0.5	1	-	-	0.5	-	1	0.5	-	-	-	-
Twin-engine bomber	-	-	1	-	1	-	-	1	-	1.5	-	-	2
Sea reconnaissance	1.5	1	0.5	-	0.5	0.5	1	-	-	-	0.5	0.5	-
Small flying-boat	0.5	-	-	-	-	-	-	-	-	-	-	-	-
Large flying-boat	-	-	1	2	-	-	1	-	1	-	-	-	-
Land training	0.5	2.5	-	-	-	-	-	-	-	-	-	-	-
Sea training	0.5	2.5	-	-	-	-	-	-	-	-	-	-	-
Research	1	0.5	-	-	-	-	-	-	-	-	-	-	-
Total	7	7.5	5	2	2	1	2	4	3	2.5	0.5	0.5	2

Total = 39 squadrons

Imperial Naval Air Service – China, 1937

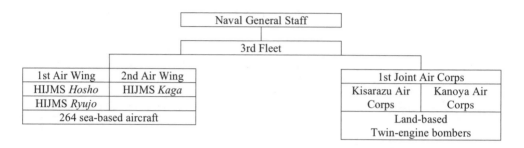

Naval General Staff

3rd Fleet

1st Air Wing	2nd Air Wing		1st Joint Air Corps	
HIJMS *Hosho*	HIJMS *Kaga*		Kisarazu Air Corps	Kanoya Air Corps
HIJMS *Ryujo*				
264 sea-based aircraft			Land-based Twin-engine bombers	

Imperial Army Air Corps – Sino-Japanese War, 1937, China

Provisional Air Command, China

Direct Command Air Unit	
3rd Squadron	1 Squadron Type 93 Heavy bomber
4th Squadron	1 Squadron Type 94 Reconnaissance
6th Squadron	1 Squadron Type 94 Reconnaissance
9th Squadron	1 Squadron Type 95 Fighter
10th Squadron	1 Squadron Type 95 Fighter
14th Squadron	1 Squadron Type 93 Light bomber
15th Squadron	1 Squadron Type 93 Heavy bomber

Summary:-

	Dcau	Air Bttlns	Totals
Fighter	2	4	6
Reconnaissance	2	6	8
Light bomber	1	4	5
Heavy bomber	2	2	4
Squadrons	7	16	23

1st Air Bttlns	2nd Air Bttlns	3rd Air Bttlns	5th Air Bttlns	6th Air Bttlns	7th Air Bttlns	8th Air Bttlns	9th Air Bttlns
1	1	1	1	1	1	1	1
2	2	2	2	2	2	2	2
2 Sqns	2 Sqns	2 Sqns	2 Sqns	2 Sqns	2 Sqns	2 Sqns	2 Sqns
Reconnaissance	Fighter	Reconnaissance	Light	Heavy	Reconnaissance bomber	Fighter-bomber	Light bomber
Type 92/94	Type 95	Type 92	Type 93	Type 93	Type 92	Type 95	Type 93

Imperial Army Air Corps Reorganisation, 1937/38

Hiko Sentai				= 40 or 30 A/C = 1 Wing
1	2	3	4	

3/4 Chutai	= 3/4 Squadrons

Basic organisation consists of:
1) 40 fighter planes or
2) 30 reconnaissance a/c or
3) 30 bombers

New Combines Sentai	Direct Command and Squadrons	Air Battalions
25	10	
27		1
31		5
44		3
45	4, 5 & 6	7
60		6
64	9	2
75		7
77		8
90		9
98	15 & 3	
59		8

Imperial Naval Air Service Expansion Programmes, 1937–45

	3rd Exp Programme 1937–40		4th Exp Programme 1939–44		5th Exp Programme 1941 to Surrender	
	Squadrons	A/C	Squadrons	A/C	Squadrons	A/C
Operational units						
Carrier-based fighters	5.0	80	6.5	104	14.0	336
Carrier-based dive-bombers	1.0	16			10.0	240
Carrier-based torpedo-bombers	4.5	72	2.0	32	2.0	48
	7.5	107	18.5	296	2.5	40
Twin-engine bombers	5.0	60	5.5	90		
Sea reconnaissance	6.0	28	2.0	44	18.0	216
Flying-boats (med/large)	1.5	18				
Flying-boats (small)						
Total	30.5	381	34.5	566	46.5	880
Sea-fighters					12	288
Anti-submarine aircraft					6.5	104
Transport aircraft					2	48
Total 1)	30.5	381	34.5	566	67.0	1320
Training units						
Carrier-based fighters	2.0	32				
Carrier-based dive-bombers	Nil	Nil				
Carrier-based torpedo-bombers	3.5	56				
	2.0	32				
Carrier-based reconnaissance	2.0	24				
Sea reconnaissance	0.5	6				
Flying-boats (small)	4.0	120	7.5	216	19.5	468
Primary trainers	5.0	120	15.5	372	21.5	516
Secondary trainers	2.0	32	17.5	280	52.0	1154
Advanced trainers	1.5	24		33		
Research aircraft & misc.						
Total 2)	22.5	446	40.5	901	93	2138
Land-based transports				55		417
Ship-based op units						
Carrier-based fighter		225		36		
Carrier-based dive-bombers		192		36		
Carrier-based torpedo-bombers		342		54		
Sea reconnaissance		329		48		
Total 3)		1,088		229		2,001
Grand Total (Total 1+2+3)	53	1,915	75	1,696	160	5,459

Imperial Naval Air Service – China, September 1938

Naval General Staff

Air Command HQ

2nd Joint Air Corps

Squadrons F = 5.5 Db = 1.5 R = 1.0 Ab = 2.5 Tb = 3.0	12th Air Corps	13th Air Corps	14th Air Corps	15th Air Corps	Kaohsiung Air Corps
	2.5 Type 96 F	2 Type 96 Ab	1 Type 96 F	1 Type 96 F	0.5 Type 96 Ab
	1 Type 96 Tb	1 Type 96 F	0.5 Type 96 Db	1 Type 96 Db	
	1 6 Seversky 2pa. R		1.5 Type 96 Tb	0.5 Type 96 Tb	

1st Joint Air Corps

Squadrons Ab = 3.0	Kisarazu Air Corps	Kanoya Air Corps
	2 squadrons	1 squadron
	Type 96 twin-engine bombers	Type 96 twin-engine bombers

Sea Command

Aircraft F = 36 Db = 30 Tb = 36 Sp = 41 Total = 143	1st Air Wing	2nd Air Wing		3rd Air Wing	
	HIJMS *Kaga* Ac	HIJMS *Soryu* Ac	HIJMS *Ryujo* Ac	Sea plane carriers	
	12 F	12 F	12 F	HIJMS *Kumoi*	9 aircraft
	12 Db	18 Db		HIJMS *Kaku Maru*	9 aircraft
	18 Tb	12 Tb	6 Tb	HIJMS *Notoro*	8 aircraft
				HIJMS *Kamikawa Maru*	6 aircraft
				HIJMS *Chitose*	9 aircraft

F = Fighter
Db = Dive-bomber
Tb = Torpedo-bomber
Ab = Attack-bomber
Sp = Seaplane
Ac = Aircraft carrier
R = Reconnaissance aircraft

Imperial Army Air Corps – China, 1939

Provisional Air Command
AOC: Lieutenant-General
Eiji Ebashi

1st Air Group	
10th Direct Command Squadron F 10	12th Sentai Hb 30
16th Direct Command Squadron R 10	59th Sentai F 40
18th Direct Command Squadron R 10	60th Sentai Hb 30
	98th Sentai Hb 30

3rd Air Group	
17th Direct Command Squadron R 10	45th Sentai Lb 30
	75th Sentai Lb 30
	77th Sentai F 40

4th Air Group Formosa	
31st Sentai	Lb 1 Squadron 10
27th Sentai	Lb 1 Squadron 10
64th Sentai	F 2 Squadron 20

	7 Air Group
Lb 10	27th Sentai
F 10	64th Sentai
Lb 30	90th Sentai

Aircraft	
Fighters	= 120
Recce	= 30
Light bombers	= 120
Heavy bombers	= 90
Total	= 360

F = Fighter
R = Reconnaissance
Lb = Light bomber
Hb = Heavy bomber

Appendix 1

Imperial Army Air Corps, May 1939

Provisional Air Command moved into Manchuria to cover the Nomonhan conflict
(Japanese/Manchurians v. Russians/Mongolians)

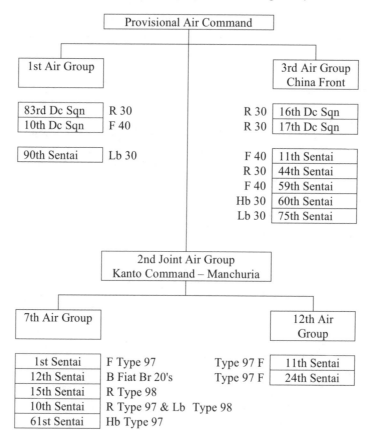

1st Air Group and 3rd Air Group	= 330 a/c
2nd Joint Air Group	= 119 a/c
Total	= 449 a/c

F = Fighter
R = Reconnaissance
Lb = Light bomber
Hb = Heavy bomber
B = Bomber

Imperial Army Air Corps, September 1939
Reorganisation of 2nd Joint Air Corps, Manchuria

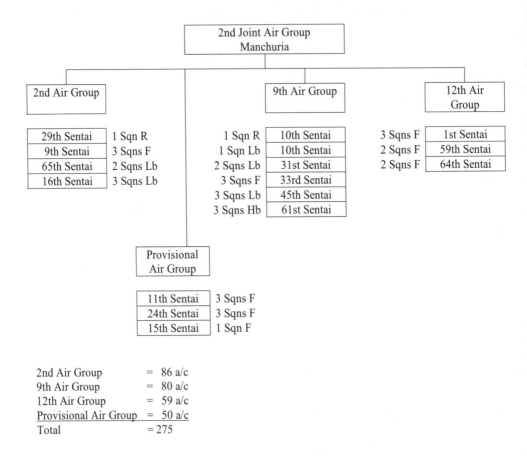

2nd Air Group	= 86 a/c
9th Air Group	= 80 a/c
12th Air Group	= 59 a/c
Provisional Air Group	= 50 a/c
Total	= 275

Sentai Analysis of the Imperial Army Air Corps, 1941

Provisional Air Command	F	R	Lb	H
1st Joint Air Group				
▪ 1st Air Group, Japan	3	-	-	-
▪ 4th Air Group, Formosa	1	1	1	1
▪ 20th Air Group, Hokkaido	1	-	1	1
2nd Joint Air Group				
▪ 2nd Air Group, Korea	1	-	1	1
▪ 7th Air Group	1	-	1	2
▪ 8th Air Group	1	3	2	1
3rd Joint Air Group – Malaya				
▪ 1st Air Group	1	1	1	-
▪ 3rd Air Group	2	1	1	1
▪ DCS	2	-	-	-
5th Joint Air Group – China				
▪ 10th Air Group	1	-	1	1
▪ 9th Air Group	1	-	1	1
▪ 12th Air Group	2	-	-	-
▪ 13th Air Group	3	2	-	-
Total	20	8	10	9 = 47

F = Fighter
R = Reconnaissance
Lb = Light bomber
H = Heavy bomber

Imperial Army Air Corps – 1941

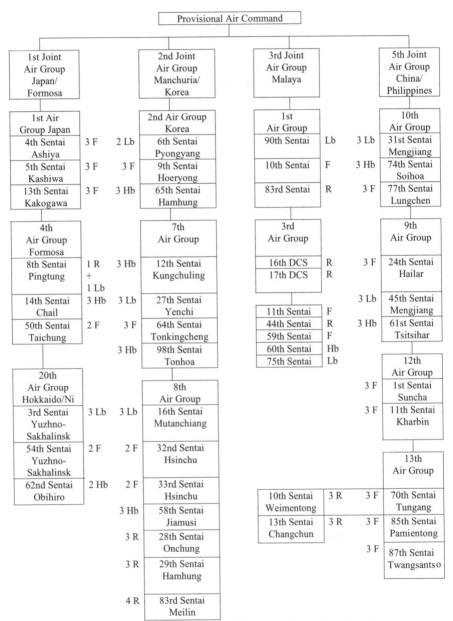

Provisional Air Command

1st Joint Air Group — Japan/Formosa

1st Air Group Japan

4th Sentai Ashiya	3 F
5th Sentai Kashiwa	3 F
13th Sentai Kakogawa	3 F

4th Air Group Formosa

8th Sentai Pingtung	1 R + 1 Lb
14th Sentai Chail	3 Hb
50th Sentai Taichung	2 F

20th Air Group Hokkaido/Ni

3rd Sentai Yuzhno-Sakhalinsk	3 Lb
54th Sentai Yuzhno-Sakhalinsk	2 F
62nd Sentai Obihiro	2 Hb

2nd Joint Air Group — Manchuria/Korea

2nd Air Group Korea

6th Sentai Pyongyang	2 Lb
9th Sentai Hoeryong	3 F
65th Sentai Hamhung	3 Hb

7th Air Group

12th Sentai Kungchuling	3 Hb
27th Sentai Yenchi	3 Lb
64th Sentai Tonkingcheng	3 F
98th Sentai Tonhoa	3 Hb

8th Air Group

16th Sentai Mutanchiang	3 Lb
32nd Sentai Hsinchu	2 F
33rd Sentai Hsinchu	2 F
58th Sentai Jiamusi	3 Hb
28th Sentai Onchung	3 R
29th Sentai Hamhung	3 R
83rd Sentai Meilin	4 R

3rd Joint Air Group — Malaya

1st Air Group

90th Sentai	Lb
10th Sentai	F
83rd Sentai	R

3rd Air Group

16th DCS	R
17th DCS	R
11th Sentai	F
44th Sentai	R
59th Sentai	F
60th Sentai	Hb
75th Sentai	Lb
10th Sentai Weimentong	3 R
13th Sentai Changchun	3 R

5th Joint Air Group — China/Philippines

10th Air Group

31st Sentai Mengjiang	3 Lb
74th Sentai Soihoa	3 Hb
77th Sentai Lungchen	3 F

9th Air Group

24th Sentai Hailar	3 F
45th Sentai Mengjiang	3 Lb
61st Sentai Tsitsihar	3 Hb

12th Air Group

1st Sentai Suncha	3 F
11th Sentai Kharbin	3 F

13th Air Group

70th Sentai Tungang	3 F
85th Sentai Pamientong	3 F
87th Sentai Twangsantso	3 F

F = Fighter Sentai, R = Reconnaissance Sentai, Lb = Light bomber Sentai, Hb = Heavy bomber Sentai, DCS = Direct Command squadron

Appendix 1

Imperial Army Air Corps, December 1941
Southern General Command

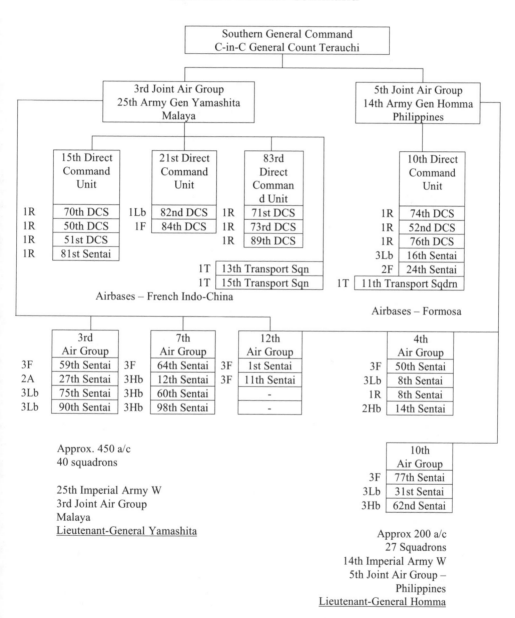

Southern General Command
C-in-C General Count Terauchi

3rd Joint Air Group	5th Joint Air Group
25th Army Gen Yamashita	14th Army Gen Homma
Malaya	Philippines

15th Direct Command Unit	21st Direct Command Unit	83rd Direct Command Unit	10th Direct Command Unit

	15th Direct Command Unit		21st Direct Command Unit		83rd Direct Command Unit		10th Direct Command Unit
1R	70th DCS	1Lb	82nd DCS	1R	71st DCS	1R	74th DCS
1R	50th DCS	1F	84th DCS	1R	73rd DCS	1R	52nd DCS
1R	51st DCS			1R	89th DCS	1R	76th DCS
1R	81st Sentai					3Lb	16th Sentai
						2F	24th Sentai
		1T	13th Transport Sqn			1T	11th Transport Sqdrn
		1T	15th Transport Sqn				

Airbases – French Indo-China

Airbases – Formosa

	3rd Air Group		7th Air Group		12th Air Group		4th Air Group
3F	59th Sentai	3F	64th Sentai	3F	1st Sentai	3F	50th Sentai
2A	27th Sentai	3Hb	12th Sentai	3F	11th Sentai	3Lb	8th Sentai
3Lb	75th Sentai	3Hb	60th Sentai		-	1R	8th Sentai
3Lb	90th Sentai	3Hb	98th Sentai		-	2Hb	14th Sentai

Approx. 450 a/c
40 squadrons

25th Imperial Army W
3rd Joint Air Group
Malaya
<u>Lieutenant-General Yamashita</u>

	10th Air Group
3F	77th Sentai
3Lb	31st Sentai
3Hb	62nd Sentai

Approx 200 a/c
27 Squadrons
14th Imperial Army W
5th Joint Air Group –
Philippines
<u>Lieutenant-General Homma</u>

Imperial Naval Air Fleet
Subsequent to the Battle of Midway, July 1942

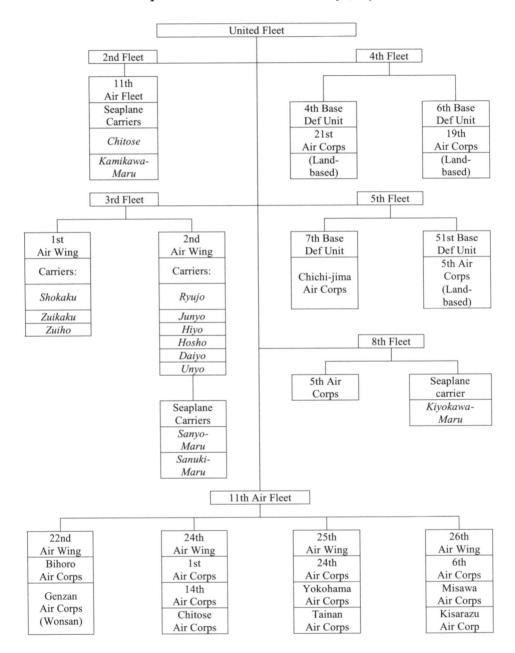

Imperial Naval Air Service
South-West District Fleet, 1942
Malaya, Indonesia, Philippines

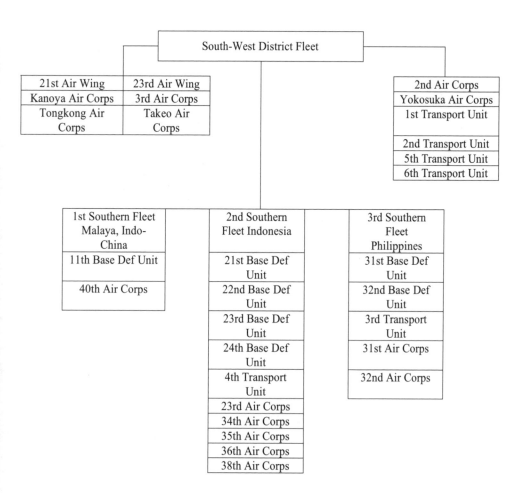

Imperial Naval Air Service, Dec 1942
South-East District Fleet
New Guinea and the Solomon Islands

		South-East District Fleet		

4th Air Fleet	11th Air Fleet			
902nd Air Corps	2nd Air Wing	24th Air Wing	26th Air Wing	25th Air Wing
952nd Air Corps	252 Air Corps	201 Air Corps	705 Air Corps	251 Air Corps
	701 Air Corps	552 Air Corps	582 Air Corps	801 Air Corps
	755 Air Corps	703 Air Corps	204 Air Corps	702 Air Corps
		752 Air Corps	802 Air Corps	

Appendix 1

Imperial Army Air Corps, July 1942

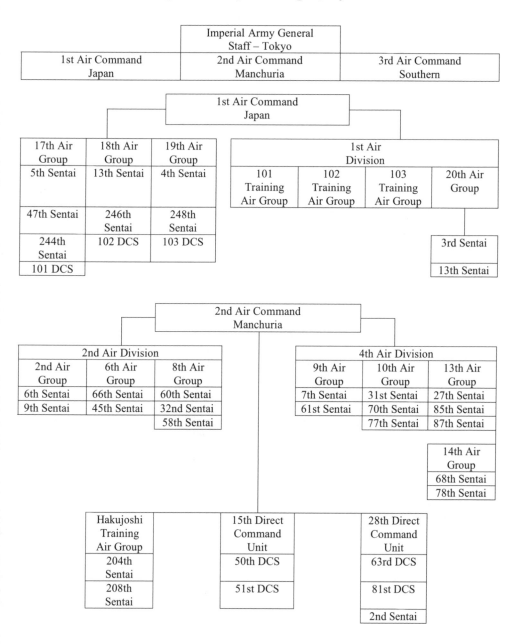

	Imperial Army General Staff – Tokyo	
1st Air Command Japan	2nd Air Command Manchuria	3rd Air Command Southern

1st Air Command Japan

17th Air Group	18th Air Group	19th Air Group	1st Air Division			
5th Sentai	13th Sentai	4th Sentai	101 Training Air Group	102 Training Air Group	103 Training Air Group	20th Air Group
47th Sentai	246th Sentai	248th Sentai				
244th Sentai	102 DCS	103 DCS				3rd Sentai
101 DCS						13th Sentai

2nd Air Command Manchuria

2nd Air Division			4th Air Division		
2nd Air Group	6th Air Group	8th Air Group	9th Air Group	10th Air Group	13th Air Group
6th Sentai	66th Sentai	60th Sentai	7th Sentai	31st Sentai	27th Sentai
9th Sentai	45th Sentai	32nd Sentai	61st Sentai	70th Sentai	85th Sentai
		58th Sentai		77th Sentai	87th Sentai

14th Air Group
68th Sentai
78th Sentai

Hakujoshi Training Air Group	15th Direct Command Unit	28th Direct Command Unit
204th Sentai	50th DCS	63rd DCS
208th Sentai	51st DCS	81st DCS
		2nd Sentai

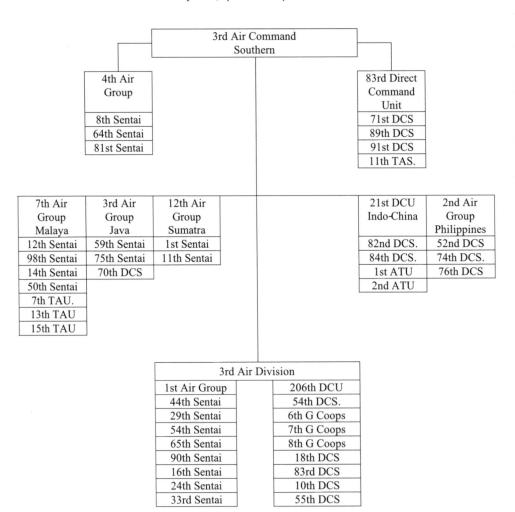

3rd Air Command Southern	

4th Air Group	83rd Direct Command Unit
8th Sentai	71st DCS
64th Sentai	89th DCS
81st Sentai	91st DCS
	11th TAS.

7th Air Group Malaya	3rd Air Group Java	12th Air Group Sumatra	21st DCU Indo-China	2nd Air Group Philippines
12th Sentai	59th Sentai	1st Sentai	82nd DCS.	52nd DCS
98th Sentai	75th Sentai	11th Sentai	84th DCS.	74th DCS.
14th Sentai	70th DCS		1st ATU	76th DCS
50th Sentai			2nd ATU	
7th TAU.				
13th TAU				
15th TAU				

3rd Air Division	
1st Air Group	206th DCU
44th Sentai	54th DCS.
29th Sentai	6th G Coops
54th Sentai	7th G Coops
65th Sentai	8th G Coops
90th Sentai	18th DCS
16th Sentai	83rd DCS
24th Sentai	10th DCS
33rd Sentai	55th DCS

TAU = Transport air unit
DCS = Direct command squadron
G Coops = Ground co-operation squadron

APPENDIX 2

Purchases by the Imperial Naval Air Service

Date	Aircraft Purchased	Country of Origin
1909	Henri Farman box-kite biplane	France
	Grade modified box-kite of Gabriel Voisin	Germany
1910	Curtiss seaplane	USA
1912	Maurice Farman Longhorn	France
	Deperdussin monoplane	France
1913	Sopwith Tabloid scout	Great Britain
	Short 184 seaplane	Great Britain
	Avro 504 biplane	Great Britain
1918	Sopwith T1 Cuckoo biplane	Great Britain
	Terrier T3 flying-boat	Great Britain
	Heinkel airframes, BMW and Mercedes engines	Germany
1919	Nieuport Nightjar or Sparrow Hawk	Great Britain
	Parnall Panther	Great Britain
	Sopwith Pup biplane	Great Britain
1920	SPAD biplane fighter	France
	Short Tractor biplane	Great Britain
	Fairey IIIF Seal Mk VI	Great Britain
1921	Felixstowe F5 flying-boat	Great Britain
	Vickers Viking flying-boat	Great Britain
	Supermarine Channel Mk II flying-boat	Great Britain
	Supermarine Seal	Great Britain
	Blackburn Swift Mk II	Great Britain
1924	Dornier Do D Flake seaplane	Germany
	Rohrbach Type R flying-boat	Germany
	Dornier Do 4 seaplane	Germany
	Dornier Wal flying-boat	Germany
1925	Dornier Komet III	Germany
1927	Blackburn T.7B	Great Britain
	Boeing F2B	USA

	Heinkel HD-23	Germany
	N3 airship	Italy
1929	Short KF1 flying-boat	Great Britain
	Supermarine Southampton MK II	Great Britain
	Bristol Bulldog	Great Britain
	Heinkel HD-56	Germany
	Vought Corsair	USA
1930	Blackburn Lincock III	Great Britain
	Short KFI flying-boat	Great Britain
1931	Boeing 100	USA
1932	Savoia S62	Italy
	Hawker Nimrod	Great Britain
	Cierva autogiro	Great Britain
1933	Northrop Gamma	USA
1934	Heinkel He 50	Germany
	Heinkel HD-66	Germany
1935	Consolidated P2Y1	USA
	Heinkel He 70	Germany
	Heinkel He 74	Germany
1936	Fiat BR20	Italy
	Fairchild A942	USA
	Heinkel He 118	Germany
	Dewoitine D510	France
1937	North American BI-9	USA
	North American BI-10	USA
	Kimmer Envoy	USA
	Douglas DF	USA
	Junkers Ju 86	Germany
1938	Bucker Jungmann	Germany
	Heinkel He 112	Germany
	Potez biplane	France
	Caudron C 600	France
	Seversky 2PA-B3	USA
	Douglas DO4	USA
1940	Junkers Ju 88	Germany
	Heinkel He 100	Germany
	Heinkel He 119 V7 and V8	Germany
	Douglas DC3	USA

Bibliography

Barnes, C.H., *Bristol Aircraft Since 1910*, Putnam, 1964

Barnes, C.H., *Short's Aircraft since 1900*, Putnam, London, 1967

Barr, Pat, *The Coming of the Barbarians*, MacMillan, 1967

Cooper, Bryan, & Batchelor, John, *Fighter*, MacDonald, London

BBC Television, Timewatch, *Sacrifice at Pearl Harbor*, 4 December 1991 (repeat)

Bergamini, David, *Japan's Imperial Conspiracy*, Heinemann

Bowyer, Chas, *Action Profile – Mitsubishi 0 Zero*, Profile Publications Ltd, 1972

British Aircraft Corporation, Correspondence. Ref. PU/NAB/WP, 15 March 1976

Burke, John, *Winged Legend. The Story of Amelia Earhart*, Arthur Barker Ltd, London

Costello, John, *The Pacific War 1941–1945*, Collins, 1981

Crampton, J., 'Herbert Smith, Aircraft Designer', *Air Pictorial* magazine, June 1975

The Daily Telegraph, 7 June 1977, Obituary, 'Sir John Masterman, Mastermind of MI5. D-Day Double Cross'

The Daily Telegraph, Reuter report, 23 September 1978

Francillon, René J., *Japanese Aircraft of the Pacific War*, Anova Books, 1987

Francillon, René J., *Profile No. 210. Mitsubishi G4M Betty and the Ohka Bomb*, Profile Publications Ltd, 1971

Garlinski, Josef, *Intercept – Secrets of the Enigma War*, Dent, London, 1979

Goerner, Fred, *The Search for Amelia Earhart*, Bodley Head, London

Hardwick, Michael, *Discovery of Japan*, Hamlyn, London

Jackson, A.J., *Blackburn Aircraft Since 1909*, Putnam, 1968

James, Derek N., *Hawker – an Aircraft Album*, Ian Allan

Jane's All the World's Aircraft, 1919

Jane's All the World's Aircraft, 1920

Jane's All the World's Aircraft, 1922

Jane's All the World's Aircraft, 1926

Jane's All the World's Aircraft, 1928

Jane's All the World's Aircraft, 1931

Jane's All the World's Aircraft, 1934

Jane's All the World's Aircraft, 1937

Jane's All the World's Aircraft, 1938

Kirby, Maj Gen S. Woodburn, *War Against Japan. Volume One. The Loss of Singapore.* HMSO, London, 1957

Mars, Alastair, *British Submarines at War 1939–1945*, William Kimbo Ltd, London, 1971

Morison, Samuel Eliot, *History of U.S. Naval Operations in World War II. Volume III. The Rising Sun in the Pacific*, Little, Brown & Company, Boston, 1968

Munson, K.G., *Aircraft of World War II*, Ian Allan, 1972

Munson, K.G., *Japanese And Russian Aircraft of World War II*, Ian Allan

Nish, Ian, *Story Of Japan*, Faber & Faber

Okumiya, Masatoke, & Horikoshi, Jiro, *Zero Fighter*, Cassell, London

Penrose, Harald, *British Aviation. The Adventuring Years 1929–29*, Putnam, 1973

Polmar, Norman, *Aircraft Carriers*, Macdonald, London, 1969

Sekigawa, Eiichiro, *Japanese Military Aviation*, Ian Allan, 1974

The Sunday Telegraph, 22 and 29 November 1981, 'Why America Was Taken By Surprise'

Taylor, John W.R. (Ed.), *Aircraft 1973*, Ian Allan, London, 1972

Toland, John, *The Rising Sun, the decline & fall of the Japanese Empire 1936–1945*, Cassell & Company, 1971

Watts, Anthony J., *Japanese Warships of World War II*, Ian Allan

Watts, Anthony J., *Pictorial History of the Royal Navy. Volume One. 1816–1880*, Ian Allan, London, 1970

Royal Aeronautical Society Proceedings. 12th Meeting, 59th Session, 3 April 1924, RAF Museum, Hendon

Semphill, Colonel the Master of, AFC, AFRAeS, 'British Aviation Mission to the Imperial Japanese Navy', Royal Aeronautical Society Proceedings, Final Meeting 1923/1924

Index

Page numbers in *italics* refer to maps.